Epilepsy Fundamentals

Zulfi Haneef
Editor

Kamakshi Patel • Jay R. Gavvala
Associate Editors

Epilepsy Fundamentals

A Concise Clinical Guide

Springer

Editor
Zulfi Haneef
Neurology
Baylor College of Medicine
Houston, TX, USA

Associate Editors
Kamakshi Patel
Associate Professor in Neurology
Baylor College of Medicine
Houston, TX, USA

Michael E. DeBakey VA Medical Center
Houston, TX, USA

Jay R. Gavvala
Associate Professor of Neurology
Texas Comprehensive Epilepsy Program
University of Texas Health Science Center at Houston
Houston, TX, USA

ISBN 978-3-031-77743-1 ISBN 978-3-031-77741-7 (eBook)
https://doi.org/10.1007/978-3-031-77741-7

This Springer imprint is published by the registered company Springer Nature Switzerland AG
The registered company address is: Gewerbestrasse 11, 6330 Cham, Switzerland

Preface

We are excited to introduce the first edition of *Epilepsy Fundamentals*. This book will serve as a comprehensive yet accessible guide for students and professionals eager to deepen their understanding of epilepsy. As the field of epilepsy rapidly evolves, there is a significant expansion in the body of literature. Given that epilepsy touches on numerous disciplines, including electrophysiology, neuroimaging, pharmacology, and genetics, it can be challenging for educators and authors to present the subject in a format that is both concise and thorough, without sacrificing clarity or rigor.

With trainees in mind, this book aims to be a valuable resource for Epilepsy fellows, Neurophysiology fellows, Neurology residents, and even more experienced professionals looking for a quick refresher. *Epilepsy Fundamentals* is not meant to replace comprehensive textbooks but rather serve as an introduction and bridge to the broader field. While more concise than existing texts, we have carefully curated content that is especially relevant for trainees during their education. The book also features numerous illustrations and tables to simplify complex concepts.

Each chapter concludes with a "pearls" section, offering key insights and additional preparation for the ABPN Epilepsy board. This book can be explored in any order, though it may be helpful to begin with Chap. 1, which outlines the most recent classification of seizures and epilepsy from the International League Against Epilepsy (ILAE). We have strived to balance traditional and modern terminology, recognizing that older terms remain in use.

In addition to sections on epilepsy classification, diagnosis, and treatment, the book includes chapters on EEG interpretation with carefully chosen examples. We hope this resource meets your needs, and we welcome any feedback that can help improve future editions.

Houston, TX, USA Zulfi Haneef

Contents

Classification of Seizures and Epilepsy 1

Francisca H. Ahn

A **seizure** is a transient occurrence of signs and/or symptoms due to abnormal excessive or synchronous neuronal activity in the brain [1]. Seizures are characterized by their *onset* into the following three categories by the International League Against Epilepsy (ILAE) [2, 3] (Figs. 1.1 and 1.2):

- **Focal:** arising from one part of the brain, further divided into:
  - Focal aware seizure (FAS): without an alteration of awareness (formerly “simple partial”).
  - Focal impaired awareness seizure (FIAS): with an alteration of awareness and/or consciousness (formerly “complex partial”/“petit mal”).
  - Focal to bilateral tonic-clonic (FTBTC): secondarily generalized event (formerly “generalized tonic-clonic”/“grand mal”).
- **Generalized:** arising from one point and rapidly engaging bilateral hemispheres.
  - Absence seizure: Characterized by behavioral arrest associated with generalized spike-wave discharges (GSW).
    Typical: 3–4 Hz GSW sudden onset/offset.
    Atypical: 1–2.5 Hz GSW may have a gradual onset/offset.
  - Myoclonic seizure: Characterized by sudden jerks associated with GSW.
  - Clonic: Characterized by repetitive rhythmic jerks of extremities.
  - Tonic: Characterized by contraction of antagonistic muscle groups.
  - Atonic: Characterized by sudden loss of muscle tone (“drop attacks”).
- **Unknown:** unclear, insufficient information to determine.

F. H. Ahn (✉)
Kellaway Section of Neurophysiology, Baylor College of Medicine, Epilepsy Center of Excellence, DeBakey VA Medical Center, Houston, TX, USA
e-mail: Francisca.Ahn@bcm.edu

Z. Haneef (ed.), *Epilepsy Fundamentals*,
https://doi.org/10.1007/978-3-031-77741-7_1

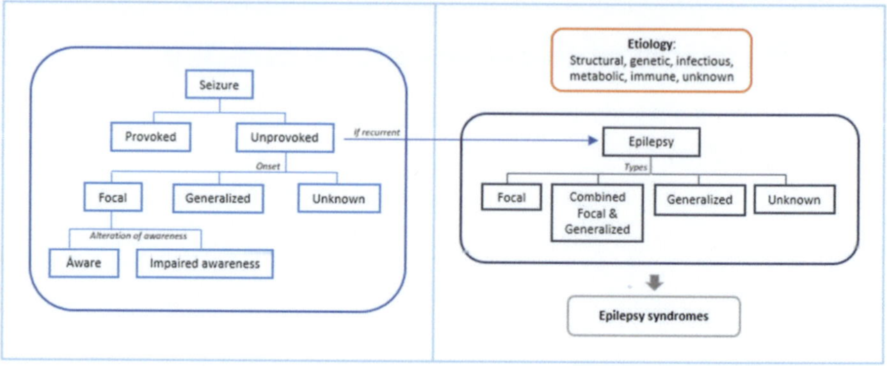

Fig. 1.1 Adapted illustration of ILAE Classifications of seizure and epilepsy types [1–3]

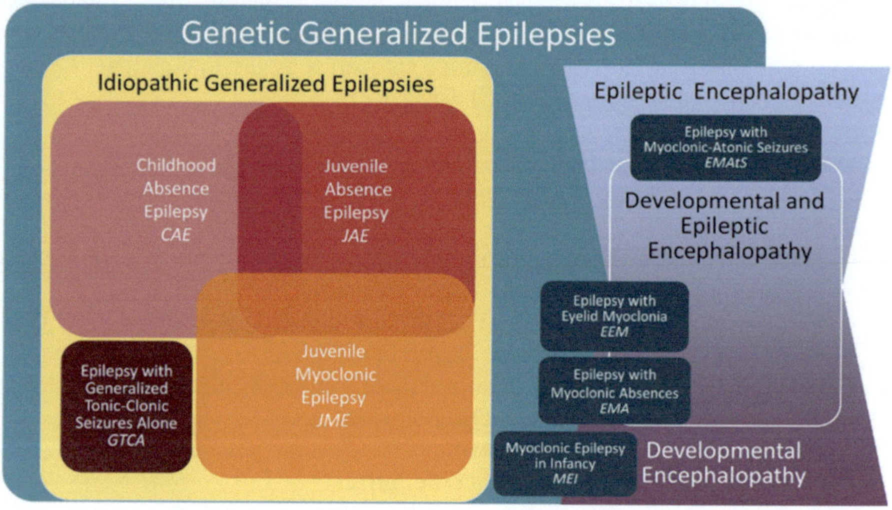

Fig. 1.2 Illustration of the concept of GGE, IGE and DEE/EE. Reproduced from Wirrell et al. (Creative Commons CC BY license) [5]

Epilepsy is defined as (1) at least two unprovoked (or reflex) seizures occurring at least 24 hours apart, (2) one unprovoked seizure with a high likelihood (>60% chance) of further unprovoked seizures, or (3) diagnosis of an epilepsy syndrome (ILAE 2014) [2]. Epilepsies are classified into four types (ILAE 2017 classification) [4], and each epilepsy type may have several seizure types.

1. **Focal**: includes unifocal and multifocal, supported by focal epileptiform discharges.
2. **Generalized**: clinical diagnosis supported by generalized epileptiform discharges. Clinical diagnosis of Genetic Generalized Epilepsies (GGEs) is drawn from previous research on inheritance patterns, and genetic testing does not need to be identified for a diagnosis.
   (a) Idiopathic generalized epilepsies (IGE) is a subgroup of GGE, which includes 4 specific epilepsy syndromes only: childhood absence epilepsy (CAE), juvenile absence epilepsy (JAE), juvenile myoclonic epilepsy (JME) and generalized tonic-clonic seizures alone (GTCA, formerly known as generalized tonic-clonic seizures on awakening) (Fig. 1.2) [5].
   (b) Developmental and epileptic encephalopathies (DEE) are described further below.
3. **Combined focal & generalized**: for example, Dravet and Lennox-Gastaut syndromes.
4. **Unknown**: definitive diagnosis of epilepsy but insufficient information present.

Developmental and epileptic encephalopathies (DEE): While DEE could be under the umbrella of GGE, depending on the clinical history, it also exists independently. The term "epileptic encephalopathy" (EE) was redefined where the epileptic activity itself contributes to severe cognitive and behavioral impairments beyond expected from other underlying pathologies [6]. This is applicable to epilepsies at all ages and can be utilized widely more than just severe epilepsies with onsets in infancy and childhood. Many epilepsies with associated encephalopathy have a genetic etiology (e.g., Infantile Epileptic Spasms Syndrome), but can also be acquired (e.g., hypoxic-ischemic encephalopathy or stroke).

The former terminology of "symptomatic generalized epilepsies" was utilized broadly and applied to a highly heterogeneous group of patients, including patients with developmental encephalopathies and epilepsy, epileptic encephalopathies, developmental and epileptic encephalopathies, and at times, generalized, and combined generalized and focal epilepsies. Given the broad range of heterogeneity, a new terminology was developed for categorizing these patient groups (Figs. 1.2 and 1.3).

Epilepsy can be categorized by its etiology, as follows: [2]

1. **Structural:** stroke, trauma, infection, or malformations of cortical development.
2. **Genetic:** either known/presumed genetic mutation in the core symptom is seizures (e.g., KCNQ2/3), suggestive of genetic etiology (previously broadly called "idiopathic") or identified pathogenic molecular mutations (e.g., SCN1A).
3. **Infectious**: meningitis, encephalitis, neurocysticercosis, tuberculosis and HIV.
4. **Metabolic**: evidence of metabolic defect with biochemical changes often overlaps genetic etiology (e.g., pyridoxine-dependent seizures).
5. **Immune**: evidence of autoimmune-mediated central nervous system inflammation (e.g., NMDA-receptor encephalitis and anti-LGI1 encephalitis).
6. **Unknown**: no identifiable focus (previously "cryptogenic").

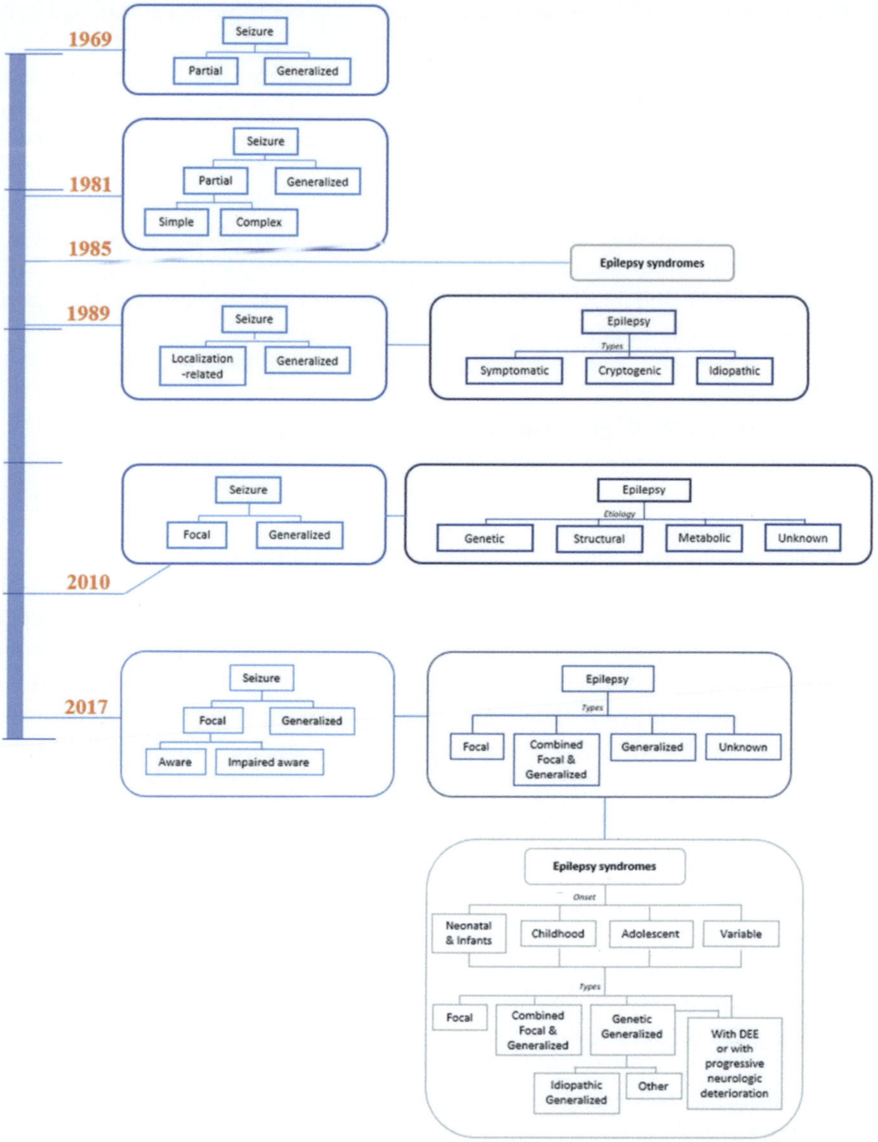

Fig. 1.3 Evolution of classifications of seizures, epilepsy, and epilepsy syndromes

1.1 Epilepsy Syndromes

The term "epilepsy syndrome" refers to a set of features, including seizure types, electrographic findings, and imaging features that tend to occur together (see Table 1.1) [5–10]. Clinical features incorporate age at onset, remission if applicable, triggers, intellectual disability, and/or psychiatric dysfunction. Identifying epilepsy syndromes can be helpful in prognostication and provide guidance in treatment plans [5].

Table 1.1 Classification of epilepsy syndromes by age of onset then by seizure typeplans [5–10]

Onset	Seizure type	Syndrome	Clinical pearls
Neonates & infants	**Focal**	Self-limited neonatal epilepsy (SeLNE)	Onset days 2–7 of life, usually remits by 6 months. Focal tonic or clonic features that may progress in a sequential pattern. Focal seizures can alternate sides and recur over hours to days. Developmentally normal. Most commonly have KCNQ2 pathogenic variants
		Self-limited infantile epilepsy (SeLIE)	Formerly known as benign infantile epilepsy. Onset 3–10 months, developmentally normal, focal seizures from posterior quadrant, seizures cluster, interictal EEG is normal. Most commonly PRRT2 pathogenic variant
		Self-limited familial neonatal-infantile epilepsy (SeLFNIE)	Same as SeLNE or SeLIE but requires family history with a broader age of onset. Autosomal dominant with high penetrance, commonly SCN2A pathogenic variant. Ictal EEG: Focal seizures, posterior region discharges
	Focal and/or generalized	Genetic epilepsy with febrile seizures plus (GEFS+)	Can start in infancy and present with febrile seizures "plus" a variety of other seizure types. Part of a spectrum that includes Dravet syndrome and Doose syndrome
	Generalized	Myoclonic epilepsy in infancy (MEI)	Spontaneous and/or reflex myoclonic jerks, ictal EEG shows GSW, interictal EEG is normal, and the prognosis is generally good
	Syndromes with DEE or with progressive neurological deterioration	Early infantile developmental and epileptic encephalopathies (EIDEE)	Includes formerly known "Ohtahara syndrome" and "early myoclonic encephalopathy." can have tonic (often upon wakening), myoclonic, epileptic spasms and sequential seizures +/− autonomic features. Both with burst suppression EEG patterns. Normal birth to onset in 0–3mo with severe psychomotor deficits and profound developmental impairments
		Epilepsy in infancy with migrating focal seizures (EIMFS)	Focal motor clonic or tonic with autonomic seizures often with secondary generalization. Ictal EEG: Multiple independent cortical sources, severe psychomotor deficits
		Infantile epileptic spasms syndrome (IESS)	Includes formerly known "west syndrome" with a triad of hypsarrhythmia, epileptic (infantile) spasms and developmental delay, and patients with epileptic spasms who may not have all three criteria
		Dravet syndrome	Formerly known as severe myoclonic epilepsy in infancy. Onset first year of life with prolonged, a/febrile seizures, focal clonic, generalized clonic seizure; poor prognosis, associated with mutations in SCN1A. Anti-seizure drugs with Na + channel-blocking properties should be avoided

Onset	Seizure type	Syndrome	Clinical pearls
Childhood	**Focal**	Self-limited epilepsy with centrotemporal spikes (SeLECTS)	Formerly known as benign epilepsy with Centro-temporal spikes (BECTS) aka Rolandic epilepsy. Focal seizures with dysarthria, sialorrhea, and dysphasia can be generalized
		Self-limited epilepsy with autonomic seizures (SeLEAS)	Formerly known as benign childhood occipital epilepsy, aka Panayiotopoulos syndrome. Onset 3–6 years with autonomic symptoms (often emesis), seizures >5 minutes typically activated by sleep. Interictal EEG: High amplitude, focal or multifocal discharges
		Childhood occipital visual epilepsy (COVE)	Formerly known as idiopathic childhood/late onset benign occipital epilepsy, Gastaut syndrome/type. Onset 8–11 years with focal sensory visual seizures with elementary visual phenomena +/− awareness, +/− motor, exclusively in wakefulness. Interictal EEG: Spikes in the occipital region only
		Photosensitive occipital lobe epilepsy (POLE)	Focal sensory visual seizures can evolve into bilateral tonic-clonic seizures. Triggered by photic stimuli. Interictal EEG: Occipital discharges with eye closure and photic stimulation
	Generalized	Epilepsy with myoclonic absences (EMA)	Absence seizures with rhythmic myoclonus time-locked with 3 Hz generalized spike-and-wave discharges. 40% of patients have remission; if they continue to have seizures, then they are associated with developmental delay
		Epilepsy with eyelid myoclonia (EEM)	Formerly known as "Jeavons syndrome." brief (<6 s), repetitive and often rhythmic 3–6 Hz myoclonic jerks of eyelids with a simultaneous upward deviation of eyeballs and head extension. Interictal EEG: 3–6 Hz generalized discharges
	Syndromes with DEE or with progressive neurological deterioration	Epilepsy with myoclonic–atonic seizures (EMAtS)	Formerly known as "Doose syndrome." onset with myoclonic and/or atonic seizures, without tonic seizures; previously normal, prognosis variable. Interictal EEG: Generalized 2–6 Hz spike-wave or polyspike-wave discharge
		Lennox–Gastaut syndrome (LGS)	Triad of multiple seizure types: Must have tonic seizures and any other type, developmental delay and EEG showing paroxysmal fast activity in sleep and slow <2.5 Hz spike-wave seizures
		Developmental and epileptic encephalopathies (DEE) or EE with spike-and-wave activation in sleep (DEE/EE-SWAS) Subtype: Landau Kleffner syndrome (LKS)	Formerly known as epileptic encephalopathy with continuous spike-and-wave in sleep and atypical benign partial epilepsy (pseudo-Lennox syndrome). Partial or generalized seizures during sleep; atypical absences when awake. EEG: Slow 1.5–2 Hz spike and wave during N-REM sleep. Previously, sleep EEG patterns were known as electrical status epilepticus (ESES). Cognitive, behavioral or motor regression related to SWAS on EEG LKS is a specific subtype of EE-SWAS that results in regression of language with acquired auditory agnosia
		Febrile infection-related epilepsy syndrome (FIRES)	History of nonspecific febrile illness within the 2 weeks prior to seizure onset. Focal and multifocal seizures that evolve into BTC seizures, which then leads to super-refractory status epilepticus
		Hemiconvulsion–hemiplegia–epilepsy (HHE)	Acute stage: Febrile, hemiclonic SE resulting in permanent hemiparesis. Chronic stage: Variable but <3 years after onset, develop unilateral focal motor or FTBTC seizures
	Idiopathic generalized epilepsies (IGE)	Childhood absence epilepsy (CAE)	Formerly known as "pyknolepsy" and "petit mal." seizures with behavioral arrest and 3 Hz generalized spike-wave, very frequent with several/day, exacerbated by hyperventilation, prognosis generally good
Adolescent		Juvenile absence epilepsy (JAE)	Absence seizures and GTCS may have myoclonic jerks and life-long seizures, but they are generally well-controlled with medication
		Juvenile myoclonic epilepsy (JME)	Myoclonic jerks, GTCS and mild absence seizures, polyspike and wave discharges on EEG, life-long but generally improve
		Epilepsy with generalized tonic-clonic seizures alone	Generalized tonic-clonic seizures occur within 1–2 hours of awakening; lifelong treatment is generally necessary

Onset	Seizure type	Syndrome	Clinical pearls
Variable	**Focal**	Mesial temporal lobe epilepsy with hippocampal sclerosis (MTLE-HS)	Most common form of epilepsy, often with a history of febrile seizures. Presents with FAS/FIAS with onset semiology consistent with medial temporal lobe network. MRI shows hippocampal sclerosis. Often medication resistant but excellent prognosis with anterior temporal lobectomy
		Familial mesial temporal lobe epilepsy (FMTLE)	Focal cognitive (particularly déjà vu), sensory, or autonomic seizures. Family history of focal seizures from mesial temporal lobe. Associated with DEPDC5, complex inheritance pattern.
		Sleep-related hypermotor (hyperkinetic) epilepsy (SHE)	Brief focal motor seizures with hyperkinetic or asymmetric tonic/dystonic features predominantly from sleep. Associated with CHRNA4/2, CHRNB2, DEPDC5, KCNT1, NPRL2/3, PRIMA1 genes.
		Familial focal epilepsy with variable foci (FFEVF)	Focal onset seizures. MRI brain may have focal cortical dysplasia. Family history of individuals with focal seizures that arise from cortical regions that differ between family members. Associated with TSC1/2, DEPDC5, NPRL2/3 genes.
		Epilepsy with auditory features (EAF)	Also called "familial lateral temporal lobe epilepsy," associated with mutations in LGI1. Focal sensory, auditory seizures and/or focal cognitive seizures with receptive aphasia. Excellent response to carbamazepine
	Generalized	Epilepsy with reading-induced seizures (EwRIS)	Reading−/language-related tasks trigger reflex myoclonic seizures involving orofacial muscles
	Syndromes with DEE or with progressive neurological deterioration	Rasmussen syndrome (RS)	Focal/hemispheric seizures that increase in frequency over weeks to months, drug-resistant, lead to progressive neurologic deficits. EEG with hemispheric slowing, epileptiform abnormality. MRI shows progressive hemiatrophy. The standard of care is early hemispherectomy
		Progressive myoclonus epilepsies (PME)	The umbrella term for catastrophic epilepsies with cortical myoclonus includes Unverricht-Lundborg (most common, least severe PME), Lafora body disease, and neuronal ceroid Lipofuscinosis

Acknowledgments Atul Maheshwari and Elizabeth Dragan for authoring a previous version of this chapter.

References

1. Fisher RS, van Emde Boas W, Blume W, et al. Epileptic seizures and epilepsy: definitions proposed by the international league against epilepsy (ILAE) and the International Bureau for Epilepsy (IBE). Epilepsia. 2005;46(4):470. https://doi.org/10.1111/j.0013-9580.2005.66104.x.
2. Fisher RS, Acevedo C, Arzimanoglou A, et al. ILAE official report: a practical clinical definition of epilepsy. Epilepsia. 2014;55(4):475–82. https://doi.org/10.1111/epi.12550.
3. Fisher RS, Cross JH, D'Souza C, et al. Instruction manual for the ILAE 2017 operational classification of seizure types. Epilepsia. 2017;58(4):531–42. https://doi.org/10.1111/epi.13671.
4. Scheffer IE, Berkovic S, Capovilla G, et al. ILAE classification of the epilepsies: position paper of the ILAE Commission for Classification and Terminology. Epilepsia. 2017;58(4):512–21. https://doi.org/10.1111/epi.13709.
5. Wirrell EC, Nabbout R, Scheffer IE, et al. Methodology for classification and definition of epilepsy syndromes with list of syndromes: report of the ILAE task force on nosology and definitions. Epilepsia. 2022;63(6):1333–48. https://doi.org/10.1111/epi.17237.
6. Berg AT, Berkovic SF, Brodie MJ, et al. Revised terminology and concepts for organization of seizures and epilepsies: report of the ILAE commission on classification and terminology, 2005-2009. Epilepsia. 2010;51(4):676–85. https://doi.org/10.1111/j.1528-1167.2010.02522.x.
7. Zuberi SM, Wirrell E, Yozawitz E, et al. ILAE classification and definition of epilepsy syndromes with onset in neonates and infants: position statement by the ILAE task force on nosology and definitions. Epilepsia. 2022;63(6):1349–97. https://doi.org/10.1111/epi.17239.
8. Specchio N, Wirrell EC, Scheffer IE, et al. International league against epilepsy classification and definition of epilepsy syndromes with onset in childhood: position paper by the ILAE task force on nosology and definitions. Epilepsia. 2022;63(6):1398–442. https://doi.org/10.1111/epi.17241.
9. Hirsch E, French J, Scheffer IE, et al. ILAE definition of the idiopathic generalized epilepsy syndromes: position statement by the ILAE task force on nosology and definitions. Epilepsia. 2022;63(6):1475–99. https://doi.org/10.1111/epi.17236.
10. Riney K, Bogacz A, Somerville E, et al. International league against epilepsy classification and definition of epilepsy syndromes with onset at a variable age: position statement by the ILAE task force on nosology and definitions. Epilepsia. 2022;63(6):1443–74. https://doi.org/10.1111/epi.17240.

Seizures and Epilepsy Syndromes

2

Zulfi Haneef and Joyce H. Matsumoto

The ILAE classification of seizures is shown in Fig. 2.1. Selected seizure types are discussed below.

2.1 Seizure Types

Absence seizures: Absence seizures are characterized by brief lapses of consciousness and are associated with 3 Hz generalized spike waves (GSW) in the **EEG**. Classically seen in childhood absence epilepsy (CAE), generalized spike-wave discharges can also be seen in other conditions such as juvenile myoclonic epilepsy (JME).

Atypical absence seizure: Generally seen in individuals with Lennox Gastaut syndrome (LGS), atypical absence is distinguished from typical absence by GSWs in the EEG, which are slower (1.5–2.5 Hz) than the 3 Hz frequency expected with typical childhood absence seizures. On **EEG,** the GSWs are less monomorphic (more irregular appearing), more asymmetric, and have broader and less clearly defined spikes than in typical absence. Atypical absence seizures are characteristically not triggered by hyperventilation or photic stimulation.

Myoclonic seizures: Myoclonic seizures are characterized by a brief, involuntary muscle contraction, causing a sudden jerking body movement. Each myoclonus is so brief (<200 ms) that it is often unclear if consciousness is affected. Myoclonic

Z. Haneef
Neurology, Baylor College of Medicine, Houston, TX, USA
e-mail: Zulfi.Haneef@bcm.edu

J. H. Matsumoto (✉)
Division of Pediatric Neurology, David Geffen School of Medicine at UCLA, UCLA Mattel Children's Hospital, Los Angeles, CA, USA
e-mail: JMatsumoto@mednet.ucla.edu

Z. Haneef (ed.), *Epilepsy Fundamentals*,
https://doi.org/10.1007/978-3-031-77741-7_2

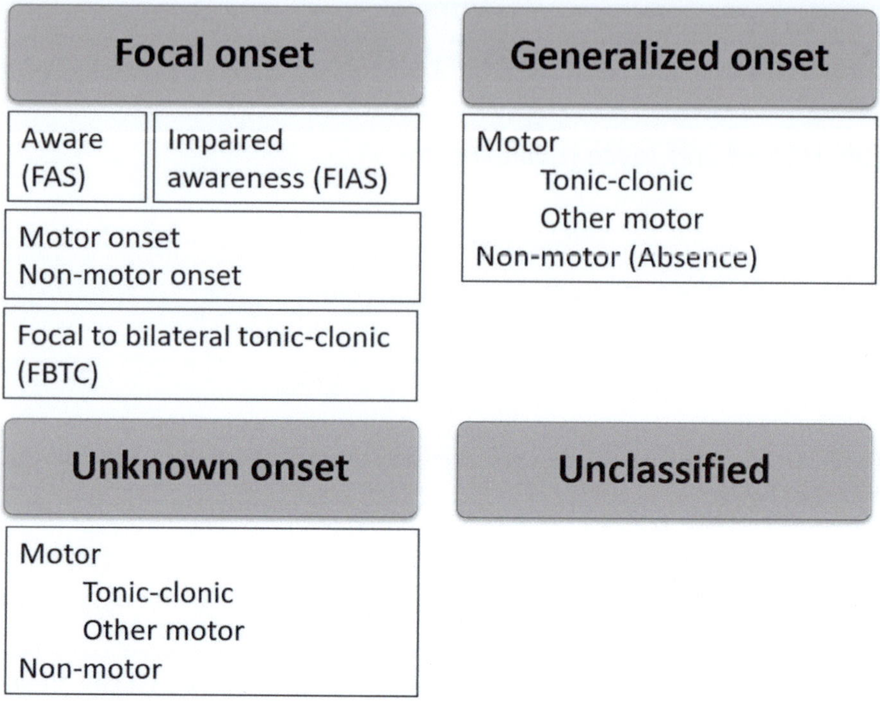

Fig. 2.1 Classification of seizures by onset (ILAE 2017). Note that focal-onset seizures can secondarily generalize

seizures are thought to be generated by a cortical/subcortical generator involving the thalamocortical/reticular projections. On **EEG,** the myoclonic jerk coincides with a generalized polyspike-wave (PSW) discharge or a bisynchronous GSW. Epileptic myoclonus can be associated with a variety of syndromes, including epilepsy with Myoclonic–Atonic seizures (EMAtS, also known as Myoclonic Astatic epilepsy (MAE) or Doose syndrome), JME, LGS, Dravet syndrome and progressive myoclonic epilepsies such as neuronal ceroid lipofuscinosis (NCL). The myoclonus associated with NCL can be associated with giant visual evoked potentials (VEPs) on photic stimulation at less than 3 Hz.

Tonic seizures: These seizures are characterized by sustained (5–20 s, up to 1 min) contraction of one or several muscle groups. Tonic seizures are a prominent feature of LGS, in which nocturnal tonic seizures (usually Stage I or II) are often considered to be a necessary clinical feature. Clinically, these can vary from uprolled eyes in isolation to the involvement of the whole body, leading to propulsive/retropulsive falls (drops) causing injury. Falls are less frequent than with atonic drop seizures but are nevertheless the most common cause of falls in children with

LGS. **EEG** shows a generalized attenuation of the background with superimposed sharply contoured 10–25 Hz fast activity, which may be of low or high amplitude. During slow-wave sleep, tonic seizures may be associated with generalized paroxysmal fast activity (GPFA).

Atonic seizures: These are generalized-onset seizures characterized by a sudden loss of postural tone leading to falls (drop seizures). Atonia can last a few seconds (drops) or minutes (akinetic seizures). The extent of involvement varies to include only the head (head nod) or the whole body (drop). Drops in atonic seizures lead to a sudden, limp collapse straight down, rather than the tonic propulsive or retropulsive drops seen in tonic seizures. Consciousness is impaired during the fall, but baseline mental status is typically restored by the time the patient stands up. Patients with atonic events often have other coexistent seizure types. **EEG** shows bursts of slow spike waves or PSW interictally. Children with EMAtS/ Doose syndrome can exhibit myoclonic astatic seizures, which are characterized by a myoclonic jerk simultaneous with an atonic event. On EEG, a generalized slow spike-wave or PSW, with the spike corresponding to a myoclonic event and the slow wave associated with atonia, is followed immediately by generalized attenuation, at times with superimposed fast activity.

Epileptic spasms: Although somewhat synonymous with infantile spasms, epileptic spasms can occur in older children and in adults that are clinically and electrographically indistinguishable. There is an initial phasic contraction of the neck, trunk, and extremities lasting less than 2 s, followed by a tonic contraction lasting 2–10 s. The contractions can be in flexion, extension, or both, ranging from a subtle head drop/shoulder shrug to a whole body movement (salaam spell/jackknife seizure). Eye deviation/nystagmus occurs in two-thirds of cases. Spasms are typical on awakening or after feeding and are less likely in sleep. They typically occur in clusters of 3–20 spasms, several times a day, which can evolve in later life into a single spasm of LGS. The exact pathophysiology is unknown. On **EEG,** each spasm movement corresponds to a moderate-to-high- amplitude generalized slow wave, often followed by a brief, 1–2 s generalized attenuation known as an electrodecremental response. In infants, the interictal EEG background is often characterized by a hypsarrhythmia pattern, which consists of a disorganized, high-voltage, chaotic background with multifocal spikes.

2.2 Epilepsy Syndromes

Chapter 1 and Fig. 2.2 give the classification of epilepsy syndromes with short descriptions. The age of onset of epilepsy can be helpful in identifying the epilepsy syndrome (Fig. 2.3). Some of the more common types are discussed in greater detail below.

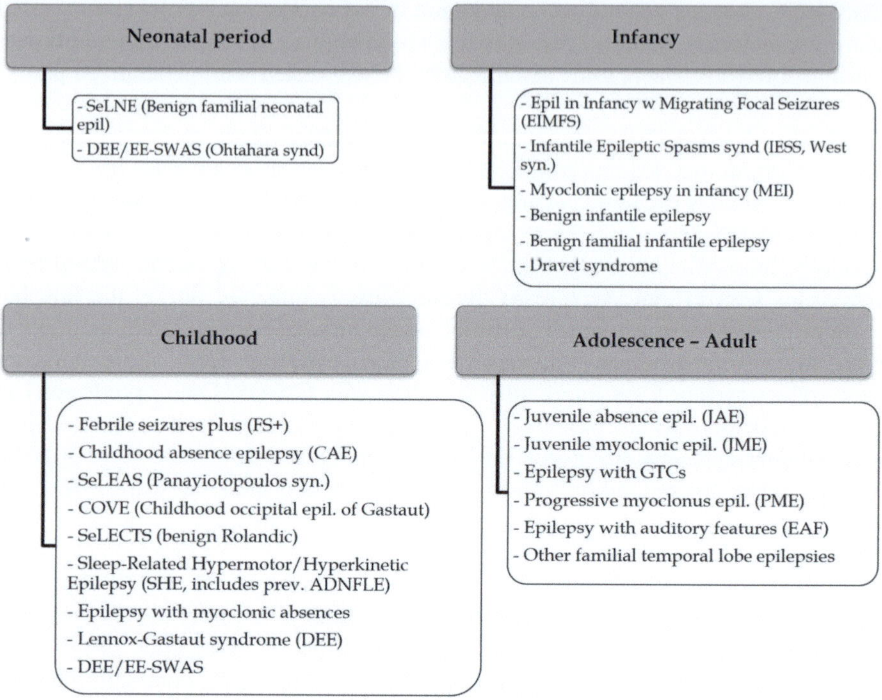

Fig. 2.2 Classification of epilepsy syndromes (ILAE 2022)

2.2.1 Neonatal

2.2.1.1 Early Infantile Developmental and Epileptic Encephalopathies (EIDEE)

Formerly Ohtahara Syndrome and Early Myoclonic Encephalopathy (EME): The earliest onset epileptic encephalopathy syndromes include Ohtahara syndrome (also known as early infantile epileptic encephalopathy with suppression-burst) and Early Myoclonic Encephalopathy (EME). These two syndromes have now been combined under the collective term EIDEE [1]. Both present very early in life (within the first 3 months, often within the first 2 weeks). Onset may be even earlier, as sometimes mothers report having felt fetal seizure activity during late pregnancy [2].

- *Etiology*: Ohtahara Syndrome and EME may be associated with severe structural brain abnormalities, such as hemimegalencephaly, agenesis of the corpus callosum, migrational disorders, and porencephaly. Alternatively, metabolic disorders

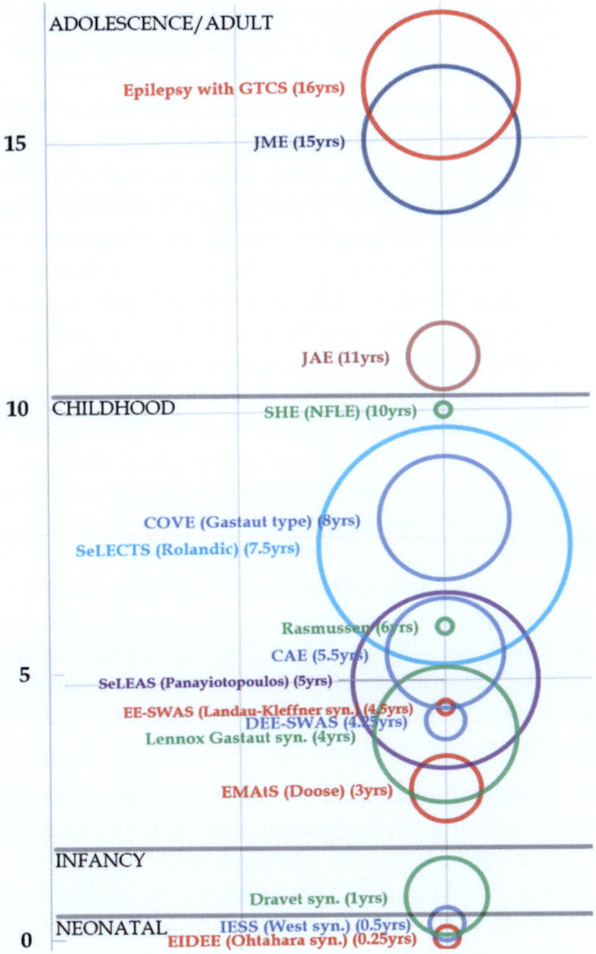

Fig. 2.3 Age of onset (y-axis) of different epilepsy syndromes (bubble size indicates frequency)

such as pyridoxine deficiency and nonketotic hyperglycinemia, along with numerous others, may be associated. Gene mutations so far associated include *ARX, SCN2A, CDKL5, STXBP1, PLB1, KCNQ2* and *KCNQ3.*

- ***Clinical Features***: In **Ohtahara Syndrome**, frequent *tonic seizures* occur throughout the day and night, occasionally in clusters, multiple times per hour up to hundreds of times per day. Other seizure types, including *focal clonic* or

hemiconvulsive seizures, may also occur. **EME** presents similarly, except that focal myoclonic jerks of the face/extremities are also prominent.
- ***Imaging*** can reveal structural abnormalities discussed previously. Focal lesions are more common in Ohtahara syndrome than in EME.
- ***EEG*** demonstrates a characteristic suppression-burst pattern with high amplitude spike/ polyspike-wave bursts separated by diffuse attenuation. In Ohtahara syndrome, this pattern persists in wakefulness and sleep, while in EME, the suppression-burst pattern may be more prominent during sleep.
- *Treatment*: Underlying pyridoxine or biotinidase deficiency responds to replenishment. In others, several ASMs and other broad-spectrum treatments such as the ketogenic diet may be tried with varying efficacy. With focal structural brain lesions, early surgical evaluation is indicated.
- *Prognosis*: Both subtypes of EIDEE have a dismal prognosis. Mortality is high before age 2, and survivors typically have severe lifelong cognitive and motor deficits. Later, many transition from tonic seizures and a suppression-burst EEG pattern into a clinical picture consistent with West Syndrome, with epileptic spasms and hypsarrhythmia.

2.2.1.2 Self-Limited Neonatal Epilepsy (SeLNE)

It is important to recognize that not all seizures in the neonatal period portend a poor prognosis. SeLNE (previously Benign Familial Neonatal Epilepsy or BFNE) typically begins within the first week of life and resolves within 1–4 months. A family history of neonatal seizures with subsequent normal development can generally be elicited.

- *Etiology*: Mutations in two potassium channel genes, KCNQ2 and KCNQ3, have traditionally been associated with SeLNE. KCNQ mutations result in a loss of function in potassium channels, which control and modulate excitability in the central and peripheral nervous systems. KCNQ 2 and 3 are the primary channels located in the brain. More recently, a syndrome of familial neonatal convulsions with later development of paroxysmal kinesogenic dyskinesia has been described in association with PRRT2 mutations.
- ***Clinical features***: Seizures are typically generalized, initially tonic with associated cyanosis, followed by clonic jerks of the entire body. Focal seizures have also been described.
- ***Imaging*** in SeLNE is unremarkable.
- ***EEG*** during seizures is characterized by a diffuse attenuation with the evolution of focal or generalized spike-wave discharges, with prolonged postictal attenuation. Interictally, the EEG background is normal.
- *Treatment*: No specific ASMs are typically recommended, though phenobarbital or carbamazepine have been successfully employed, amongst others.
- *Prognosis*: Neurocognitive development in SeLNE is normal.

KCNQ encephalopathy: A potentially treatable EIDEE variant.

Although KCNQ2 mutations have traditionally been associated with SeLNE, in recent years genetic testing of EIDEE cases has identified a subset associated with KCNQ2 mutations. Seizures in KCNQ2 encephalopathy typically begin within 1 week of life. Still, unlike SeLNE seizures, which are typically benign, interictal EEG demonstrates multifocal abnormalities with generalized attenuations and poor developmental outcomes.

Ezogabine, also known in Europe as retigabine, is an ASM whose primary mechanism of action relies on increasing current through KCNQ2 channels. This mechanism would presumably correct the pathophysiology of KCNQ2 encephalopathy, though confirmatory studies are required [3]. Unfortunately, due to limited use likely related to side effects of somnolence, urinary retention, and blue discoloration of the skin and retina/sclera, ezogabine/retigabine was withdrawn from the market. Other medications with reported efficacy for KCNQ2 encephalopathy in the case series include carbamazepine, levetiracetam, zonisamide, phenytoin, topiramate and valproic acid. Recently, there has been renewed interest in medications targeting potassium channels, with two compounds currently in development.

2.2.2 Infancy

2.2.2.1 West Syndrome (Infantile Spasms)

Infantile spasms are mostly regarded as a type of seizure that characteristically occurs during a certain neurodevelopmental period of life, typically between 2 weeks and 18 months (peak onset 4–6 m) of age. As these infants grow older, spasms typically recede and transition into other seizure types, particularly the tonic, myoclonic and tonic-clonic seizures of LGS. Occasionally, epileptic spasms can persist into later childhood and adulthood. Spasms are often associated with significant developmental deterioration, with lasting cognitive deficits into adulthood.

West syndrome triad

Epileptic spasms
Hypsarrhythmia
Developmental arrest/regression

- ***Etiology:*** Infantile spasms may have many different etiologies, including genetic anomalies such as tuberous sclerosis (~20%), Down syndrome (~15%), neurofibromatosis type 1, perinatal insults such as hypoxic-ischemic injury, hypoglycemia, metabolic conditions, or postnatal (infection, trauma) causes. The cause for about 10–15% of cases is unknown, with no identifiable structural, metabolic, or genetic etiology.

- ***Clinical features:*** Epileptic spasms (described earlier) are the hallmark of West syndrome and are part of the clinical triad (see box) required for diagnosis. The term "epileptic spasms" encompasses a similar semiology that appears beyond infancy and even in adulthood.
- ***Imaging:*** Spasms may be associated with a variety of structural abnormalities, including malformations of cortical development (cortical dysplasias, hemimegalencephaly) and perinatal stroke, showing an abnormality in most, while 10–15% show no lesion. Among these non-lesional cases, a subset termed "idiopathic West syndrome" shows normal development, with spasms stopping after a while.
- ***EEG:*** Interictal EEG shows hypsarrhythmia (Fig. 2.4), which is a high amplitude (hypsos in Greek means "mountainous") generally >200 mV, disorganized, chaotic background, with admixed multifocal spikes and spike and wave discharges, and is thought to be pathognomonic for infantile spasms. Several variants have been identified and classified as "modified hypsarrhythmia," though these variants do not appear to correlate with differences in treatment response or outcome. This interictal pattern generally subsides before 24 months of age, even when clinical spasms persist. Spasms are associated with a generalized attenuation, or "electrodecrement," on EEG.
- ***Treatment:*** A trial of folinic acid and pyridoxine for possible metabolic etiologies should be considered. The mainstay of treatment encompasses either hormonal therapy with ACTH/high-dose prednisolone or vigabatrin. Vigabatrin is particularly effective in West syndrome associated with tuberous sclerosis. Treatment should be undertaken as quickly as possible to rescue the developmental trajectory.
- ***Prognosis:*** The onset of spasms is generally associated with developmental regression. Spasms are frequently refractory to treatment, and the majority of affected children suffer lifelong cognitive impairment, often with evolution to other seizure types consistent with LGS. With shorter latency to diagnosis and prompt response to treatment, intellectual outcomes improve significantly.

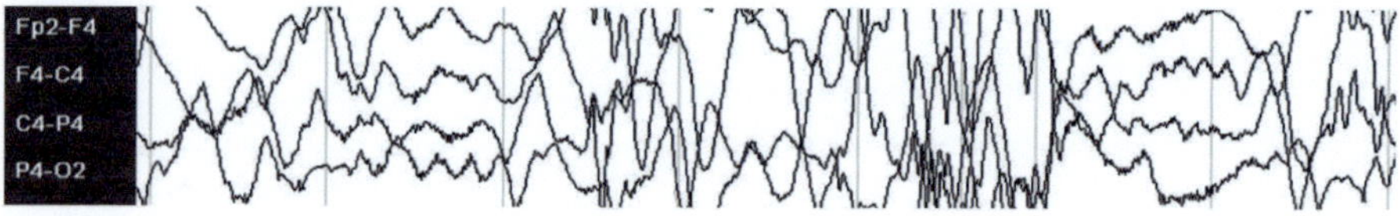

Fig. 2.4 Hypsarrhythmia in West syndrome

2.2.2.2 Dravet Syndrome (Previously Severe Myoclonic Epilepsy in Infancy, SMEI)

This is one of the most severe epilepsies in infants.

- ***Etiology***: Most patients have a de novo *SCN1A* mutation.
- ***Clinical features***: Three phases have been described in the course of the disease (1) Onset within 1 year of life with febrile seizures, consisting of generalized or unilateral clonic seizures, which are typically very prolonged (2). Between 1–4 years of age, seizures worsen. Non-febrile seizures occur, which are prolonged (>20 min) or in clusters and may evolve into status epilepticus. Myoclonus appears, which may be generalized. Absence seizures can also occur. Psychomotor delay may appear after the child learns to walk and speaks a few words. Effective language may never develop. Seizures may be triggered by changes in temperature. Hot water is a trigger commonly described in reports from Japan (3). After age 4, seizures can improve, which may represent a burnt-out phase.
- ***EEG***: The background is initially normal in year 1. A photoparoxysmal response may occur in 40%. Subsequently, there can be diffuse slowing, generalized spike-waves and focal/multifocal spikes [4].
- ***Treatment***: Sodium channel blockers (including LTG, CBZ, PHT, OXC and LCM) should be avoided as they can worsen seizures (*SCN1A* mutation can disable up to 50% of sodium channels located on inhibitory interneurons; further blockade of these channels leads to less inhibition, and therefore increased seizure propensity). VPA is the first medication often used. Benzodiazepines (such as clobazam) or TPM may be added when VPA becomes ineffective. In recent years, several medications have been approved for Dravet syndrome, including pharmaceutical-grade cannabidiol, fenfluramine (serotonergic agonist), and stiripentol (increases GABA transmission).
- ***Prognosis***: The long-term outcome is poor, with a 15% mortality and poor cognitive outcome, which may be related to seizure severity in the first 2 years of life.

2.2.2.3 Generalized Epilepsy with Febrile Seizures Plus (GEFS+)

Generalized epilepsy with febrile seizures plus (GEFS+) is an autosomal dominant condition where febrile seizures (FS) are associated with mutations in the beta or alpha subunits of the sodium channel (SCN1B in chromosome 19, or SCN1A in chromosome 2). GEFS+ is associated with FS in early childhood with persistent FS beyond 6 years of age. Various generalized afebrile seizures can develop later but typically remit by adolescence. Na + channel blockers such as CBZ should be avoided since they can exacerbate seizures.

SCN1A Mutations

SCN1A is a gene that codes for the voltage-gated sodium channel (SCN) type I alpha subunit located in the long arm of chromosome 2 (Fig. 2.5). Mutations cause malformation of sodium channels and are associated with a wide range of clinical phenotypes including:

- Simple febrile seizures.
- Generalized epilepsy with febrile seizures plus (GEFS+).
- Epilepsy with myoclonic-atonic seizures (EMAtS, also known as Doose syndrome).
- Dravet syndrome (previously Severe myoclonic epilepsy in infancy).
- Intractable childhood epilepsy with generalized tonic-clonic seizures (ICE-GTC).
- Familial hemiplegic migraine type 3 is also associated with SCN1A mutation.

2.2.3 Childhood

2.2.3.1 Childhood Absence Epilepsy (CAE)

CAE (previously pyknolepsy) is characterized by absence seizures and 3 Hz GSW on EEG. Typically seen in young children (peak 4–7 y, range 2–10 y, F:M = 2:1), these infrequently persist into adulthood.

- ***Etiology:*** A strong genetic component is noted with 85% twin concordance. Several genetic foci have been described, but monogenic inheritance is rare. GSWs are thought to be generated within a thalamocortical network involving both excitatory (spikes) and inhibitory (wave) components.
- ***Clinical features:*** **Typical absence** seizures occur associated with transient loss of awareness lasting about 10 s (range 1–45 s), without postictal confusion or, at most, brief confusion lasting 2–3 s. From 10 to 100 seizures may occur in a day. Associated automatisms or mild myoclonia can occur. Typical absences are almost universally triggered by hyperventilation for 3–5 min and in a smaller number (15%) by photic stimulation. GTCS may occur in CAE. Subtle cognitive and behavioral issues are also noted.

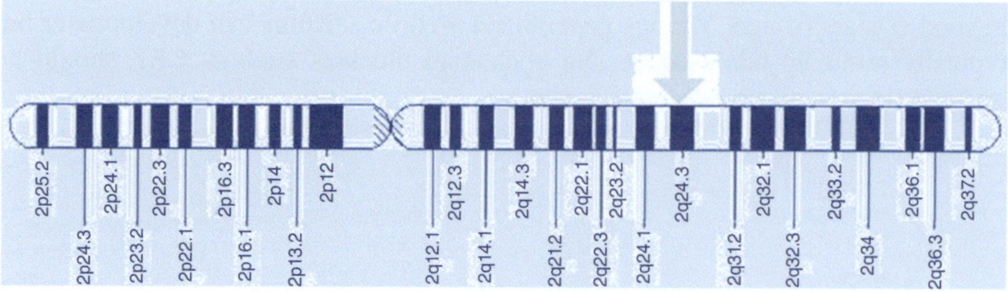

Fig. 2.5 Chromosome 2 showing location of SCN1A gene

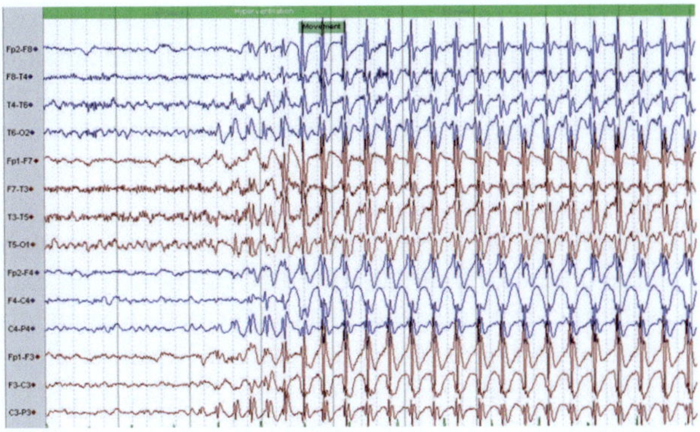

Fig. 2.6 Absence epilepsy showing 3 Hz generalized spike waves

- ***EEG*** shows monomorphic 3 Hz GSWs interictally or during seizures. In a longer run, the frequency may start faster at 3–4 Hz, and slow to 2.5–3 Hz by the end (Fig. 2.6). Polyspikes in sleep are associated with a less favorable prognosis.
- ***Imaging***: MRI is normal. Volumetry has reported an enlarged anterior thalamus. fMRI-EEG has shown thalamic activation, with cortical deactivation, during GSW discharges.
- ***Treatment***: Ethosuximide (ESX) is generally considered the first line for pure absence, though it will not treat GTCS. VPA is also effective but has more undesirable side effects than ESX [5]. LTG is better tolerated but less frequently efficacious than ESX and VPA. Seizures may be worsened by CBZ, PHT, OXC, TGB, and VGB.
- ***Prognosis:*** Spontaneous remission occurs by age 10–12 in 80% with absence alone and 30% with associated GTCS. Drug withdrawal can be considered after 2 years of seizure freedom in association with EEG resolution.

2.2.3.2 Epilepsy With Myoclonic-atonic Seizures (Also Doose Syndrome)

Epilepsy with myoclonic atonic seizures (EMAtS, previously known as myoclonic astatic epilepsy (MAE) or Doose syndrome) affects a previously healthy child (range 2–5 y, peak 3 y, M:F = 2.7:1) with several seizure types, including tonic-clonic, myoclonic, myoclonic-astatic, tonic, and absence, among others.

- ***Etiology***: Unclear. Genetic testing may demonstrate pathogenic variants in one of multiple epilepsy-related genes including SCN1A and SLC2A1, but in many cases is unrevealing.
- ***Clinical features***: Epilepsy onset may mimic other generalized epilepsy syndromes, with one predominant seizure type such as absence, followed by the emergence of different seizure types (myoclonic, atonic, tonic clonic) appearing over time. In some cases, the onset may be abrupt and explosive, with multiple seizures of various types in a day. A distinctive seizure type in EMAtS is the myoclonic-atonic seizure, where the child has a myoclonic jerk at the onset of the atonia, leading to significant head and facial injuries from falls. Protective

helmets are often required. *Absence* seizures also occur in >50%. Cognitive and motor dysfunction in the form of ataxia and apraxia can occur.

- ***EEG*** initially shows slowing and 2–6 Hz generalized spike-waves or poly spike-wave discharges. Video EEG can show myoclonic, atonic, and myoclonic-atonic seizures.
- ***Imaging:*** Generally normal.
- ***Treatment***: VPA and benzodiazepines such as clobazam may treat several seizure types. Rufinamide and FBM can be helpful, particularly for tonic and atonic seizures. The ketogenic diet (KD) can be particularly effective for EMAtS, more than medications [6, 7]. Early use of KD in refractory cases may improve time to remission and cognitive outcome. Avoid CBZ, PHT, and VGB, which can cause worsening. LTG can be helpful, though it may exacerbate myoclonic seizures.
- ***Prognosis***: Total recovery can occur in months to 3 years in most cases. Behavioral improvement is slower. However, those whose seizures do not respond to treatment may suffer developmental impairment and evolve into an LGS-like phenotype.

2.2.3.3 Childhood Occipital Epilepsy (COE, subtypes SeLEAS and COVE)

COE has two variants: (a) an early onset (Self-limited Epilepsy with Autonomic Seizures [SeLEAS], aka Panayiotopoulos) type, and (b) a late onset (Childhood Occipital Visual Epilepsy [COVE] aka Gastaut) type, with distinct clinical features (Table 2.1).

(a) ***Self-limited Epilepsy with Autonomic Seizures (SeLEAS), aka Panayiotopoulos Syndrome.***

This is the more common type of COE and occurs in early childhood (peak 5 y, range 2–8 y, F > M).

- ***Clinical features:*** Nocturnal seizures with prominent vomiting, followed by tonic eye deviation and occasional head/extremity involvement is seen. All patients have seizures in sleep, while one-third also have awake events.
- ***EEG*** shows high amplitude bilateral occipital spikes (Fig. 2.7) exhibiting "fixation-off sensitivity" (attenuation with eye-opening and induction by darkness or visual occlusion). Extraoccipital spikes may also occur.
- ***Treatment***: Most children experience only a few, and in some cases only one, seizure. Therefore, they do not need ASMs and can be managed by instructing parents to administer rectal diazepam in the event of seizure recurrence.
- ***Prognosis*** is good in early onset COE (SeLEAS) with remission in 1–2 years.

(b) ***Childhood Occipital Visual Epilepsy (COVE), aka Childhood Occipital Epilepsy of Gastaut.***

This is the late-onset type (peak 8 y, range 3–15 y, F = M).

- ***Clinical features:*** Visual symptoms (amaurosis/hallucinations) predominate in this variant. These visual hallucinations may occur multiple times daily with preserved awareness. Occasionally, the visual hallucinations will progress to larger seizures. Postictal and interictal headache, nausea and vomiting occur.

Table 2.1 Clinical features of the childhood occipital epilepsy subtypes

	SeLEAS (Panayiotopoulos)	COVE (Gastaut)
Peak age of onset	4–5 years old	8–10 years old
Seizure frequency	Rare	Frequent
Seizure medications	Daily medication generally not required Rescue medications recommended	Daily medications and rescue medications generally required
Prominent emesis	Frequent	Rare
Visual hallucinations	Rare	Frequent
Migraine headaches	Rare	Frequent

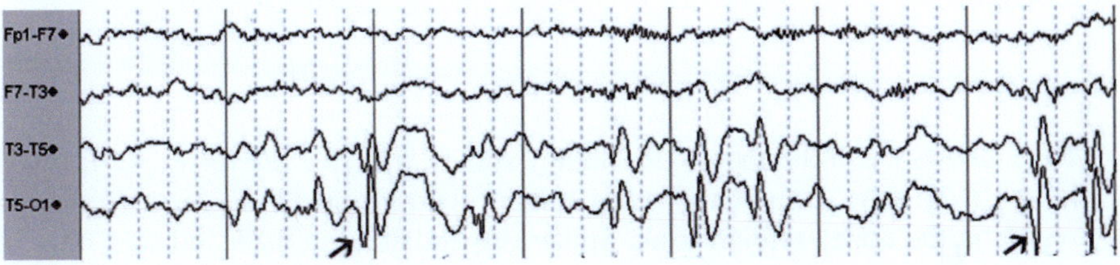

Fig. 2.7 SeLEAS (Panayiotopoulos syndrome) showing occipital spikes

- ***EEG*** is similar in showing bilateral occipital spikes. Extraoccipital spikes are less common than in the early-onset Panayiotopoulos type.
- ***Treatment***: Most patients experience simple partial visual seizures multiple times per day, with more frequent progression to complex partial and generalized tonic-clonic seizures. Treatment with ASMs is therefore indicated for the COVE (late-onset COE, Gastaut type.
- **Prognosis:** less clear in the late-onset COE than in the early-onset type.

2.2.3.4 Self-Limited Epilepsy with Centrotemporal Spikes (SeLECTS)

Also called Benign Epilepsy with Centrotemporal Spikes (BECTS) or Benign Rolandic Epilepsy (BRE), SeLECTS is the most common focal epilepsy below age 15, accounting for 10–15% of all seizures (peak 7–8 y, range 2–13 y, M:F = 3:2).

- ***Etiology***: Inheritance is autosomal dominant with heterogeneous genetics.
- ***Clinical features***: Affected children are neurologically normal. Seizures occur typically at sleep onset and are infrequent (once every 1–2 weeks). Seizures during wakefulness are focal aware (sensory or motor of the lower face, dysarthria, or drooling), while nocturnal seizures can be focal (hemifacial spasm ± FIAS), or be secondary generalized with a postictal Todd's paralysis. In some patients, neuropsychological examination reveals disturbances in auditory-verbal memory, learning and some executive function, particularly in those cases where the spikes are prominently activated by sleep (approaching or fulfilling ESES criteria). These patients have been described as "atypical Rolandic epilepsy." [8] The cognitive and behavioral deficits typically improve once clinical seizures and electrographic abnormalities resolve.

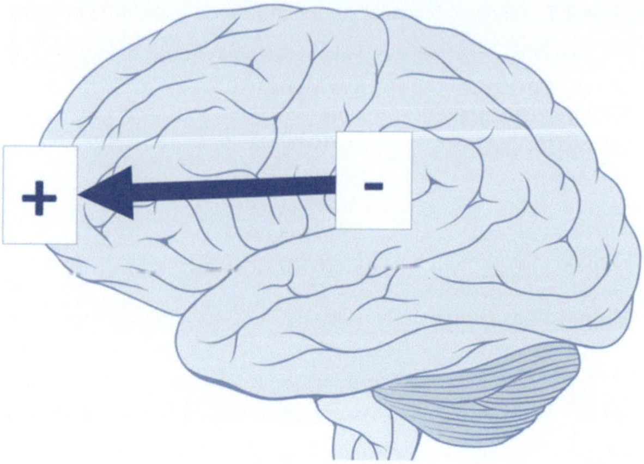

Fig. 2.8 SeLECTS with horizontal dipole (negative centrally, positive frontally)

- ***EEG*** shows centrotemporal (Rolandic) spikes with a *transverse (horizontal) dipole* (Fig. 2.8*)*, consistent with a seizure focus from the lower perirolandic region in the upper sylvian bank. Spikes are unilateral in 60% and seen only in sleep in 30%. Family members may have a similar EEG appearance. Only 10% of subjects with this EEG picture have seizures; therefore, isolated EEG findings without clinical correlation are not diagnostic of SeLECTS. During seizures, the EEG can show a reversal of the transverse dipole (positive centrally, negative frontally). EEG evidence of Rolandic spikes in siblings can help diagnose atypical cases.
- ***Imaging:*** reveals no brain lesions.
- ***Treatment*:** If seizures are primarily focal aware, clinical observation without ASMs is common. With frequent secondary generalization or frequent seizures in the daytime, ASMs are helpful.
- ***Prognosis:*** excellent with spontaneous remission in 95% by age 16, without long-term sequelae. The cognitive dysfunction also resolves as seizures remit. Remission occurs sooner with older onset age.

2.2.3.5 Sleep-Related Hypermotor (Hyperkinetic) Epilepsy (SHE)

This condition is characterized by brief nocturnal hyperkinetic or tonic seizures and encompasses several similar prior epilepsy syndromes, including autosomal dominant nocturnal frontal lobe epilepsy (ADNFLE) and nocturnal frontal lobe epilepsy.

- ***Etiology***: Both genetic and structural etiologies exist. Mutations in *CHRNA4*, *CHRNB2*, or *CHRNA2*, which encode subunits for the neuronal nicotinic acetylcholine receptor, are seen in 20% with a family history but only in <5% of those without. KCNT1 mutations and DEPDC5 have also been described. Occasionally, focal cortical dysplasia or an acquired structural lesion may present as SHE.
- ***Clinical features***: Seizure onset can vary between infancy and adulthood (mostly first 2 decades, peak 10 y) and persist throughout life. Clusters of stereotyped nocturnal motor seizures lasting 5 s–5 min, varying from simple arousals to bizarre, hyperkinetic events with bicycling, pelvic thrusting, ballism (flinging/throwing arm movement), and tonic/dystonic features occur. Consciousness is

preserved and fear of falling asleep may be expressed. These happen usually in non-REM and most commonly in stage 2 sleep.

- ***EEG***: is often normal but can show infrequent frontal epileptiform discharges. Ictal EEG is often obscured by movement but can show an 8–11 Hz spike pattern.
- ***Treatment***: Low-dose CBZ causes remission in 70%. About 30% are drug-resistant and may require multiple drug trials or VNS. Ser284Leu mutation confers a response to ZNS.

2.2.3.6 Lennox Gastaut Syndrome (LGS)

LGS is a common seizure disorder of childhood (peak 3–5 y, range 2–10 y) characterized by multiple seizure types, including nocturnal tonic seizures and drops, medical intractability, and cognitive/behavioral deficits. LGS is currently categorized within the broader DEE spectrum.

> LGS seizure types may show characteristic EEG changes:
>
> ***Tonic***: bilateral fast rhythms or attenuation.
> ***Atypical absence***: 2–2.5 Hz slow spike waves.
> ***Atonic***: slow spike wave, polyspike wave, electrodecrement.

- ***Etiology***: LGS can evolve from pre-existing brain injury or seizures such as infantile spasms ("symptomatic" LGS) or appear in otherwise normal children ("cryptogenic" LGS).
- ***Clinical features:*** The hallmark of LGS is the concurrence of multiple seizure types, including nocturnal tonic seizures. Other coexisting seizure types can include absence, atypical absence, myoclonic, atonic (astatic/akinetic) drops, and generalized tonic-clonic events. Patients frequently require helmets to prevent injury from falls resulting from both tonic and atonic events. Often, there is associated developmental/cognitive impairment.
- ***EEG*** shows generalized "slow" 2–2.5 Hz (slower than typical 3 Hz of CAE) GSWs and diffuse slowing with poor organization. Focal/multifocal spikes and polyspikes can be seen. Generalized paroxysmal fast activity in sleep is characteristic (Fig. 2.9). **Secondary bilateral synchrony** may be responsible for some of the EEG and clinical manifestations in LGS.
- ***Treatment:*** Seizures are often drug resistant. Drugs used include VPA, FBM, and benzodiazepines, including CLB. LTG and VNS can be helpful. Pharmaceutical-grade cannabidiol and fenfluramine have recently received approval from LGS for an orphan drug. VPA/ESX or VPA/LTG combinations are synergistic. RUF, VNS and callosotomy may be helpful for atonic and tonic drop seizures. VNS may also benefit other seizure types. A ketogenic diet may also be considered. Avoid CBZ, PHT, TGB, and clonazepam, which can induce refractory seizures/status epilepticus in this condition.
- ***Prognosis***: Seizures in LGS are often quite refractory to treatment, often with daily seizures in spite of multiple medications. Early onset (<3y), symptomatic etiology, high seizure frequency and multifocal EEG patterns all confer an unfavorable prognosis.

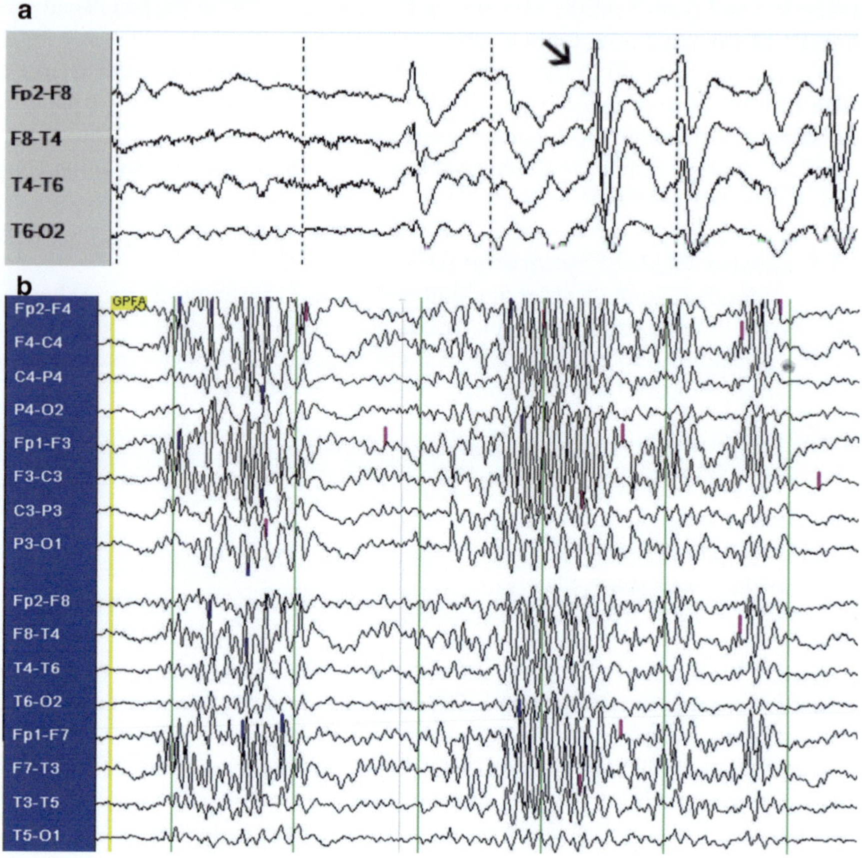

Fig. 2.9 LGS showing (**a**) "slow" spike and wave and (**b**) paroxysmal fast activity (PFA)

2.2.3.7 Developmental and Epileptic Encephalopathy with Spike-Wave Activation in Sleep (DEE-SWAS) and Epileptic Encephalopathy with Spike-Wave Activation in Sleep (EE-SWAS)

DEE-SWAS and EE-SWAS refer to a group of syndromes previously including Continuous Spike Waves during Sleep (CSWS), Landau-Kleffner syndrome (LKS) and atypical Rolandic epilepsy (aka pseudo-Lennox syndrome) [9]. DEE-SWAS occurs in children with pre-existing neurodevelopmental delay and encompasses patients who were previously described as CSWS. EE-SWAS describes children with previously normal neurologic development. LKS and atypical Rolandic epilepsy are two of the subtypes included in EE-SWAS. DEE-SWAS and EE-SWAS are characterized by (1) progressive neuropsychological/behavioral deterioration affecting global cognition or selective functions; (2) motor deficits such as ataxia or unilateral weakness; (3) significant activation of epileptiform discharges in sleep; and (4) epilepsy, including nocturnal focal or generalized seizures, drop seizures (50%), or absence seizures associated with spike-waves.

EEG background changes in which >85% of slow wave sleep is occupied by spikes are termed Electrographic Status Epilepticus in slow wave Sleep (ESES). Although 85% is used as the EEG criteria for ESES, lower percentages may also be associated with neurodevelopmental regression.

- ***Etiology:*** Because DEE-SWAS and EE-SWAS encompass a spectrum of conditions, a variety of etiologies, both genetic and structural, may be associated.

DEE/EE-SWAS Diag. Criteria
Neurobehavioral change
Motor deficits
ESES in sleep (85%)
Epilepsy

Diverse brain lesions can lead to DEE-SWAS. Focal childhood epilepsy syndromes such as SeLECTS and SeLEAS, may have significant sleep-activation of spike-wave discharges and, if associated with neurodevelopmental difficulties, represent a mild transient form of EE-SWAS.

- ***Clinical features:*** Seizures can be focal motor, hemiclonic, complex partial, absence and secondary generalized. Interictal cognitive deficits include multisystem cognitive decline and may include aphasia.
- ***EEG:*** Interictal EEG shows frontal spikes and spike waves. ESES usually consists of 2 Hz (range 1.5–5 Hz) spike-waves in frontal regions activated by slow sleep and interrupted by REM or wakefulness. Amitriptyline can induce slow-wave sleep and enhance EEG findings.
- ***Imaging***: CSWS may have MRI changes such as unilateral atrophy, pachygyria, porencephaly, hydrocephalus, or multilobar polymicrogyria.
- ***Treatment*****:** Standard ASMs may control seizures but not cognitive impairment. Medications with spike-suppressive properties such as LEV, clobazam, and high-dose diazepam may be effective. CBZ may decrease seizures while increasing spike-wave burden and should be avoided. ACTH, corticosteroids, and perhaps IVIG may be effective in improving cognitive outcomes. Multiple subpial transections and VNS have some efficacy. Specific management should also address behavioral changes.
- ***Prognosis*****:** Seizures usually remit in the teenage years. However, long-term neuropsychological/behavioral sequelae persist proportional to the duration of the active condition.

2.2.3.8 Landau-Kleffner Syndrome (LKS)

LKS, also known as "acquired epileptic aphasia," is a subtype of EE-SWAS, in that progressive language dysfunction, along with cognitive and behavioral deterioration in a previously neurotypical child, are associated with EEG findings of ESES.

- ***Etiology:*** Etiology is less often evident in LKS than in DEE-SWAS. Repetitive electrical discharges in the posterior temporal region may produce auditory agnosia, which later leads to acquired aphasia.

- ***Clinical features:*** In LKS, language develops normally in the first 2 years of life. Progressive language regression (peak onset age 3–7, range 2–14, M:F 2:1) is the hallmark, with initial loss of auditory comprehension, described as an "auditory agnosia," followed by paraphasias and phonological errors. Progressive language deficits ensue, often ending in mutism. Seizures may be subtle (eye blinking, head drops, minor automatisms, rare generalization) or absent in 20–30% cases.
- ***EEG*** shows slow waves, spikes, and spike-waves, predominating in the posterior temporal regions. Dipole mapping and MEG reveal that the discharges may be localized to the Heschl's gyrus (superior posterior temporal lobe) on one side, with secondary contralateral involvement. ESES occurs in a lower percentage of LKS than in CSWS.
- ***Imaging*** is typically normal.
- ***Treatment*:** Similar to DEE-SWAS and EE-SWAS above. Speech therapy is also needed.
- ***Prognosis***: Seizures remit by teenage years, although significant language impairments persist, especially if aphasia was present for more than 2–3 years.

2.2.4 Adolescence/Adulthood

2.2.4.1 Juvenile Absence Epilepsy (JAE)

JAE is best understood as an epilepsy subtype having aspects of both CAE and JME- not just in name but also in clinical features. While most patients have onset around puberty (peak 10–12 y, range 10–17 y), another peak may be present at an earlier age, around 6–7 years.

- ***Clinical features***: While the predominant seizure type is absence, GTCS can be seen in 50–80% and myoclonic jerks in 10–15%. The frequency of absence seizures in JAE is lower than in CAE. Absence seizures in JAE are typically not induced by hyperventilation.
- ***EEG***: The GSW in JAE is slightly faster than in classic absence, about 4 Hz, and is less monomorphic.
- ***Treatment***: Broad-spectrum medications that can treat both absence and GTCS are favored. VPA is effective, although side effects can be an issue (e.g., teratogenic and weight gain effects in females). LTG and clobazam are also helpful. LEV can be tried when myoclonus is present. CBZ/OXC can aggravate absence seizures in JAE.
- ***Prognosis***: A good response to VPA occurs, but seizure-free rates are lower than with CAE.

2.2.4.2 Juvenile Myoclonic Epilepsy (JME)

JME is a primary generalized epilepsy syndrome with onset in adolescence ("juvenile" age, median 15 y, F > M).

- ***Etiology***: Often familial, there is a presumed channel defect. Various studies have found associations with calcium channel (*EFHC1*, *CACNB4*), chloride channel (*CLCN2*) and GABA receptor (*GABRA1*) genes.

- ***Clinical features***: Myoclonic jerks (single or multiple jerks), typically on awakening and involving the arms/shoulders, are characteristic of the disease. Specific questioning may reveal unintentional "throwing" of items (e.g., brush, breakfast) and/or early morning clumsiness due to mild myoclonus. Early morning tonic-clonic seizures may also occur. Absence seizures can also occur in JME. Myoclonus is often triggered by sleep deprivation, alcohol use, or photic stimulation.
- ***EEG*** shows "fast" (faster than the typical 3 Hz of CAE) GSW (4–6 Hz, Fig. 2.10), at times provoked by photic stimulation. Such GSW or polyspike waves occur with myoclonic seizures.
- ***Imaging***: MRI is normal. Volumetry has revealed the thickened gray matter of the mesial frontal cortex. Metabolic changes have also been shown in the frontal cortex using PET and MRS.
- ***Treatment:*** JME generally demonstrates an excellent response to VPA. Other medication options, including LEV, BRV, TPM and ZNS, should be considered. LTG can also be effective, though this may exacerbate myoclonus in some patients. Other Na^+ channel blockers and GABAergic ASMs can also exacerbate seizures. Myoclonus has a good response with VPA, LEV, BRV and clonazepam. Good sleep habits and avoidance of alcohol may also be essential for seizure control. Polarized glasses may be helpful in those with photosensitivity.
- ***Prognosis:*** The prognosis for seizure control is excellent. However, drug withdrawal is successful only in 10–20% of patients, and therefore continued lifelong therapy is often required.

2.2.4.3 Epilepsy with GTCS

This condition occurs in later childhood or early adulthood (10–25 y).

- ***Clinical features***: Generalized tonic-clonic seizures without an aura, usually in the early morning or following a nap, are characteristic.
- ***Etiology***: GTCS on awakening has shown linkage to the EJM1 locus on chromosome 6.
- **EEG** is normal or has less well-defined GSWs. Photosensitivity occurs in 13%.
- ***Treatment*** is with VPA, LTG, LEV, BRV, TPM, CBZ, OXC or PHT. Avoid CBZ, PHT, and OXC if patients have myoclonic jerks or absences.
- ***Prognosis*** is good with fairly easy seizure control in about 75%. Remission is variable. Recurrence on stopping treatment is likely in GTCS on awakening.

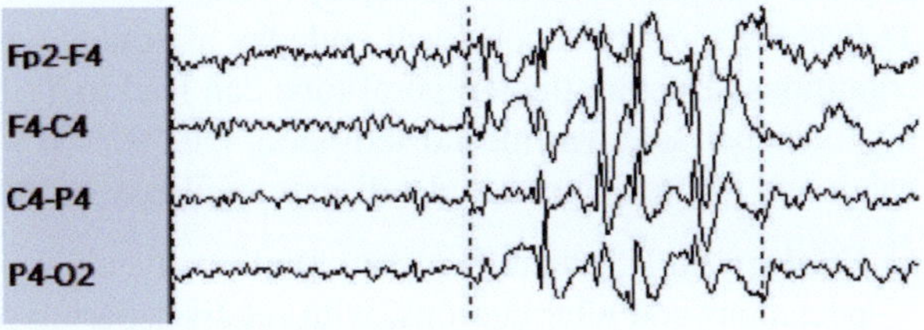

Fig. 2.10 JME showing "fast" spike waves

2.2.5 Distinctive Constellations

2.2.5.1 Rasmussen Encephalitis

Rasmussen Encephalitis is a childhood onset (median 7 y, range 1–14 y, M = F) progressive drug-resistant epilepsy with hemispheric atrophy, intellectual decline and hemiparesis.

- ***Etiology***: Generally idiopathic, about 50% report an infectious or inflammatory episode approximately 6 months prior to onset.
- ***Clinical features***: Seizures are the first manifestation in a previously healthy child. Progressive neurological deficits commence within a year. Seizures are typically focal motor, suggesting a *fronto-central onset*. Epilepsia partialis continua (EPC), in which persistent twitching of one side of the face or one extremity with preserved consciousness, is common and occurs in more than 50%. Complex partial and secondary generalized seizures can also occur. Later development of sensory (somatosensory, visual, auditory) seizures suggests a *posterior migration* of the disease from the frontocentral regions. RE has been classified into 3 stages: (1) seizure onset, (2) fixed hemiparesis, and (3) "burnt-out" residual stage after completion of neurologic deterioration.
- ***Etiology:*** Pathology shows inflammatory changes, including perivascular lymphocytic cuffing, microglial nodules, neuronal loss, and gliosis, suggesting a potential inflammatory origin.
- ***EEG*** shows focal slowing, attenuation and epileptiform discharges. Seizures can appear non-localizable, lateralized, or generalized. Seizures may not have a clear scalp EEG correlation, particularly with EPC. However, a combination of EEG and clinical features can sufficiently lateralize the abnormal hemisphere in 90%.
- ***Imaging*** is initially normal. The neuroimaging hallmark of Rasmussen Encephalitis is progressive atrophy of one hemisphere, involving both gray and white matter, starting in the temporal regions. Fronto-insular or fronto-central regions follow. Basal ganglia involvement occurs, particularly in the caudate and putamen. SPECT and PET may show concordant changes. MR spectroscopy shows reduced NAA and increased choline, lactate, myoinositol and glutamine. Blood and CSF may inconsistently show anti-GluR3 antibodies.
- ***Treatment*** is difficult as ASMs are often ineffective. Limited transient benefit may be provided by immunosuppression with IVIG, plasmapheresis, or high-dose steroids. Anatomic or functional hemispherectomy is the only definitive treatment that provides longstanding improvement or prevents progression.

2.2.5.2 Temporal Lobe Epilepsy (TLE)

TLE is the most common focal epilepsy affecting adolescents and adults. Several conditions affecting the temporal lobe can lead to TLE. Mesial TLE (mTLE) with hippocampal sclerosis/mesial temporal sclerosis (HS/MTS) is the most common cause and is often considered a discrete epilepsy syndrome (MTLE-HS).

- ***Etiology***: In HS, there is preferential involvement of the CA1 region of the hippocampus and hilar neurons, with relative sparing of CA2 neurons. Mossy fiber sprouting is typically seen in HS. Prolonged febrile seizures in early childhood predispose to MTLE-HS. Cortical dysplasia has been recognized more recently as a significant cause of TLE, on its own or in association with HS. Low-grade gliomas such as gangliocytomas have a predisposition for the temporal lobe and may cause TLE. With persistent seizures for many years, involvement of the contralateral hippocampus can occur, leading to independent epileptogenic foci from bilateral temporal regions.

- ***Clinical features***: Focal impaired awareness seizures (FIAS, aka "complex partial" or dyscognitive) are the most common seizure type, often with premonitory auras with preserved awareness. Progression to secondary generalized convulsive seizures can occur (see Chap. 8 for details). Though responsive to ASMs in a majority of cases, the clinical course can be stuttering with the development of pharmacoresistance as the course progresses. Mesial temporal involvement characteristically leads to memory deficits, although a variety of neurobehavioral changes have been noted because TLE is a network disease.
- ***EEG***: The interictal EEG shows anterior temporal sharp waves/spikes. Bilateral discharges are seen to be up to 30%. Ictal EEG shows an evolution of characteristic 5 Hz or faster (theta/alpha) rhythmic activity beginning in one temporal region [10].
- ***Imaging***: Hippocampal sclerosis is diagnosed by findings of focal hippocampal atrophy (best seen in T1 sequences) and signal hyperintensity/loss of internal architecture (best seen in T2/FLAIR sequences). Secondary atrophy along the limbic circuit (fornix, mammillary body, and thalamus) may also be seen. Isolated dilatation of the ipsilateral temporal horn is not significant without these other changes. PET scans show ipsilateral hippocampal hypometabolism. SPECT and MEG may also be used to diagnose during the presurgical workup. The MRI and PET appearance of MTS are shown in Chap. 10.
- ***Treatment/prognosis***: Several of the available ASMs are effective in treatment, though 20–30% of cases are medication refractory. For drug-resistant cases, anterior temporal lobectomy offers a remission rate of 60–70% [11]. Post-surgical seizure control with HS in MRI and concordant ictal EEG may approach 90%. Bilateral EEG discharges, bilateral PET findings, normal MRI, and discordance between functional and imaging data are all associated with less optimal outcomes.

2.3 Other Epilepsies

2.3.1 Hemiconvulsion-Hemiplegia-Epilepsy Syndrome

This is a condition characterized by three phases: (1) hemiconvulsions in the setting of a febrile illness in the very young age (6 m–2 y), which can last from 1 h to days; (2) persistent hemiplegia lasting more than 7 days (compared with a typical Todd's paralysis, which is <48 h) and residual persistent weakness; and (3) development of focal seizures.

2.3.2 Febrile Seizures (FS)

FS are benign seizures associated with fever without evidence of intracranial pathology, most (90%) occurring between the ages of 6 m–3 y (range 3 mo–5 y, F > M).

- ***Etiology***: FS are 2–3 times more common with affected family members (risk for siblings> offspring> nieces/nephews). Asians have a higher risk than Westerners. The fever may be caused by any childhood illnesses, particularly with Shigellosis or with DPT and measles immunization. Linkage studies show an association with several sodium channel genes. *FEB3* mutations have a higher risk of later epilepsy.
- ***Clinical features***: Seizures typically occur during the rising phase of fever early in the course. However, temperature is not directly correlated to the incidence of FS. **Simple febrile seizures** (80–90%) last less than 15 min and lack focality. **Complex febrile seizures** are associated with the following features: duration >15 min, focal semiology, recurrence within 24 h, abnormal neurological exam, and afebrile seizures in first-order relatives. The risk for later afebrile seizures is greater with complex FS (6% with 2 or more risk factors). Febrile status epilepticus is the most common cause of status epilepticus in children.
- ***EEG*** shows paroxysmal changes and slowing within 24 h. EEG lacks prognostic significance.
- ***Investigations*** are not always needed with simple FS. Age under 6 months should prompt CSF examination to rule out underlying bacterial meningitis. Complex FS are more likely to need investigations such as CSF or neuroimaging. Febrile status epilepticus appears to increase the risk of developing hippocampal sclerosis later in life.
- ***Treatment***: Parents are counseled about the prompt use of antipyretics and tepid sponging to control fever. Prophylactic ASMs are not typically recommended, as there is a low incidence of morbidity/mortality, and prevention of recurrent FS is not known to reduce the risk of later epilepsy. Prolonged episodes can be terminated by rectal diazepam administration at home or lorazepam/phenobarbital by medical personnel.
- ***Prognosis*:** The recurrence risk for FS is 33%, which typically occurs within 1 year of the initial event. Age is the most important predictor of recurrence, with a higher risk associated with onset age less than 1 year (50% risk). Multiple risk factors lead to a higher recurrence risk (30% with 2 risk factors, 60% with 3 risk factors). Between 2–5% have later afebrile seizures, particularly in complex FS with risk factors.

2.3.3 Epilepsia Partialis Continua (EPC) (Kojevnikov Syndrome)

EPC is a partial motor status epilepticus consisting of myoclonic jerks in a localized body region. EPC can be non-lesional or associated with a brain lesion, as seen characteristically with Rasmussen syndrome. Development of EPC should prompt investigation for an underlying lesion. **EEG** can show regional spiking. Ictal scalp

EEG can also be normal because the region of the cerebral cortex recruited during the seizure is too small. At least 6 square centimeters [2] of the cortex must be involved for the resulting EEG changes to be visible on scalp recordings.

2.3.4 Supplementary Sensori-Motor Area (SSMA) Seizures

SSMA seizures are brief (10–40 s), associated with rapid asymmetric tonic posturing of extremities with preserved consciousness and no postictal confusion. The SSMA is the secondary zone for motor and sensory representation for the body and is often the "symptomatogenic zone" in parasagittal cortex seizures. The involvement of the sensory regions can cause an aura of tension/movement or the desire to move the contralateral body. Motor involvement can lead to a variety of bilateral posturing/movements, the most well-known, albeit not very common, of which is the "fencing" posture with contralateral arm extension, ipsilateral arm flexion, and head deviation contralaterally as if looking toward the extended arm. It has also been called the "M2e" to describe the contralateral tonic abduction and external rotation of the shoulder with flexion of the elbow, with or without head turning. Bilateral complex leg movements in the form of "bicycling" can also occur. Spread to primary motor areas can cause clonic movements of the contralateral body. **EEG:** Midline interictal epileptiform discharges (IEDs) may be seen, which may be quite difficult to distinguish from vertex waves. Clues to differentiate midline IEDs from vertex waves include (a) the presence of after-going slow waves, (b) polyspikes, or (c) consistent lateralization of the vertex-like wave. Considerable EMG artifacts often obscure SSMA seizures.

2.4 Traditional Classification

The traditional classification of epilepsy syndromes prior to the 2010 classification is shown in Fig. 2.11. Syndromes associated with a disturbance of brain function (cognition, behavior, motor, language etc.) thought to result from the epileptiform abnormality are sometimes called "epileptic encephalopathies." The prototype is continuous spike-wave in sleep/Landau Kleffner syndrome (CSWS/LKS). Others include LGS, Dravet syndrome, Rasmussen's encephalitis, and the Hemiconvulsion-Hemiplegia-Epilepsy syndrome.

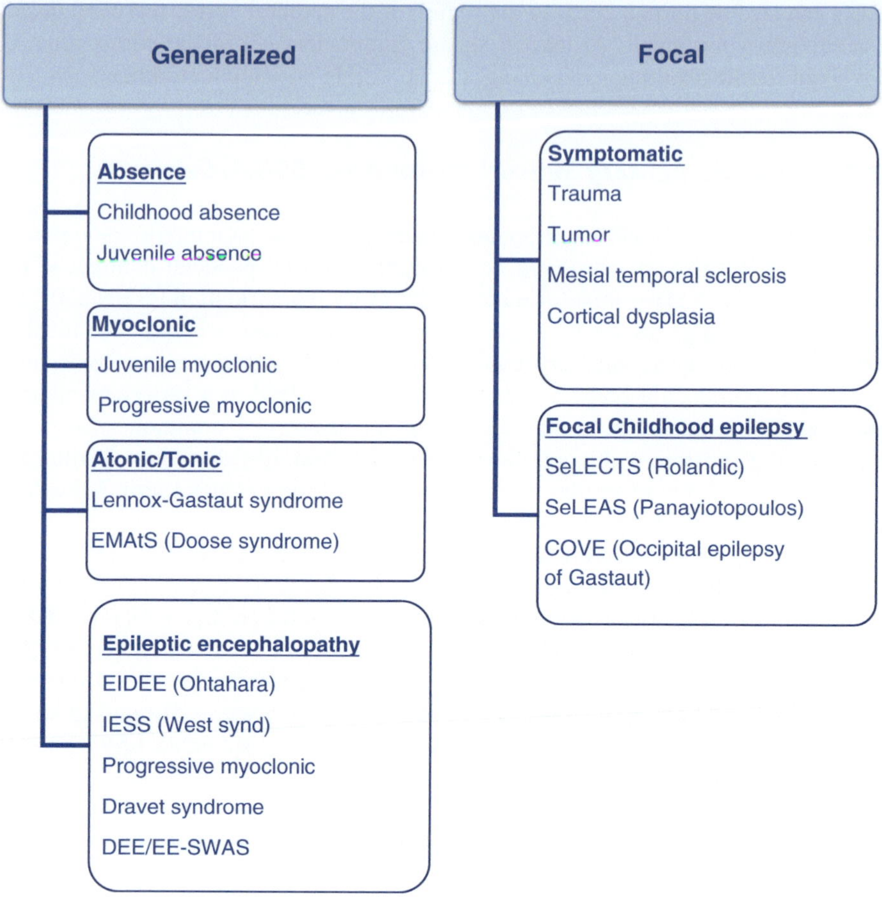

Fig. 2.11 Traditional classification of epilepsy syndromes

2.5 Seizure Mimics

Many paroxysmal events can mimic seizures (Fig. 2.12). Based on video-EEG monitoring and clues from the semiology of the events, one can usually distinguish between epileptic and nonepileptic events. The etiology of nonepileptic events can be further subdivided into physiologic and psychogenic based on history, semiology, and diagnostic tests.

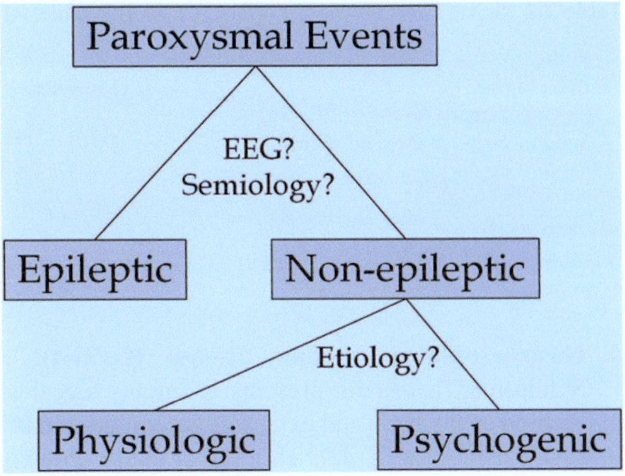

Fig. 2.12 Algorithm for evaluation of paroxysmal events

2.6 Physiologic Nonepileptic Events

1. **Syncope:** Loss of consciousness occurs due to reduced blood flow to the brain. In contrast to epilepsy, the trigger is often identifiable, there is a feeling of lightheadedness (pre-syncope) prior to the event, and the loss of consciousness is typically brief (<20 s) with rapid return to baseline [12]. Convulsions, if present, are generally brief, lasting seconds. The scoring system in Table 2.2 can help differentiate seizures from syncope.
2. **Acute confusional migraine: P**atients describe a typical migraine aura followed by headache and confusion. They often do not recur and are usually present in patients with a history of migraine with aura.
3. **Breath-holding spells (BHS):** These occur typically between 6 months and 6 years old. With prolonged breath-holding, the child may lose consciousness and fall, with subsequent tremors or convulsions. They are associated with bradycardia and cerebral hypoperfusion. *Cyanotic BHS* is more common. In response to an emotional trigger, the child will cry and suddenly stop breathing, become cyanotic, and lose consciousness, followed by unresponsiveness for 1–2 min with rapid resumption of consciousness afterward. *Pallid BHS* is less common, following minor trauma or surprise. There is loss of consciousness and convulsions associated with profound bradycardia or asystole. Serum ferritin levels should be assessed, as BHS may be associated with iron-deficiency anemia. Iron supplementation may decrease the frequency of breath-holding events, even with normal ferritin levels. In most cases, parental reassurance and behavioral modification provide sufficient treatment. Events generally resolve by 5–6 years old.

Table 2.2 Seizure versus syncope (score ≥ 1 likely seizure, <1 likely syncope) [13]

Feature	Points
Tongue biting	2
Déjà vu or jamais vu	1
Emotional stress associated with LOC	1
Head turning during a spell	1
Lightheaded spells	−2
Sweating before spell	−2
Spell associated with prolonged sitting or standing	−2

4. **Gastroesophageal reflux disease (GERD):** Also known as "Sandifer Syndrome," it usually presents in infants less than one-year-old with episodic arching of the back and extension/lateral flexion of the head, often with feeding. It can also occur in older children or adults with cognitive impairment.
5. **Self-stimulatory behavior:** Common in neurologically impaired children, this includes rhythmic hand shaking, body rocking, and head swaying, while remaining relatively unresponsive. **Rett Syndrome** should be considered in young girls with repetitive "hand washing" movements. Masturbatory behavior in infants, including rocking back and forth, may also mimic seizures.
6. **Shuddering attacks:** These start as early as 3–4 months but gradually decrease by 10–11 months. They resemble shivering, occurring both during wake and sleep states, described as a rapid tremor involving the head, arms, trunk or lower extremities.
7. **Benign myoclonus of early infancy:** Resembles infantile spasms but occurs only during the awake state, and, in contrast to West syndrome, the developmental milestones and EEG are normal.

2.7 Psychogenic Nonepileptic Seizures (PNES)

PNES is further discussed in the chapter on neuro-psychiatric co-morbidities.

PNES mimics: Just as PNES mimics seizures, there are a few seizures that have atypical movements suggesting PNES to the untrained eye. Examples include frontal lobe epilepsy and supplementary sensori-motor area (SSMA) seizures, which may have asymmetric movements (see above) with preserved consciousness and little, if any, postictal state.

Pearls: Epilepsy Syndromes

Absence seizures are often provoked by hyperventilation. If a typical staring event can be induced by 2 min of hyperventilation, one can probably start treatment even before obtaining a confirmatory EEG.

In SeLEAS (**Doose** syndrome), a characteristic parietal theta rhythm (4–7 Hz) has been described, which persists in wakefulness and can persevere after seizures remit.

Parents may only witness the aftermath of a focal seizure from **SeLECTS**. They may describe only unilateral face ± arm weakness ± drooling. Think of SeLECTS in your differential for a possible transient ischemic attack in a child!

The American Academy of Pediatrics recommends no imaging or EEG in children who present with a simple **febrile seizure** (generalized seizure in the setting of a fever, lasting less than 15 min without recurrence in 24 h.

It has been said that GTCS in **Dravet** syndrome may respond to bromides when all else fails. One of the first ASMs developed, bromides, is still used in veterinary medicine and may be available from the veterinary pharmacy.

A "**teddy bear sign**" has been described, where the presence of a stuffed animal in the video EEG suite suggests the diagnosis of NES. However, this finding has been subsequently challenged [14].

Perisylvian polymicrogyria syndrome is a distinct syndrome where patients often have swallowing difficulties along with the expected developmental delay and seizures.

References

1. Zuberi SM, et al. ILAE classification and definition of epilepsy syndromes with onset in neonates and infants: position statement by the ILAE task force on nosology and definitions. Epilepsia. 2022;63:1349–97.
2. Beal JC, Cherian K, Moshe SL. Early-onset epileptic encephalopathies: Ohtahara syndrome and early myoclonic encephalopathy. Pediatr Neurol. 2012;47:317–23.
3. Gunthorpe MJ, Large CH, Sankar R. The mechanism of action of retigabine (ezogabine), a first-in-class K+ channel opener for the treatment of epilepsy. Epilepsia. 2012;53:412–24.
4. Lee H-F, Chi C-S, Tsai C-R, Chen C-H, Wang C-C. Electroencephalographic features of patients with SCN1A-positive Dravet syndrome. Brain and Development. 2015;37:599–611.
5. Glauser TA, et al. Ethosuximide, valproic acid, and lamotrigine in childhood absence epilepsy: initial monotherapy outcomes at 12 months. Epilepsia. 2013;54:141–55.
6. Nickels K, Kossoff E, Eschbach K, Joshi C. Epilepsy with myoclonic-atonic seizures (Doose syndrome): clarification of diagnosis and treatment options through a large retrospective multicenter cohort. Epilepsia. 2021;62:120.
7. Li B, Tong L, Jia G, Sun R. Effects of ketogenic diet on the clinical and electroencephalographic features of children with drug therapy-resistant epilepsy. Exp Ther Med. 2013;5:611–5.
8. Verrotti A, et al. Typical and atypical rolandic epilepsy in childhood: a follow-up study. Pediatr Neurol. 2002;26:26–9.

9. Specchio N, et al. International league against epilepsy classification and definition of epilepsy syndromes with onset in childhood: position paper by the ILAE task force on nosology and definitions. Epilepsia. 2022;63:1398.
10. Risinger MW, Engel J, Van Ness PC, Henry TR, Crandall PH. Ictal localization of temporal lobe seizures with scalp/sphenoidal recordings. Neurology. 1989;39:1288–93.
11. Wiebe S, Blume WT, Girvin JP, Eliasziw M. A randomized, controlled trial of surgery for temporal-lobe epilepsy. N Engl J Med. 2001;345:311–8.
12. Anon. Epilepsy: a comprehensive textbook. Philadelphia: Wolters Kluwer Health/Lippincott Williams & Wilkins; 2008.
13. McKeon A, Vaughan C, Delanty N. Seizure versus syncope. Lancet Neurol. 2006;5:171–80.
14. Cervenka MC, et al. Does the teddy bear sign predict psychogenic nonepileptic seizures? Epilepsy Behav. 2013;28:217–20.

Normal EEG

3

Mostafa Hotait, Atul Maheshwari, and Zulfi Haneef

The normal adult awake EEG is defined by a frequency-amplitude gradient (faster low amplitude rhythm anteriorly and slower high amplitude rhythm posteriorly) and a posterior dominant rhythm (PDR) between 8–12 Hz, best seen with eye closure (Fig. 3.1).

Age	PDR (Hz)
3 months	3
12 months	6
3 years	8
9–10 years	9

In a bipolar montage, eye closure creates a downward deflection in anterior leads due to the relatively electropositive cornea moving closer to the frontopolar leads (*Bell's* phenomenon). The *squeak* rhythm is where the PDR appears faster with eye closure, as seen below.

Starting at 3 months, the PDR normally increases, starting at 3–4 Hz and then gradually increasing up to 9–11 Hz, reaching a peak between 13–19 years old.

M. Hotait
Louisiana State University Health Shreveport, Shreveport, LA, USA

A. Maheshwari
Department of Neurology, Baylor College of Medicine, Houston, TX, USA

Z. Haneef (✉)
Department of Neurology, Baylor College of Medicine & Neurology Care Line, Houston VA Medical Center, Houston, TX, USA
e-mail: Zulfi.Haneef@bcm.edu

Z. Haneef (ed.), *Epilepsy Fundamentals*,
https://doi.org/10.1007/978-3-031-77741-7_3

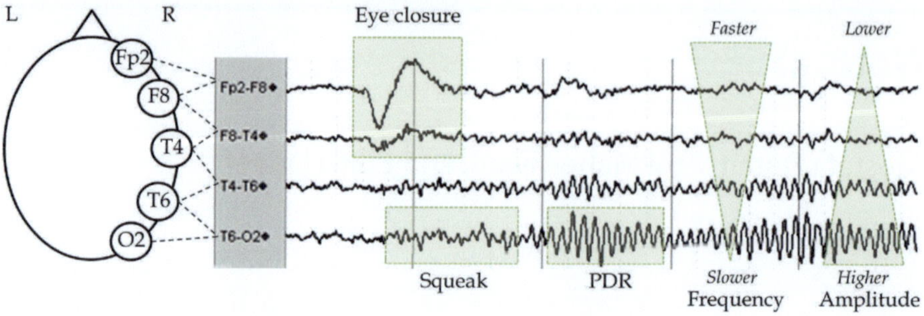

Fig. 3.1 The normal awake EEG includes high amplitude deflections in the anterior leads consistent with eye closure, a posterior dominant rhythm (PDR) and a frequency-amplitude gradient best appreciated when the patient's eyes are closed

Table 3.1 EEG features in different stages of sleep

	Stage 1	Stage 2	Slow-Wave Sleep (SWS)	Rapid Eye Movement (REM)
EEG Features	POSTS Vertex waves	POSTS Vertex waves Sleep spindles K-complexes	Delta waves Occupy >20% of the background rhythm	Rapid eye movements; loss of muscle tone; "Saw-tooth" waves

3.1 The Drowsy and Sleep EEG

During drowsiness in adults, the posterior dominant (alpha) rhythm breaks up and then disappears; there can be an increase in beta activity and enhancement of generalized theta activity. **Sleep architecture is best visualized in the midline leads; positive occipital sharp transients of sleep (POSTS) are seen best in the posterior leads** (Table 3.1 and Figs. 3.2, 3.3, 3.4 and 3.5).

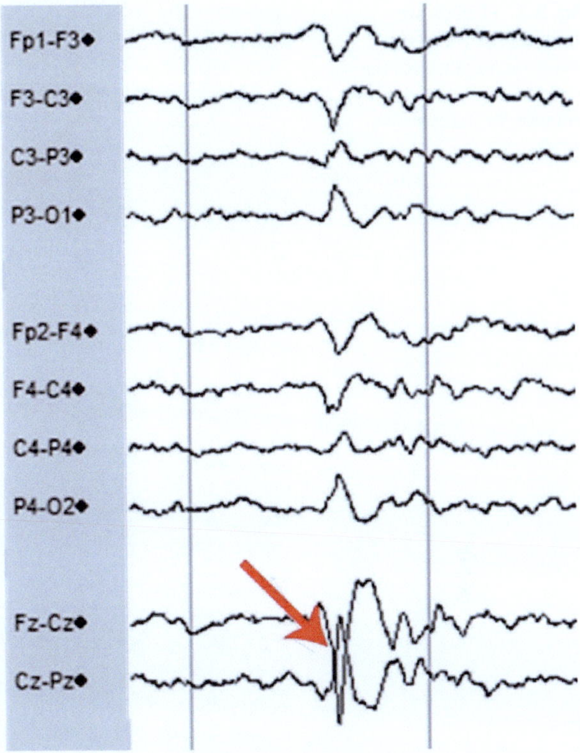

Fig. 3.2 Seen in Stage 1 sleep, vertex waves are 200 msec diphasic sharp transients. Lateralization toward one side can be abnormal

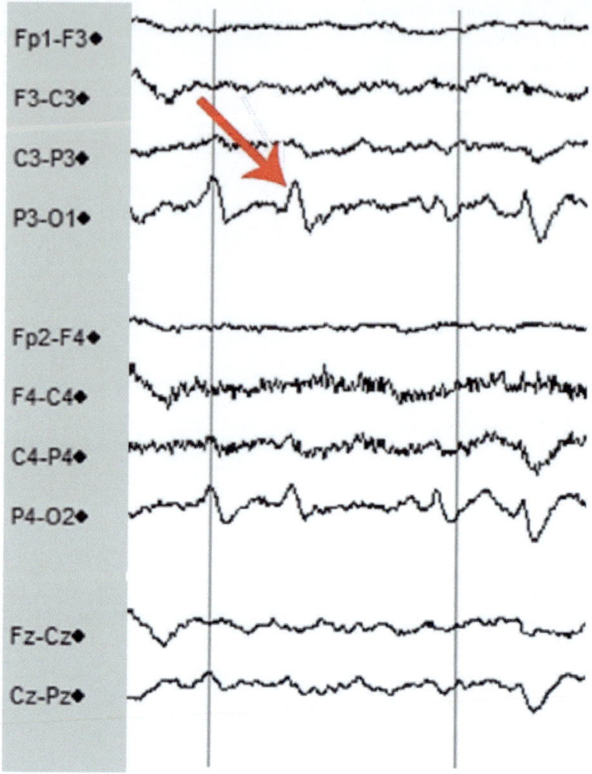

Fig. 3.3 POSTS are surface-positive sharp waves occurring over the occipital regions either in isolation or in repetitive bursts

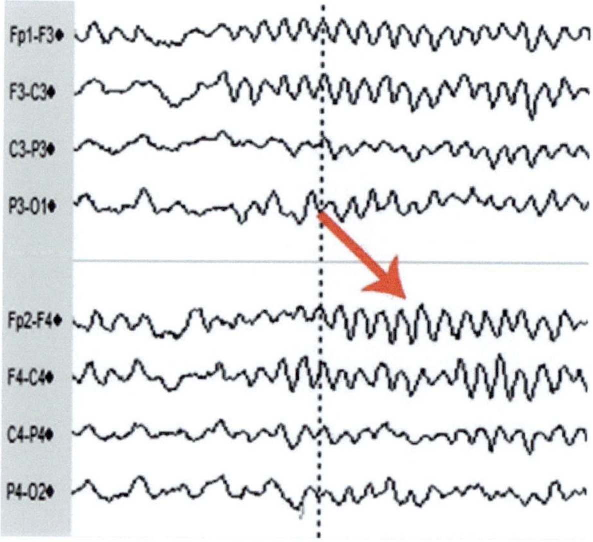

Fig. 3.4 Sleep spindles are transient 1–2 s sinusoidal waveforms with a frequency of 12–14 Hz centrally

Fig. 3.5 A K-complex is a high-amplitude diphasic wave, can occur with a sleep spindle, and can be provoked by an auditory stimulus

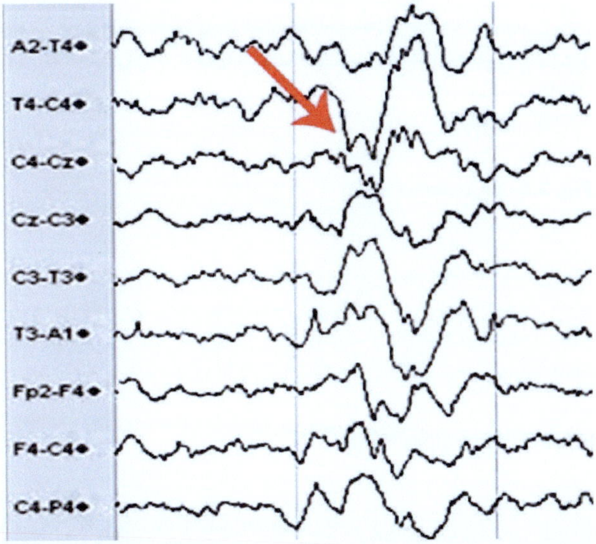

3.2 Normal/Benign Variants

Normal or benign variants are EEG patterns that appear abnormal but are within the range of normal variation. The electroencephalographer should be vigilant about identifying normal/benign variants because misinterpretation as an abnormality can lead to inappropriate patient management. These are described below:

Fast Alpha variant (Fig. 3.6): Here, the posterior leads show a supraharmonic frequency double the true PDR (arrow in Figure). This may periodically blend with the true PDR (seen to the right of the image), giving a notched or bifurcated appearance. The amplitude generally decreases, which can seem to make the frequency and amplitude gradient disappear.

Slow Alpha variant (Fig. 3.7): In contrast to the pattern above, in the slow alpha variant, the posterior leads show a subharmonic frequency of the true PDR (arrow), which is typically half the normal PDR (seen to the right of the image). Like the fast alpha variant, the appearance can be notched. Both alpha variants should remain reactive to eye-opening and closure.

Rhythmic Mid-Temporal Theta of Drowsiness (*RMTD,* Fig. 3.8): Previously called the "psychomotor variant," this is currently believed to be a benign pattern of sharply contoured theta activity that may mimic seizure activity. However, these are brief and without spatial or temporal evolution. RMTDs can occur bilaterally, simultaneously or alternating in the left and right temporal regions.

Midline Theta Rhythm (*or Ciganek rhythm,* Fig. 3.9): Originally described by Ciganek in patients with temporal lobe epilepsy, these are now considered benign. The pattern consists of waxing and waning 4–7 Hz theta rhythm expressed over the vertex/midline region. Typically, this is seen in the awake and drowsy states without any clinical correlate.

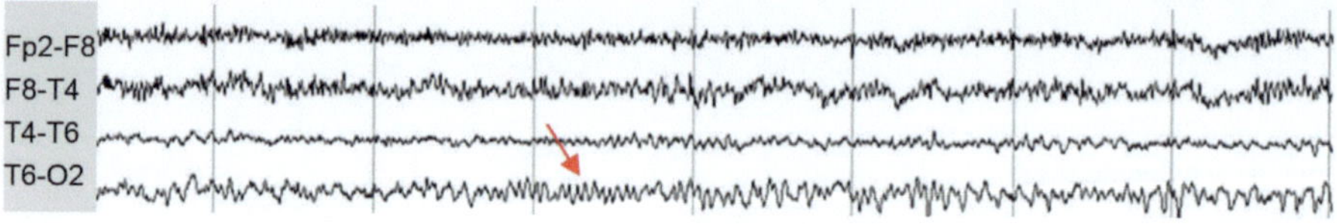

Fig. 3.6 Fast alpha variant

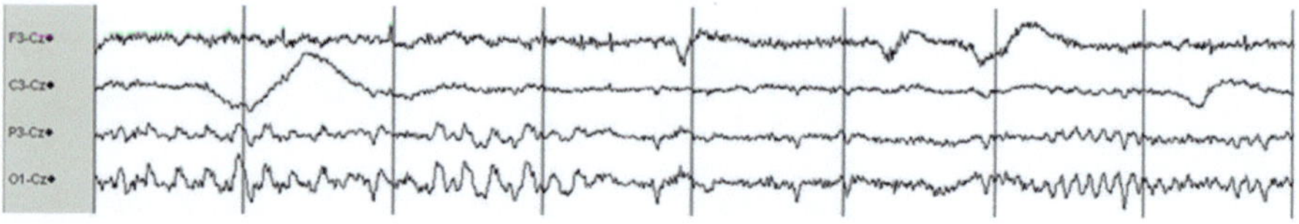

Fig. 3.7 Slow alpha variant

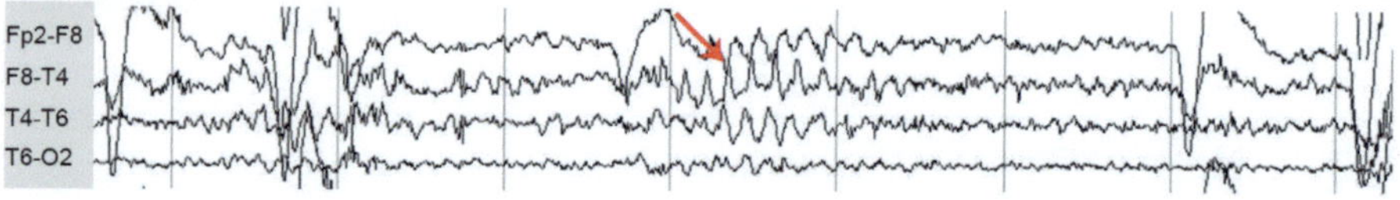

Fig. 3.8 Rhythmic Mid-Temporal Theta of Drowsiness (RMTD)

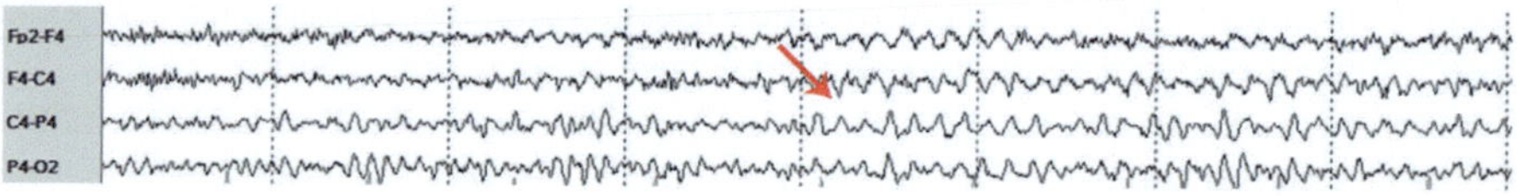

Fig. 3.9 Midline Theta Rhythm of Ciganek

14-and 6-Hz positive bursts (Fig. 3.10): Also called "ctenoids" (Greek: comb), these bursts consist of a spindle-shaped architecture over the posterior temporal regions, lasting 0.5–1 s. Most bursts commonly have 14 Hz and/or 6 Hz sharply contoured waveforms, forming the appearance of a "comb." The 14 & 6 bursts are almost exclusively seen in children (8–14y) during drowsiness and light sleep. They are absent in wakefulness and deeper sleep.

Benign Epileptiform Transients of Sleep (*BETS*) (Fig. 3.11)*:* Also called small sharp spikes (SSS), these waves are small (low voltage, <50 μV) and sharp (short duration, <50 msec). They disappear in wakefulness and slow-wave sleep states.

Wickets (Fig. 3.12): These are seen most commonly in adults over 30 years old, usually occur with a high amplitude (up to 200 μV), have a frequency of 6–11 Hz, and can be unilateral or bilateral and independent. The morphology comprises sharply contoured waves in singlets or brief bursts seen over the temporal regions during drowsiness and light sleep.

Lambda waves (Fig. 3.13): Like POSTS (see above), these waveforms occur in the posterior/occipital regions and have a similar morphology. However, lambda

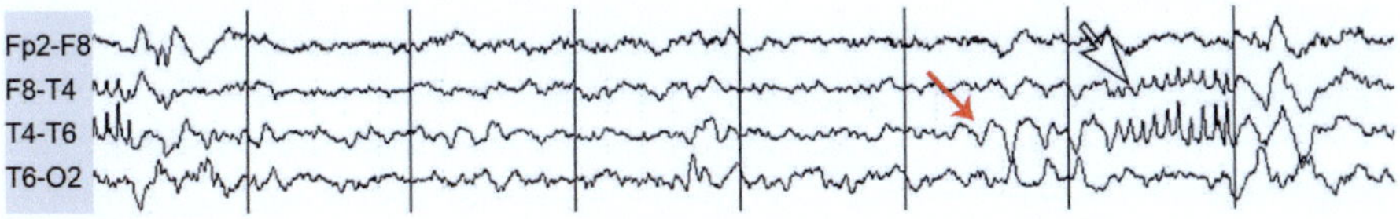

Fig. 3.10 14-and 6-Hz positive bursts. White arrow shows 14 Hz and red arrow shows 6 Hzcomponents

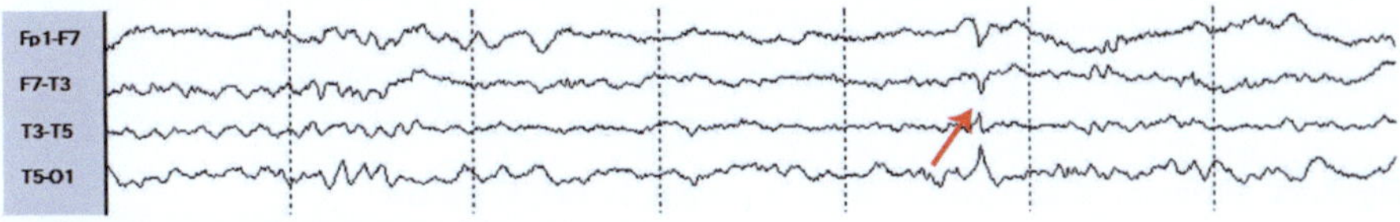

Fig. 3.11 Benign epileptiform transients of sleep

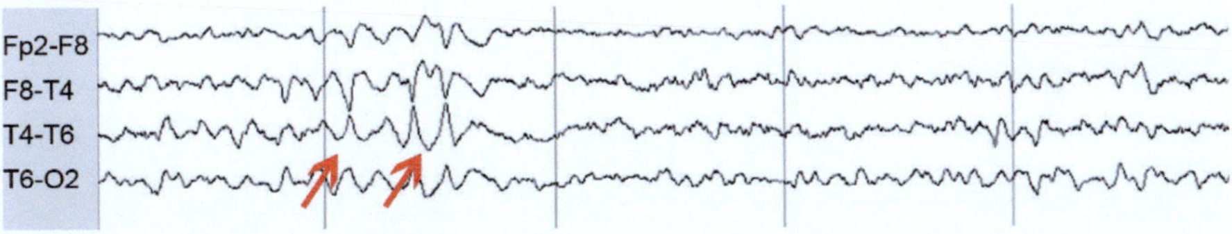

Fig. 3.12 Wickets

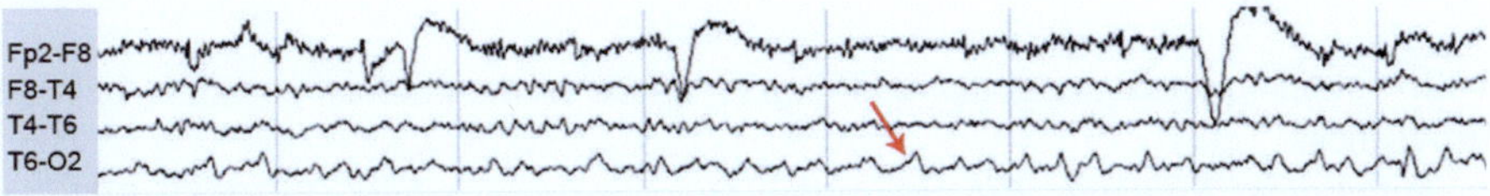

Fig. 3.13 Lambda waves

waves occur in the awake state and correlate with visual scanning. The eye blinks in the frontal leads (see image) also indicate that this patient is in the awake state.

6-Hz phantom spike-wave (Fig. 3.14): This pattern occurs in adolescents/adults during relaxed wakefulness and stage I sleep and disappears during deeper sleep (as opposed to pathological spike-waves, which increase in sleep). They are bisynchronous bursts lasting 1–2 s and consist of a small spike, with an average amplitude of 25 uV and duration <30 ms (spike may be difficult to see) followed by a larger slow wave. Two patterns have been described: WHAM (wakefulness, high amplitude, anterior predominance, males) and FOLD (females, occipital, low amplitude, drowsiness). Of these, WHAM is more likely to be epileptiform and FOLD tends to be benign [1].

Hypnagogic hypersynchrony (Fig. 3.15): Generally seen in children, this pattern consists of 3–5 Hz centrally predominant paroxysmal bursts in drowsiness, often

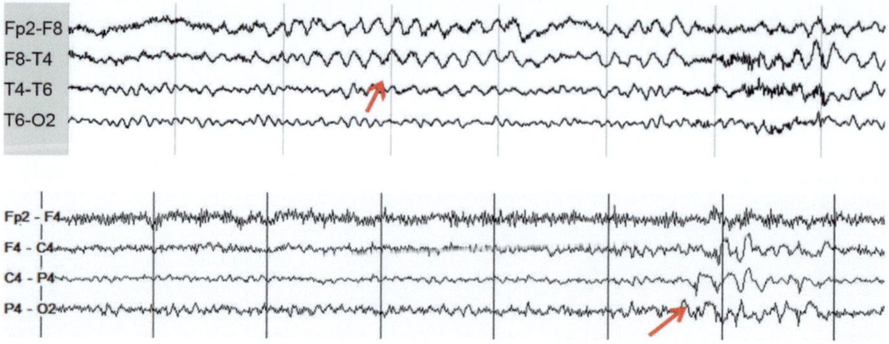

Fig. 3.14 Phantom spike-waves A. WHAM pattern; B. FOLD pattern

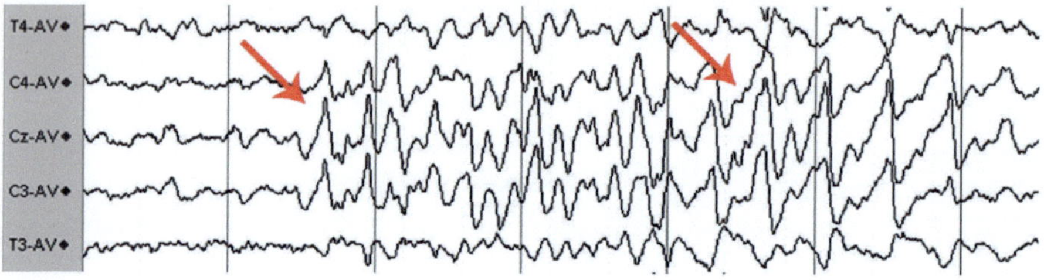

Fig. 3.15 Hypnagogic hypersynchrony

with voltage approaching up to 350 μV. There is no clinical correlate beyond drowsiness.

Subclinical Rhythmic Electrographic Discharges in Adults (*SREDA*): SREDA appear similar to an epileptic rhythm with the progression of frequencies in the delta-theta range, distributed broadly across parietal/posterior temporal regions, and can last 1–5 min, but without clinical correlate. SREDA is most common in patients over age 50 and is usually seen in wakefulness.

3.3 Artifacts

Much of what a beginner in EEG learns is how to differentiate the abnormal EEG from abnormal -appearing EEG findings caused by artifacts. Several typical EEG artifacts are described below (Figs. 3.16, 3.17, 3.18, 3.19, 3.20, 3.21 and 3.22).

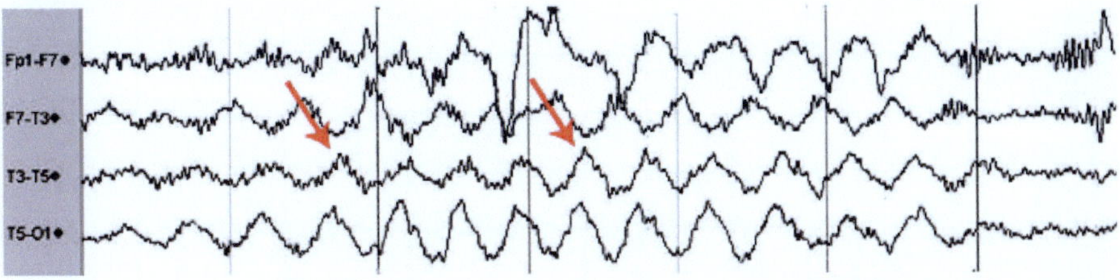

Fig. 3.16 Movement artifact (shaking legs in the air in this example) simulating seizure: multiple phase reversals (at T3 and F7 here), sudden onset and offset, and lack of post-ictal slowing can sometimes help to differentiate between movement artifact from electrographic seizure activity

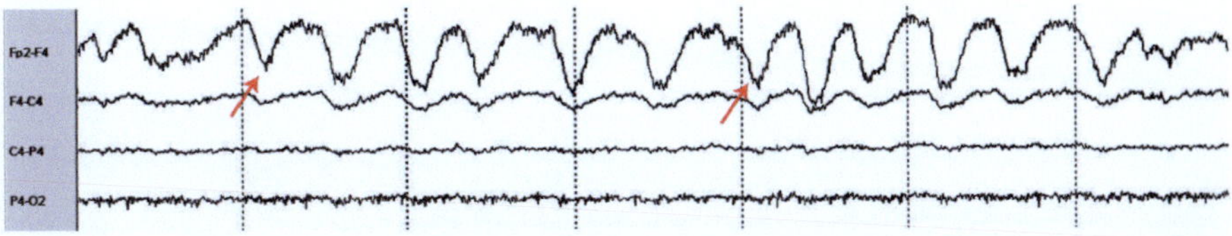

Fig. 3.17 Eyelid flutter: note the high-amplitude 2–3 Hz oscillations that are largely restricted to the frontopolar leads. These are generally bilaterally symmetric, created by the Bell's Phenomenon where the eyeballs move up with eye closure leading to a positive deflection in the anterior leads as the cornea is relatively positive compared to the retina

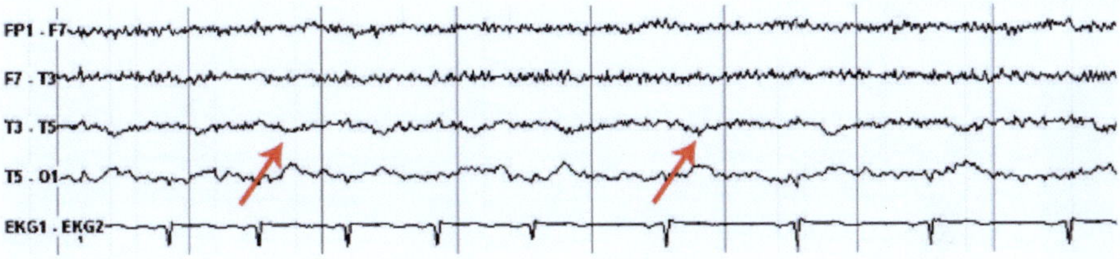

Fig. 3.18 Pulse artifact: the mechanical movement of a lead near an artery can produce slow waves mimicking focal cortical slowing. See here how they immediately follow the EKG potentials

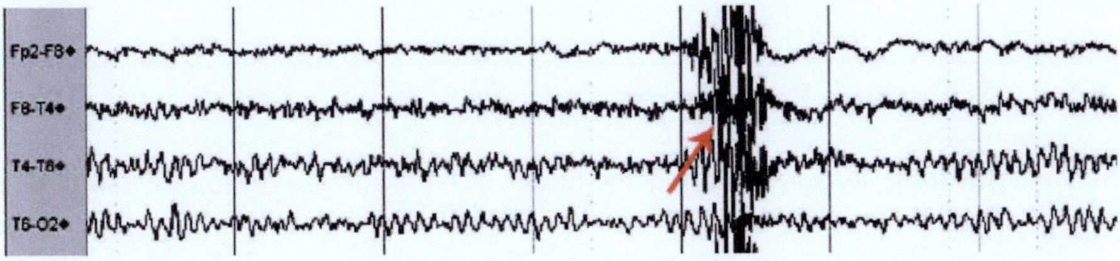

Fig. 3.19 Swallowing artifact: This artifact is generally widespread over many leads and is composed of a mix of fast and slow waves. It is created by a combination of pharyngeal muscle myogenic potentials (EMG artifact, fast activity) as well as tongue movement artifact due to the tongue's inherent dipole (glossokinetic artifact, slow activity)

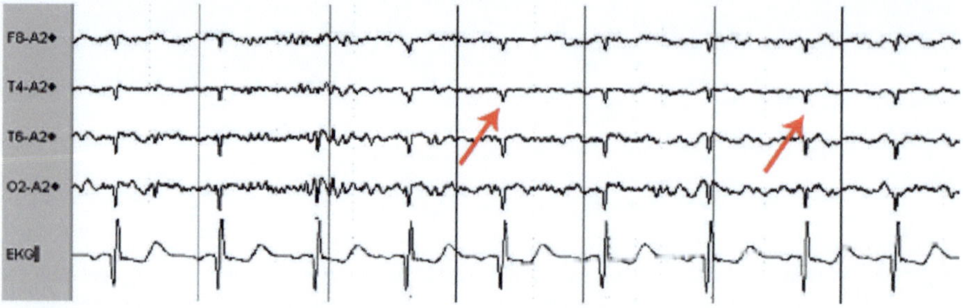

Fig. 3.20 EKG artifact: periodic sharp waves that adhere to a strict frequency should be checked against the EKG lead to exclude contamination by the EKG's far-field potential

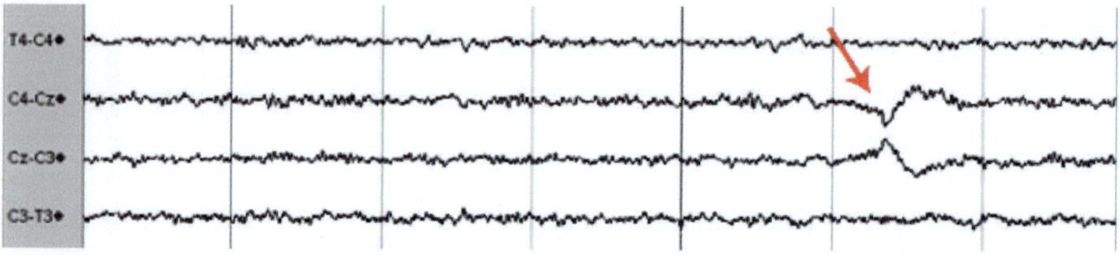

Fig. 3.21 Electrode artifact (also called lead artifact or electrode "pop"): This artifact has a characteristic mirror image" appearance and often worsens with time until the offending electrode is secured to the scalp. There should be no involvement of adjacent channels (no electrographic field) in contrast to epileptiform sharp waves

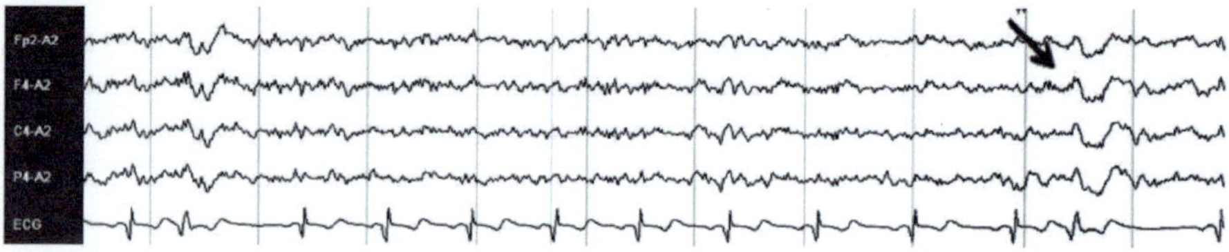

Fig. 3.22 Extrasystole (see EKG) can cause intermittent pulse artifact (one marked with arrow) seen here contaminating the ipsilateral ear lead

3.4 Activation Procedures

Sleep deprivation, hyperventilation and photic stimulation are commonly employed to help elicit epileptiform activity [2].

Hyperventilation: with adequate rapid, deep breathing for 3–5 min, frontally-predominant slow (theta>delta) activity is normally enhanced. Epileptiform activity is rarely enhanced in focal epilepsy but can be activated in up to 80% of generalized epilepsies, especially in absence epilepsy. Focal background slowing can also be enhanced by hyperventilation. As it causes vasoconstriction, hyperventilation is avoided in patients with cardiovascular risk/elderly, sickle cell anemia/trait, and pregnancy (Fig. 3.23).

Photic stimulation: Flickering light at 1–30 Hz frequencies normally causes rhythmic evoked potentials in the occipital regions, called *photic driving* (Fig. 3.24). Photic driving is best elicited by frequencies near the subject's PDR. *Photomyoclonic/photomyogenic responses* are normal myogenic artifacts that occur from rhythmic reflex contraction of frontalis or orbicularis oculi muscles.

A *photoparoxysmal* (also called *photoconvulsive*) *response* (Fig. 3.25) is an abnormal burst of generalized spike-or polyspike-and-wave complexes, occurring most frequently at 15–20 Hz stimulation, correlated most strongly with a diagnosis of idiopathic generalized epilepsy, particularly juvenile myoclonic epilepsy (Fig. 3.26).

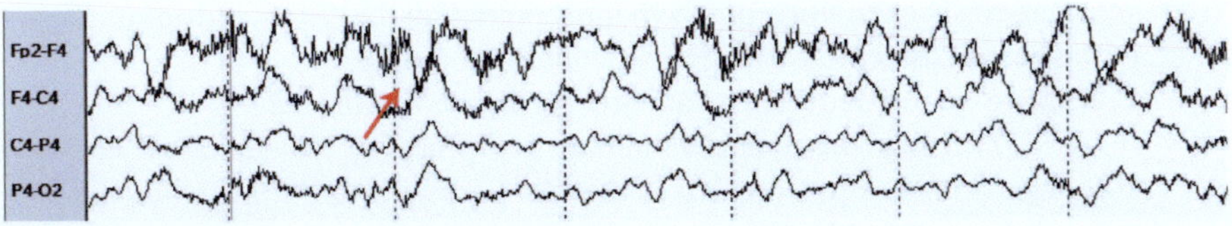

Fig. 3.23 Hyperventilation-induced slowing

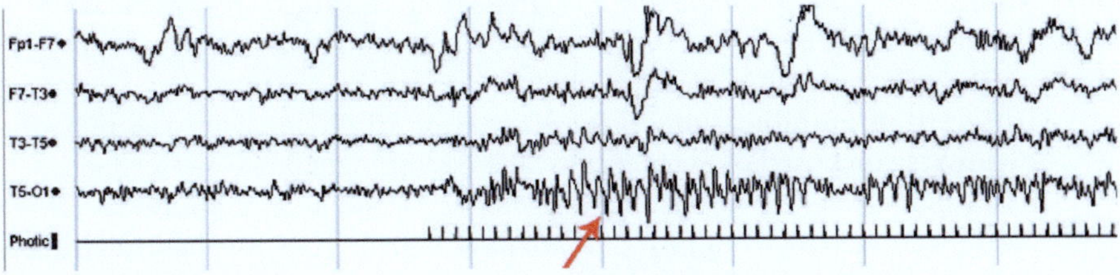

Fig. 3.24 Photic driving

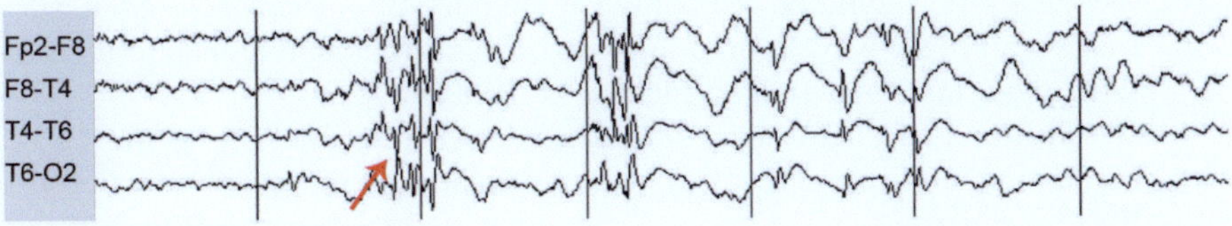

Fig. 3.25 Photoparoxysmal response

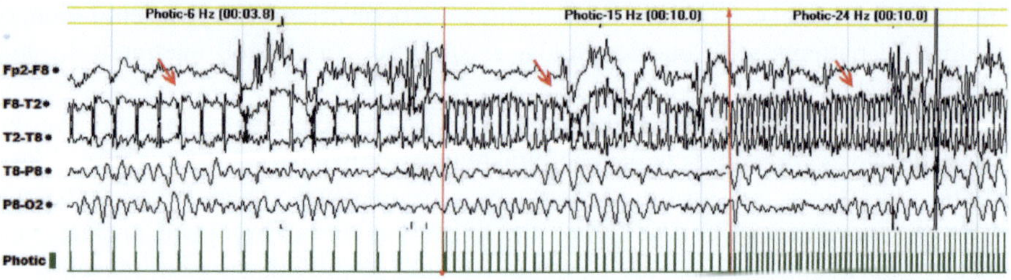

Fig. 3.26 Photo-electric artifact. Photic stimulation at 6 Hz, 15 Hz and 24 Hz producing the corresponding artifact time-locked to the photic stimulus in the T2 channel (arrows). This is caused by direct stimulation of light/photons ("photo") on the electrode causing a voltage change ("electric")

Pearls: Normal EEG

- A majority of the signal responsible for creating the EEG arises from the summation of excitatory post-synaptic potentials (EPSPs) and inhibitory post-synaptic potentials (IPSPs), not action potentials.
- **Electrode placement:**The standard is the *International 10–20 system* (Fig. 3.27) using anatomical landmarks to place electrodes at 10% and 20% across the scalp. Note that in the *10–10 system* (*Modified combinatorial nomenclature* [3]), there are more electrodes placed closer together. Electrode impedance (resistance) should be maintained *between 100 and 5000 Ω.*
- **Montages:** Most commonly, EEG signals from each lead are referenced to a nearby lead (*bipolar montage*), a single common lead (*referential montage*), or the average of all leads (*average reference montage*).
- **Differential amplification: (Fig. 3.28)** Raw EEG signal (see A in the Figure) is composed of many electrical sources, including electrical activity of interest and noise. Subtracting the signal found at a different reference electrode (B) from this, a *differential* signal (A-B) is obtained with the noise that is common to both electrodes is removed. This signal is then *amplified* $(A\text{-}B)^x$ to give the trace that is seen on routine EEG.
- **Nyquist theorem:** To reliably record a waveform, the sampling rate (Nyquist rate) should be at least twice the maximum frequency (Nyquist frequency) of the waveform of interest. Lower sampling rates can result in falsely low frequency (undersampling or "aliasing").
- **Notch filter:** Electrical noise (60 Hz in some countries such as the US, 50 Hz in others such as Canada) in the EEG can be removed by selectively filtering a narrow band of frequency ranges called a *notch* filter (Fig. 3.28).

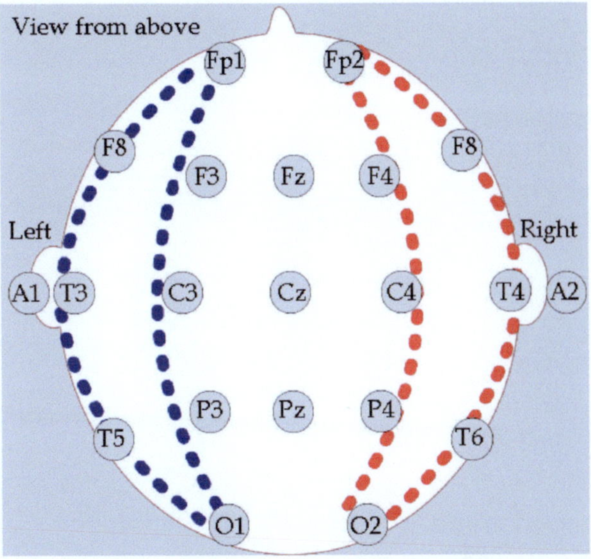

Fig. 3.27 10–20 system of electrode placement with the "double banana" montage shown (dotted lines)

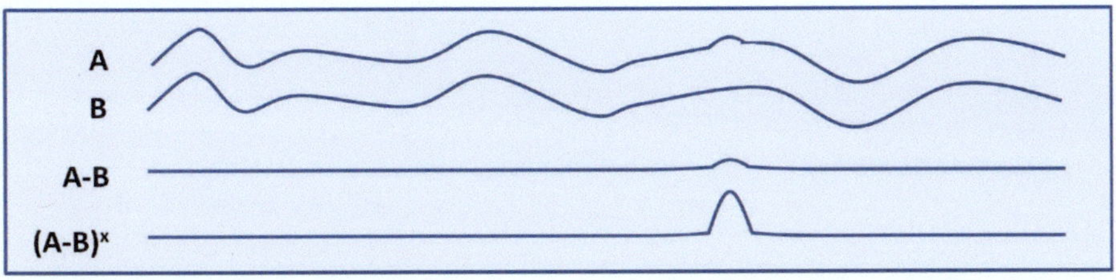

Fig. 3.28 Differential amplification

References

1. Hughes JR. Two forms of the 6/sec spike and wave complex. Electroencephalogr Clin Neurophysiol. 1980;48:535–50.
2. Edwards JC, Kutluay. Patterns of unclear significance. In: Schomer D, Lopes da Silva F, editors. Niedermeyer's electroencephalography: basic principles, clinical applications and related fields. Baltimore: Williams and Wilkins; 2011. p. 267–80.
3. Jurcak V, Tsuzuki D, Dan I. 10/20, 10/10, and 10/5 systems revisited: their validity as relative head-surface-based positioning systems. NeuroImage. 2007;34:1600–11.

Non-epileptiform EEG Abnormalities

4

Zulfi Haneef and Mostafa Hotait

This chapter covers non-epileptiform EEG abnormalities, including EEG abnormalities with focal lesions, encephalopathy/coma (including periodic patterns), brain death, aging and dementia, and drug effects. Much of the terminology has been revised and clarified with the widespread adoption of the American Clinical Neurophysiology Society (ACNS) Guideline on Standardized Critical Care EEG Terminology which have been followed in this Chapter [1, 2].

4.1 Focal EEG Changes

Before the advent of modern neuroimaging (CT and MRI), EEG was used widely to aid the clinical localization of focal cerebral lesions. Nowadays, focal EEG abnormalities provide ancillary information when neuroimaging is difficult to obtain or suboptimal. Some of these are described below.

Focal attenuation: Superficial lesions cause focal amplitude attenuation of the EEG (**A**bove = **A**ttenuation). See Fig. 4.1.

Focal slowing: Deeper lesions cause slowing (**S**ubcortical = **S**lowing). Focal slowing can be polymorphic or rhythmic. Continuous polymorphic (arrhythmic) delta activity (PDA) usually happens with localized structural lesions such as tumors, stroke, or abscesses (Fig. 4.2). While isolated subcortical (white matter) involvement is correlated with focal slowing, associated involvement of the superficial cortex can cause superimposed focal attenuation, which adds to the localizing

Z. Haneef (✉)
Department of Neurology, Baylor College of Medicine & Neurology Care Line, Houston VA Medical Center, Houston, TX, USA
e-mail: Zulfi.Haneef@bcm.edu

M. Hotait
Louisiana State University Health Shreveport, Shreveport, LA, USA

Z. Haneef (ed.), *Epilepsy Fundamentals*,
https://doi.org/10.1007/978-3-031-77741-7_4

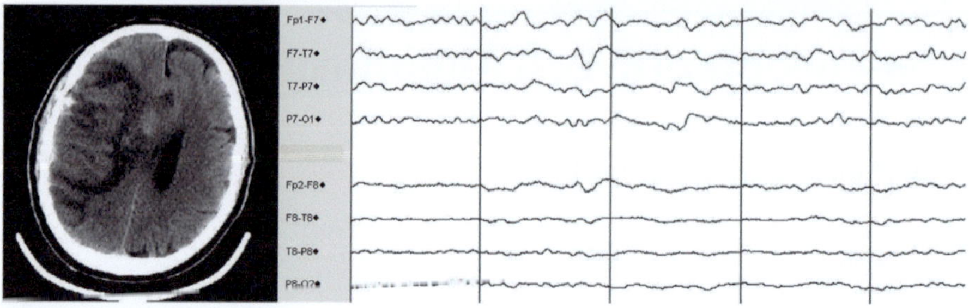

Fig. 4.1 EEG showing right attenuation in a patient with a right MCA infarct

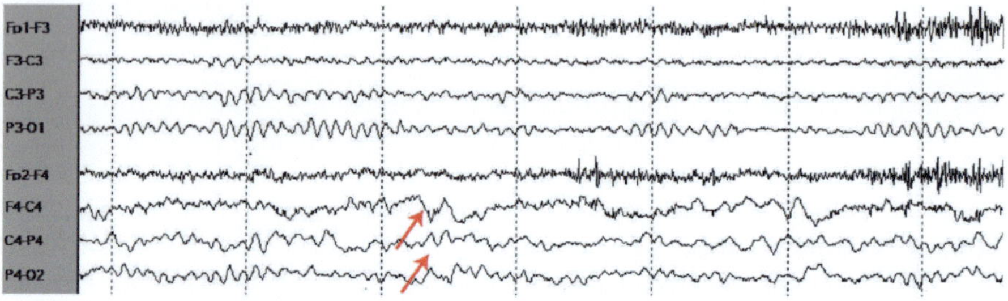

Fig. 4.2 Polymorphic delta activity

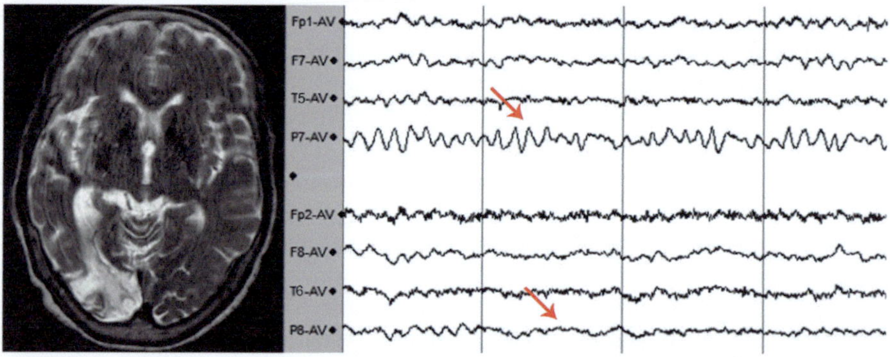

Fig. 4.3 Alpha asymmetry

value. The typical amplitude of focal PDA is 100–150 μV, although this will be lower with sufficient destruction of the overlying cortex. Deep lesions cause less defined focal changes and can cause bilateral diffuse slowing.

Focal beta asymmetry is defined as a 35% difference in beta voltage between sides. Structural pathology (CVA, tumor) or supra-cortical collections can cause focal suppression of beta activity.

Focal alpha rhythm asymmetry is either a (a) frequency asymmetry greater than 1 Hz between sides, (b) loss of alpha reactivity (attenuation) with eye-opening (also called the Bancaud's phenomenon), or (c) voltage attenuation. The most common cause of alpha asymmetry is a parieto-occipital lesion (Fig. 4.3). Voltage asymmetry

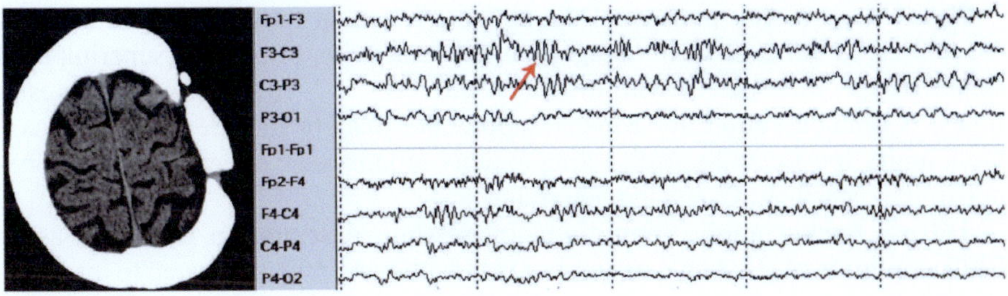

Fig. 4.4 Craniotomy (breach) defect showing faster frequency higher amplitude waveforms

is defined as amplitude asymmetry of 50% between both sides in referential montage.

Focal enhancement: A focal increase in EEG amplitudes, particularly faster frequencies that appear sharper (Fig. 4.4) is typical with craniotomy defects (skull breach).

4.2 Encephalopathies, Coma and Brain Death

Diffuse brain dysfunction due to various reasons (encephalopathies) can be metabolic (hepatic, renal, electrolyte imbalance), toxic, anoxic, endocrine (hypo/hyper thyroid, adrenal, calcemia), nutritional (B1, niacin, B6, B12 deficiency), infectious (meningitis, encephalitis, prion disease) or from other causes. The EEG changes can vary from mild to severe and include the following stages.

4.2.1 EEG Patterns in Early Encephalopathy

Background slowing with slowing of the posterior dominant rhythm indicates the earliest stages of encephalopathy. As an encephalopathy worsens, the following changes are seen.

Diffuse slowing: Initially, there is an increase in the amount of theta activity followed by an increase in delta activity. Diffuse slowing is commonly reported as mild, moderate or severe by taking the following characteristics into account (1) **mild:** slowing of the posterior dominant rhythm (PDR) with occasional diffuse theta; (2) **moderate:** loss of the PDR, diffuse intermixed theta and delta activity, background variability (spontaneous) and reactivity (to external stimulation) may be present; and (3) **severe:** diffuse predominant delta activity associated with an invariant, non-reactive background.

4.2.2 EEG Patterns Seen in Deeper Coma

Diffuse continuous slowing is a continuous arrhythmic, non-reactive delta activity. Cortical deafferentation may play a major role in its development. Polymorphic delta activity (PDA) is said to require deafferentation by involvement of subcortical white matter.

Burst suppression pattern: This pattern consists of bursts of synchronous or asynchronous bilateral high voltage delta/theta activity, often with superimposed spikes/sharps, interspersed with interburst intervals of 2–10 s (or even minutes at times). Common causes for a burst suppression pattern include hypoxic/anoxic brain injury (post-cardiac arrest), drug toxicity or iatrogenic effects (e.g., with phenobarbital, anesthetics), and hypothermia.

Diffuse background attenuation/suppression: Attenuation is defined as an EEG amplitude of 10–20 uV or amplitude reduction by 50% compared to the background. Background suppression occurs when the voltage is less than 10 μV. This EEG pattern is typically non-reactive to external stimulation.

Rhythmic delta activity (RDA): Repetition of a waveform (0.5 to ≤4 Hz), for at least 6 cycles, with relatively uniform morphology and duration and without an interval between the consecutive waveforms. Generalized rhythmic delta activity (GRDA) with frontal predominance, GRDA-Frontal, was previously known as frontal intermittent rhythmic delta activity, FIRDA. GRDA is seen with diffuse encephalopathies of various causes (Fig. 4.5). GRDA-Frontal is also seen in mesial frontal/parietal lesions or deeper lesions such as third ventricle tumors or raised intracranial pressure. GRDA with occipital predominance (GRDA-Occipital, previously OIRDA) is similarly present in children below 10 years of age and has been described in children with Childhood Absence Epilepsy (CAE). IRDA is thought to be due to abnormal interactions between cortical and subcortical (thalamic) structures.

Generalized periodic discharges with Triphasic morphology, GPDs-Triphasic (Triphasic waves) (Fig. 4.6) occur typically in metabolic encephalopathies. The typical morphology consists of an initial negative wave, a larger positive wave, and another small negative wave. However, monophasic, diphasic, or

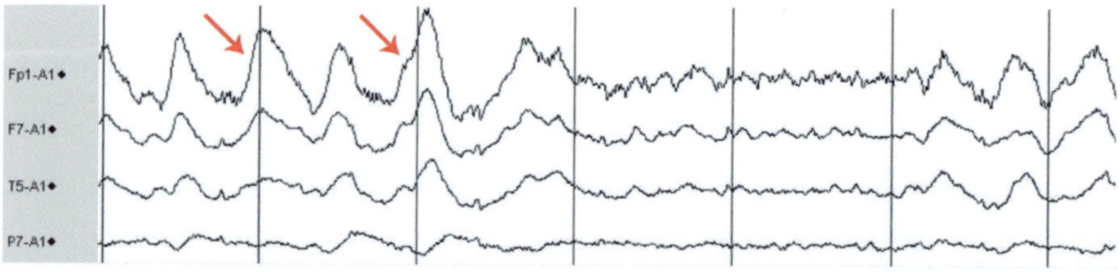

Fig. 4.5 Generalized rhythmic delta activity with frontal predominance (GRDA-frontal)

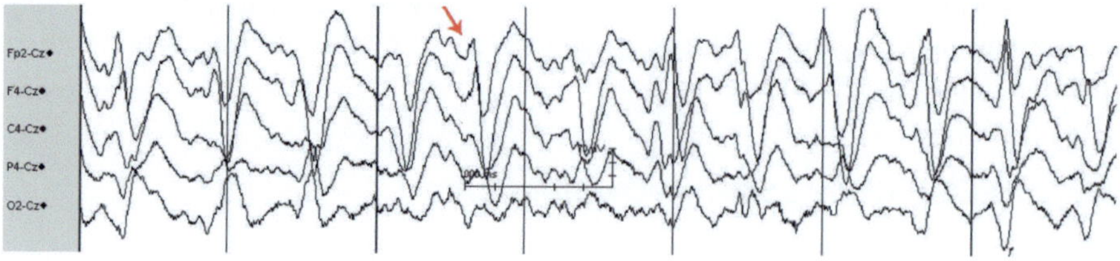

Fig. 4.6 Triphasic waves

quadriphasic waves have also been described. The usual frequency is about 2 Hz (originally termed "2 per second blunt spike and wave"). The distribution is anterior in 60% and posterior or diffuse in 40%. An anteroposterior gradient of the large positive wave is characteristic where the anterior (frontal) waves appear to precede the more posterior waves. This occurs in 75% of cases. The usual causes of triphasic waves are metabolic encephalopathies, characteristically in hepatic encephalopathy but also in renal dysfunction and severe hyponatremia. Other causes are valproate- or topiramate-induced hyperammonemia, hypercalcemia, anoxia, myxedema, hypothermia, CNS depressant drug effect, and lithium toxicity. Triphasic waves are not described in children.

Periodic patterns: Non-epileptic periodic patterns occurring at a frequency of once every 1–2 s are seen in Creutzfeldt-Jakob disease (Fig. 4.7) or lipoidosis in children. Periodic complexes occur once every 4–10 s in subacute sclerosing panencephalitis (SSPE). Other periodic patterns are described with epileptiform changes in Chap. 5.

Special patterns in coma have been described as follows:

- **Alpha coma:** An alpha frequency activity predominates in the EEG, typically in a frontal distribution (Fig. 4.8). The usual cause is post-anoxic. The prognosis is extremely poor (80% mortality). A less frequent cause is a pontomesencephalic lesion (cortical deafferentation).
- **Spindle coma:** Spindle activity (similar to stage II sleep) is seen diffusely. This is seen in post-traumatic or post-encephalitis encephalopathy. This pattern carries a good prognosis for recovery.
- **Beta coma:** Seen following drug intoxication. Prognosis is good.

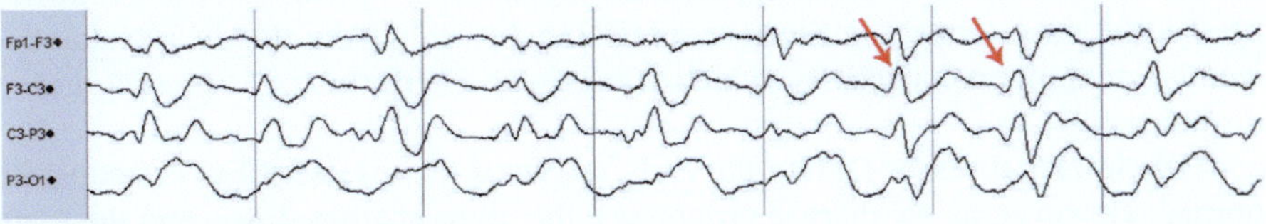

Fig. 4.7 Creutzfeldt-Jakob disease

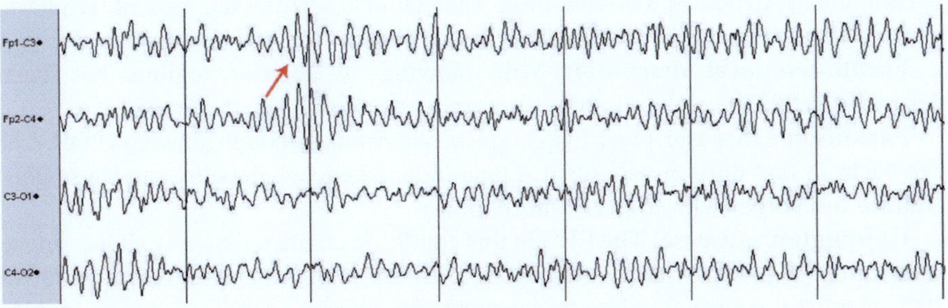

Fig. 4.8 Alpha coma (double distance montage)

- **Delta/ theta coma:** This is a non-specific finding in severe encephalopathy from many causes.
- **Theta coma:** This is a pure theta pattern seen frontally and associated with a very high mortality.

4.2.3 EEG in Brain Death

EEG for brain death determination requires the demonstration of electrocerebral inactivity (ECI). For this, several pre-requisites should be met, including (1) the exclusion of confounding factors (hypothermia <32° C, drug intoxication, metabolic/endocrine derangements); (2) a full set (21or more) of electrodes is ideal; (3) the sensitivity of at least 2 μV/mm to be used; (4) the use of double distance electrodes with inter-electrode distance of ≥10 cm; (5) Inter-electrode impedance should be between 100 and 10,000 Ω; (6) Record for at least 30 minutes; (7) Repeat EEG if there is doubt about electro-cerebral inactivity [3].

EEG should be done at least 8 h following circulatory arrest and should be repeated after 6 h in case of doubt.

4.3 EEG with Aging and Dementia

It is unclear if there is a normal slowing of the EEG background with aging. Previous evidence pointed to a slower PDR in the seventh and eighth decades.

Temporal slowing in the elderly has been described in up to 40% of people over age 60, with temporal theta and delta frequency slowing. Such slowing is more common on the left side. Temporal theta lasting up to 10–15% of recording time has been considered normal [4]. **Sleep onset** GRDA-Frontal, in the elderly, is characterized by rhythmic delta activity, which is maximal anteriorly and is present during drowsiness.

Alzheimer's disease: During the early stages of the disease, EEG is frequently normal. However, as the disease progresses, the alpha rhythm slows, and there are diffuse irregular theta and random delta waves. A correlation between the severity of cognitive decline and EEG changes has been described. Epileptiform discharges are uncommon in Alzheimer's disease even when clinical seizures are present. Triphasic waves with a posterior distribution have been described in Alzheimer's disease.

Fronto-temporal dementia: Mild slowing in anterior regions has been described, while the alpha rhythm is preserved even with late dementia.

Parkinson's disease: The EEG is typically normal, although slowing (PDR < 8) can occur in one-third of patients. L-Dopa-induced encephalopathy can cause generalized discharges with triphasic morphology.

Huntington's disease: The EEG in this condition characteristically shows severe attenuation (Fig. 4.9), with an amplitude of <20 μV in 60% of patients (compared to 10% of normal subjects), while background suppression of <10 μV occurs in 33% patients (does not occur in normal subjects).

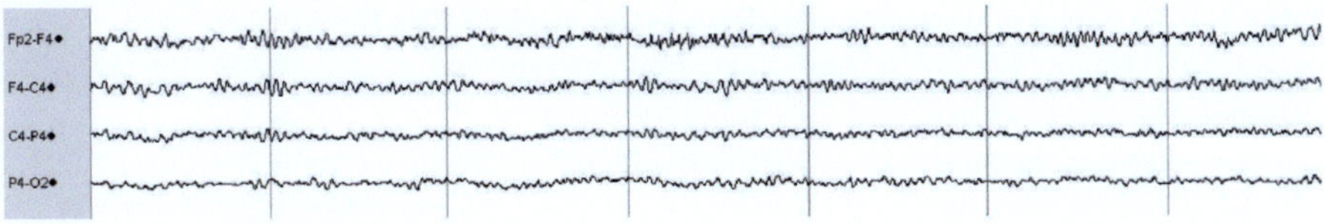

Fig. 4.9 Huntington's disease

4.4 Drug Effects

Drugs can have many effects on the EEG. The most common pattern seen is diffuse slowing.

Anti-seizure medications (ASMs): Ictal and interictal changes can be suppressed by benzodiazepines. Suppression of interictal changes with benzodiazepines is thought to be a good prognostic sign of eventual control [5]. *Benzodiazepines* (e.g., diazepam, lorazepam) and *barbiturates* (e.g., phenobarbital and primidone) cause an increase in rhythmic fast (beta) activity. Benzodiazepine-induced fast activity can be present for up to 2 weeks following drug ingestion. Higher doses cause diffuse slowing. Later stages show burst suppression and electrocerebral inactivity with barbiturates. Barbiturate withdrawal can cause high-voltage spike-and-wave abnormalities.

All of the first-generation ASMs cause slowing of the posterior dominant rhythm and diffuse slowing. Of the newer ASMs, levetiracetam and zonisamide appear to produce fast activity in some patients.

Phenytoin causes slowing of the posterior dominant rhythm at levels >20 μg/mL. *Phenytoin and carbamazepine* cause no changes to epileptiform discharges. *Valproate* has no effect on focal IEDs but reduces both generalized spike-waves (this effect lasts up to 3 months after drug discontinuation) and photoparoxysmal discharges.

Induction of epileptiform discharges by ASMs (see Table 4.1**): Sodium channel blockers are known to induce/activate the frequency of generalized spike and wave discharges (seen in genetic generalized).** Generalized spikes can be induced by *vigabatrin*. *Tiagabine* has been reported to cause nonconvulsive status epilepticus. *Carbamazepine* can exacerbate absence, atonic, and myoclonic seizures. ASMs can cause paradoxical worsening of seizures: risk factors include young age, cognitive impairment, prominent IEDs, high seizure frequency, and ASM polytherapy, e.g., Lennox-Gastaut syndrome [6]. However, care must be taken not to mistake the normal variability in seizure frequency for paradoxical seizure exacerbation due to an ASM.

Neuroleptic agents can reduce the seizure threshold. The risk is highest with clozapine, which causes EEG changes in 16–74% of patients, including slowing and IEDs (bilateral anterior spike-waves). Clozapine-induced EEG changes are associated with improvement in psychosis. This is the converse to forced normalization with ASMs, where clinical improvement of seizures is associated with the induction of psychotic features [8].

Table 4.1 Examples of worsening of epilepsy syndromes by ASMs [7]

Epilepsy syndrome	ASMs implied	Effect
Childhood absence epilepsy	CBZ, PHT, PB, VGB	Increases absences
Juvenile myoclonic epilepsy	CBZ, VGB, GBP	Increases myoclonia
Benign “Rolandic” epilepsy	CBZ	ESES-like drop attacks
Progressive myoclonic epilepsy	CBZ	Increases myoclonus
Idiopathic generalized epilepsy	ESX, PHT	Provokes GTCS
Lennox-Gastaut syndrome	Benzodiazepines	Serial tonic seizures

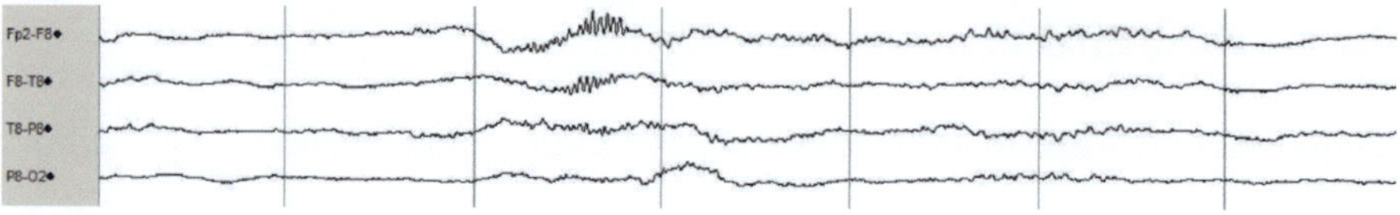

Fig. 4.10 Propofol-induced frontally predominant delta with superimposed beta- a common pattern in the intensive care unit

Lithium causes EEG changes in 50% of patients. In addition to background slowing, a GRDA-frontal (previously FIRDA)-like pattern, increasing with hyperventilation, occurs. Intoxication causes patterns resembling triphasic waves or periodic complexes, as in Creutzfeldt-Jakob disease. Lithium also accentuates pre-existing IEDs.

Antidepressants: Tricyclic antidepressants can increase beta/theta frequencies, with slowing of the alpha rhythm. Among newer antidepressants, maprotiline and bupropion induce seizures, while SSRIs are less likely to do so.

Anesthetic agents: Many anesthetic agents produce EEG changes similar to barbiturates (increased beta, loss of alpha, diffuse slowing, burst suppression). Etomidate can increase IEDs. Propofol causes a characteristic frontally predominant delta with superimposed beta (Fig. 4.10) that resembles delta brushes seen in infancy.

Halogenated anesthetics (e.g., desflurane, halothane) cause alpha activity in the anterior head regions resembling sleep spindles.

Alcohol: Alcohol use or cessation is associated with a variety of EEG changes. Acute alcohol intoxication causes mild slowing of the alpha rhythm. Higher doses cause generalized slowing. Chronic alcohol abuse causes EEG attenuation (voltages <25 μV). Alcohol withdrawal can cause IEDs and seizures. Photosensitivity (photomyogenic, photoconvulsive) has been described with alcohol withdrawal, although this has been contested.

Drugs of abuse:

- *CNS stimulants (amphetamines, cocaine)* increase beta and alpha activity with attenuation of the background EEG.
- *Cannabis* causes no specific EEG changes. THC, its active ingredient, causes increased alpha and decreased beta and theta activity.
- *PCP* causes a characteristic EEG signature of non-reactive rhythmic theta with intermittent delta activity.

References

1. Hirsch LJ, LaRoche SM, Gaspard N, et al. American clinical neurophysiology Society's standardized critical care EEG terminology: 2012 version. J Clin Neurophysiol. 2013;30:1–27.
2. Hirsch LJ, Brenner RP, Drislane FW, et al. The ACNS subcommittee on research terminology for continuous EEG monitoring: proposed standardized terminology for rhythmic and periodic EEG patterns encountered in critically ill patients. J Clin Neurophysiol. 2005;22:128–35.
3. Guideline 3: Minimum Technical Standards for EEG Recording in Suspected Cerebral Death. [online]. American Clinical Neurophysiology Society. http://www.acns.org/pdf/guidelines/Guideline-3.pdf. Accessed 13 Dec 2015.
4. Pedley TA, Miller JA. Clinical neurophysiology of aging and dementia. Adv Neurol. 1983;38:31–49.
5. Ebersole JS, Pedley TA. Current practice of clinical electroencephalography. New York: Lippincott Williams & Wilkins; 2003.
6. Bauer J. Seizure-inducing effects of antiepileptic drugs: a review. Acta Neurol Scand. 1996;94:367–77.
7. Genton P, McMenamin J. Summary. Aggravation of seizures by antiepileptic drugs: what to do in clinical practice. Epilepsia. 1998;39(Suppl 3):S26–9.
8. Pakalnis A, Drake MEJ, John K, Kellum JB, Forced normalization. Acute psychosis after seizure control in seven patients. Arch Neurol. 1987;44:289–92.

Epileptiform EEG Abnormalities

5

Zulfi Haneef and Mostafa Hotait

EEG is important in the diagnosis of epilepsy when it demonstrates epileptiform abnormalities. However, epilepsy is a clinical diagnosis and cannot be excluded by normal EEGs. Around 10% of patients with epilepsy never show epileptiform changes. On the contrary, 0.5% of healthy adults without known seizures have epileptiform changes in the EEG [1]. Between 30–70% of patients with a clinical diagnosis of epilepsy have interictal epileptiform discharges (IEDs) on a single EEG. Multiple EEGs may enhance the yield, and up to 4 EEGs have been reported to provide a yield of 92%, with further EEGs having minimal benefit [2]. Although the sensitivity is low with a single EEG, the specificity of EEG is better at 78–98% [1]. The type of epilepsy changes this yield—for example, untreated absence epilepsy is typically almost always associated with an abnormal routine EEG. The diagnostic yield of EEG may also be increased by prolonged monitoring, including sleep, or by activation maneuvers of hyperventilation, photic stimulation, or sleep deprivation. Furthermore, EEG done within 24 h of a seizure has a higher likelihood of showing IEDs (51%) compared to a later study (34%) [3]. Prolonged EEG sampling can increase the yield by 20% [1].

Epileptiform EEG changes are divided into focal and generalized patterns.

Z. Haneef (✉)
Department of Neurology, Baylor College of Medicine, Houston VA Medical Center, Houston, TX, USA
e-mail: Zulfi.Haneef@bcm.edu

M. Hotait
Louisiana State University Health Shreveport, Shreveport, LA, USA

Z. Haneef (ed.), *Epilepsy Fundamentals*,
https://doi.org/10.1007/978-3-031-77741-7_5

5.1 Focal Epileptiform Patterns

Focal seizures are typically associated with localized sharps/spikes. An epileptiform spike is defined as a waveform with a duration ≥20 ms and less than 70 ms, while a sharp wave lasts between 70 and 200 ms. The term "sharply contoured activity" is sometimes used to describe waveforms with a sharp morphology of >200 ms (too long to qualify for a sharp wave). The clinical significance of sharply contoured activity is unclear and may represent non-epileptiform features such as focal slowing. IEDs are suggestive of an epileptic focus, although epileptiform discharges can also occur in the absence of seizures. As discussed in Chap. 3, normal EEG changes can sometimes be mistaken for epileptiform abnormality.

Typical IEDs are oriented radially (deep from the cortical surface) as the electrical activity spreads deep from the cortical surface following the radial direction of axons from the cortical surface neurons. This leads to surface electronegativity. When the cortical surface configuration leads to the generation of a horizontal dipole oriented parallel to the cortical surface, an electropositive component of the spike is seen at a different location from the electronegative component. Classically described in benign Rolandic epilepsy (Fig. 5.1), it is also seen in occipital lobe epilepsy. Surface-positive spikes can also be seen with cortical distortion, which occurs following trauma.

An epileptiform discharge is thought to correspond to **"paroxysmal depolarizing shifts"** in the membrane potential of a group of neurons. Pacemaker neurons called **"epileptic neurons"** may trigger these changes. Such depolarizing shifts are limited by hyperpolarizations, except in the case of seizures when they spread. A minimum of 6 cm^2 of cortical surface needs to be activated before it can be detected on a scalp EEG electrode, although typically 10 cm^2 or more is involved [4, 5]. Focal IEDs in refractory epilepsy have been associated with cognitive deficits, independent of and additive to lesional effects or drug use [6].

Secondary bilateral synchrony: A focal epileptiform discharge can have rapid transcallosal propagation such that it appears to be simultaneously present bilaterally, simulating a generalized epileptiform discharge (Fig. 5.2). Mesial frontal lesions are more likely to cause secondary bilateral synchrony, although temporal lesions can also be causative. Electrographic distinction from a generalized

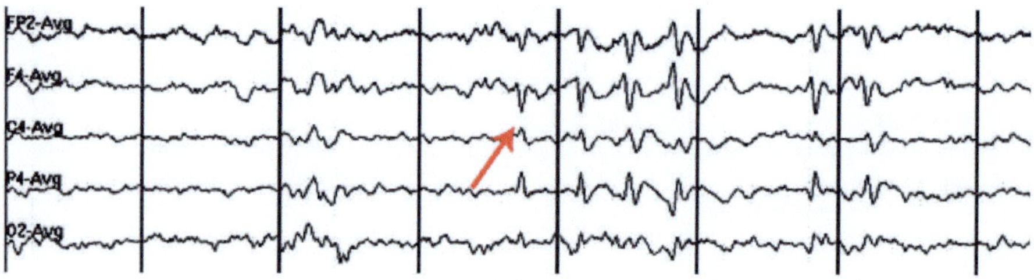

Fig. 5.1 Horizontal dipole in benign Rolandic epilepsy. Note the negative end of the dipole in parietal (P4) and positive phase in frontal (F4/Fp2) leads in this referential (average) montage

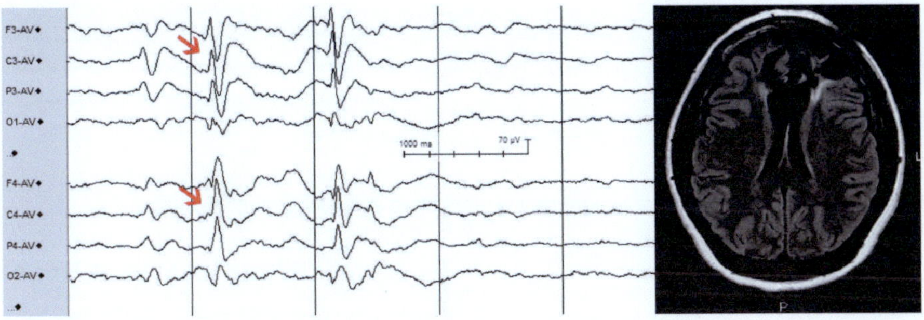

Fig. 5.2 Secondary bilateral synchrony in a patient with post-traumatic bi-frontal lesions

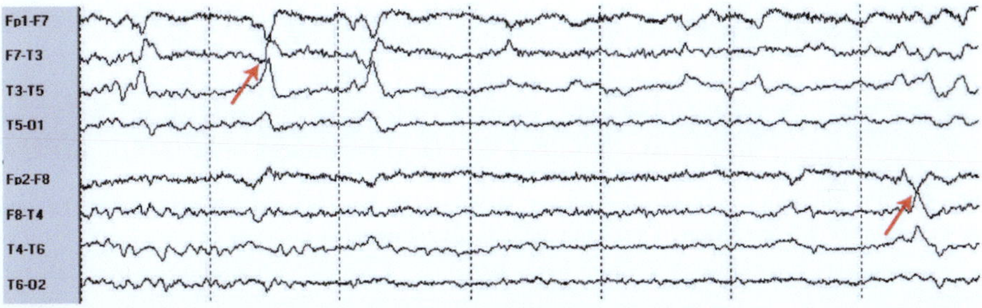

Fig. 5.3 Bilateral temporal foci

epileptiform discharge is aided if a phase reversal can be demonstrated in a transverse montage or a minor phase lag or voltage asymmetry can be visualized between the two sides. Although these discharges may appear in trains of spike-waves simulating generalized discharges further, the spike waves show morphological variation (polymorphic) and are usually less than 2.5 Hz when rhythmic.

Secondary and Mirror foci: Repeated input from distant epileptogenic foci may lead to secondary cortical epileptogenicity. Such an independent epileptogenic focus that develops in a previously normal cortex is called a **secondary focus**. If this secondary focus is in the homotopic contralateral cortex (e.g., left TLE leading to right temporal lobe discharges, Fig. 5.3), it is called a **mirror focus**. The development of a secondary focus has been likened to the development of epileptogenicity in rat brains by repeated electrical stimulation, a phenomenon termed **"kindling."**

Non-epileptic conditions can also cause focal epileptiform spikes in EEG. Examples include occipital spikes in early onset blindness, posterior spikes in basilar migraine, and very high amplitude occipital spikes with slow photic stimulation seen in infantile cerebral lipidoses (e.g., Bielshowsky-Jansky disease). **Non-epileptiform mimics (benign variants)** of focal epileptiform discharges include RMTD, SSS (BETS), wickets and midline theta (Ciganek) rhythms. See Chap. 3 for further details. As opposed to a benign variant simulating IED, an IED has the following characteristics: (1) the predominant waveform is electronegative on the surface (except for horizontal dipoles); and (2) there is disruption of the background rhythm by an after-going slow wave (Fig. 5.4).

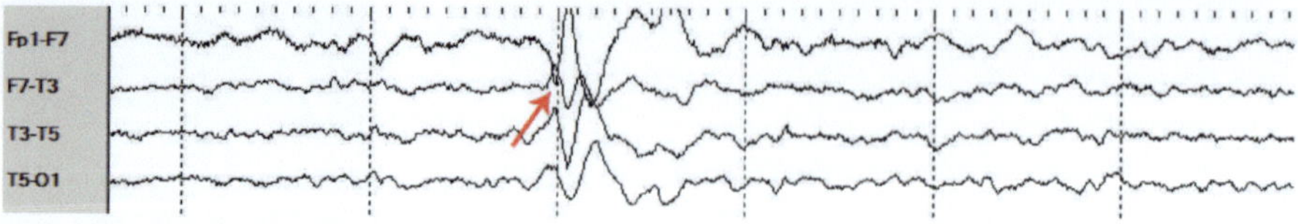

Fig. 5.4 Epileptiform discharge, after-going slow wave and background disruption

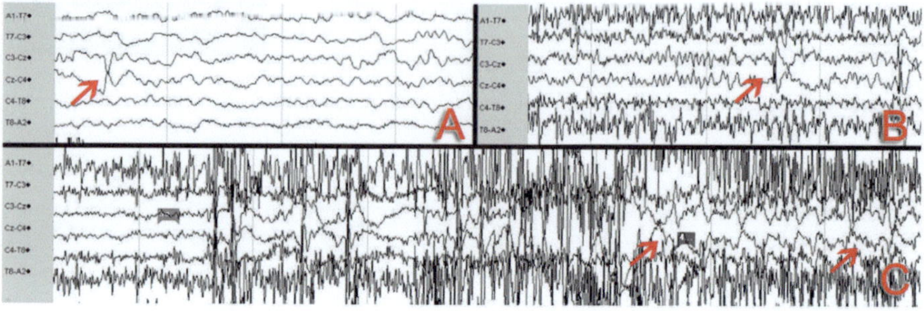

Fig. 5.5 Patient with a midline seizure focus showing Cz sharp wave discharges in sleep (**a**) and in wakefulness (**b**). Evolution of a seizure at Cz is also shown (**c**)

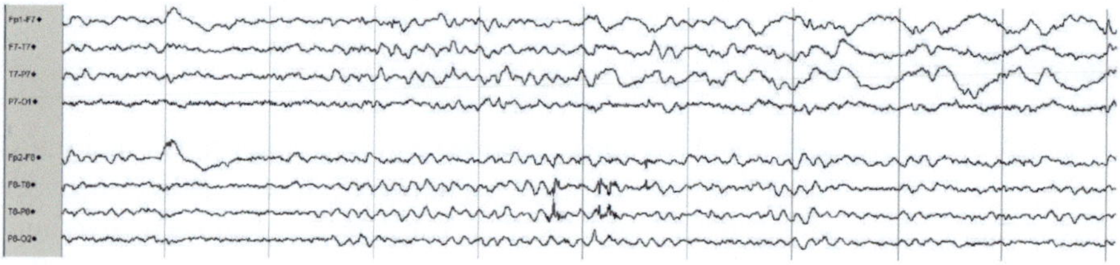

Fig. 5.6 Lateralized Rhythmic Delta Activity (LRDA, temporal) or Temporal Intermittent Rhythmic Delta Activity (TIRDA)

Occasionally distinguishing epileptiform activity from a benign waveform can be challenging, as with differentiating a vertex wave from an IED at the vertex (Fig. 5.5).

Adults usually have only one (or a few) stereotyped interictal/ictal changes, while infants/children are more variable and several interictal/ictal patterns may coexist. Lesions in the temporal and frontal lobes are more likely to cause epileptogenic foci than lesions in the occipital and parietal lobes

Lateralized Rhythmic Delta Activity with temporal predominance (LRDA-Temporal, *previously TIRDA***):** LRDA—Temporal consists of rhythmic slow waves (delta rhythm) seen in an epileptogenic temporal lobe (Fig. 5.6). It carries the same significance as an epileptiform discharge and is considered to be a reliable marker of epileptogenicity.

5.2 Generalized Epileptiform Patterns

Generalized epilepsy is typically associated with monomorphic generalized spike waves (GSW) (Fig. 5.7). GSWs appear simultaneously over the homotopic cortex bilaterally and characteristically have a frontal or fronto-central predominance. They typically appear in brief runs lasting a few seconds or in sub-second "fragments." The classic 3 Hz GSW seen in the absence epilepsy is the **"typical generalized spike wave"** (Fig. 5.7). The 3 Hz GSW of absence epilepsy is characteristically activated by hyperventilation and photic stimulation. **Atypical spike wave** implies a less rhythmic and usually faster GSW seen in atypical absence and in JME (4–6 Hz). A **slow spike and wave** implies a frequency of ≤2.5 Hz (1.5–2.5 Hz), typically seen in Lennox-Gastaut syndrome. They are more polymorphic than typical GSW and are not induced by hyperventilation. Cerebral generators for GSWs are thought to include the cortex and the thalamus (cortico-reticular network). **Polyspikes** are typically seen in JME.

Generalized epilepsy type	EEG findings
Absence epilepsy	Typical 3 Hz GSW
Lennox-Gastaut syndrome	Slow spike and wave, GPFA
Juvenile myoclonic epilepsy	Fast 4–6 Hz GSW, polyspikes
Infantile spasms	Hypsarrhythmia, multifocal spikes
Atypical absence	Atypical spike wave, polyspikes
Akinetic/atonic drops	Slow spike and wave

Epileptiform activity does not have to appear exclusively as spike and wave activity. **"Hypsarrhythmia"** is a special type of generalized discharge seen in infantile spasms, characterized by high amplitude, chaotic background activity. **Generalized paroxysmal fast activity (GPFA)** is a generalized EEG rhythm characterized by sharp waveforms in the beta frequency, usually appearing in sleep, maximal in frontal/frontocentral leads and lasting 2–4 s.

Non-epileptic conditions can also lead to the generation of generalized epileptiform spikes in the EEG. Examples include cerebral anoxia, hypoglycemia, barbiturate withdrawal and metabolic encephalopathies. **Non-epileptiform mimics**

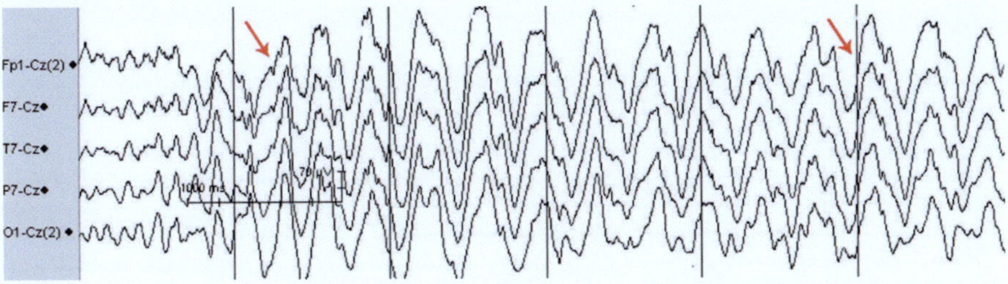

Fig. 5.7 Absence epilepsy with 3 Hz GSW. Longer bursts can start with 3.5–4 Hz and slow to 2–3 Hz by the end

(benign variants) of generalized epileptiform discharges include 6 Hz (phantom) spike and wave, 14- and 6-Hz positive spikes, hyperventilation-induced slowing, and hypnagogic hypersynchrony (see Chap. 3).

5.3 Ictal Changes

Ictal activity in focal seizures is comprised of repetitive rhythmic discharges with evolution of frequency, amplitude and distribution over the course of the seizure. Ictal activity in generalized epilepsy could be a longer repetition of the interictal activity or can change in frequency or morphology as the seizure progresses. Ictal EEG changes can assume a variety of morpholgies (Figs. 5.8, 5.9, 5.10 and 5.11). **Non-epileptiform mimics (benign variants)** of focal seizures include SREDA (see Chap. 3). Not all seizures are associated with surface EEG changes (e.g., focal aware seizures/simple partial seizures show EEG changes in <25%). Conversely, not all EEG seizures are associated with clinical seizures ("subclinical" or "electrographic" seizures) (Fig. 5.12).

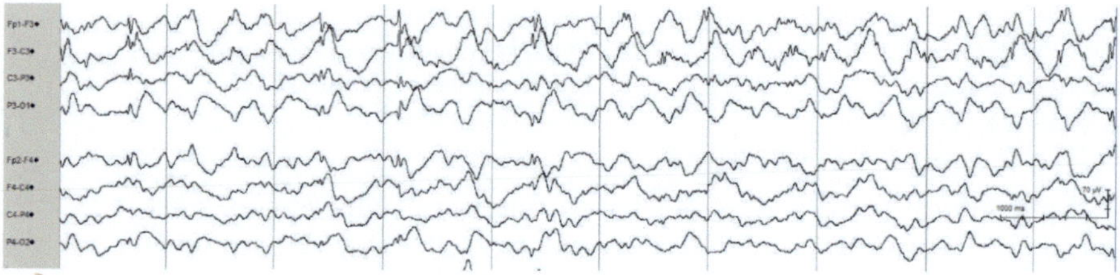

Fig. 5.8 Ongoing left frontal delta frequency seizure

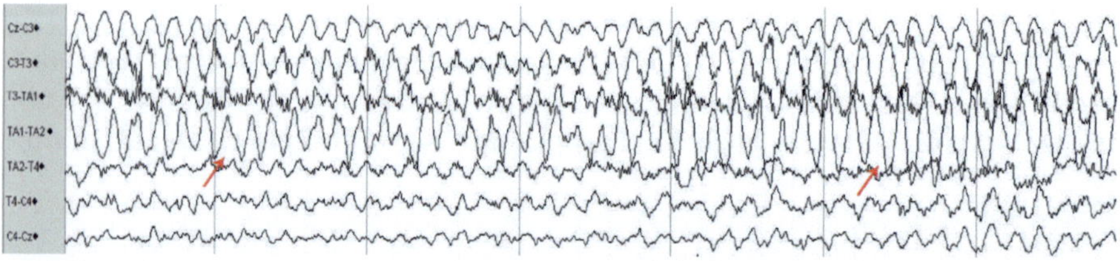

Fig. 5.9 Ongoing left temporal seizure demonstrating typical theta frequency activity

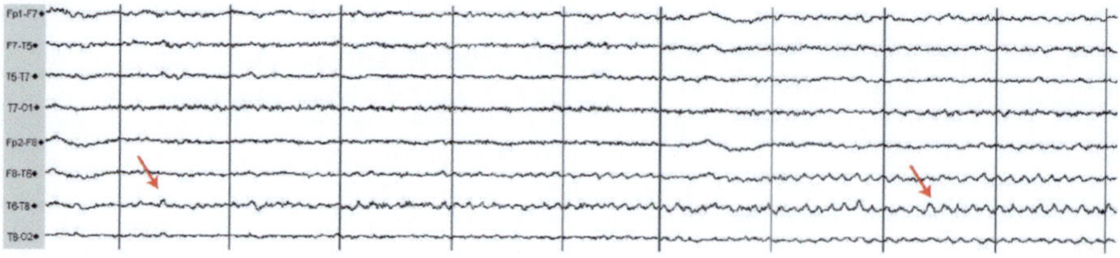

Fig. 5.10 Evolving right temporo-parietal seizure in the alpha frequency range

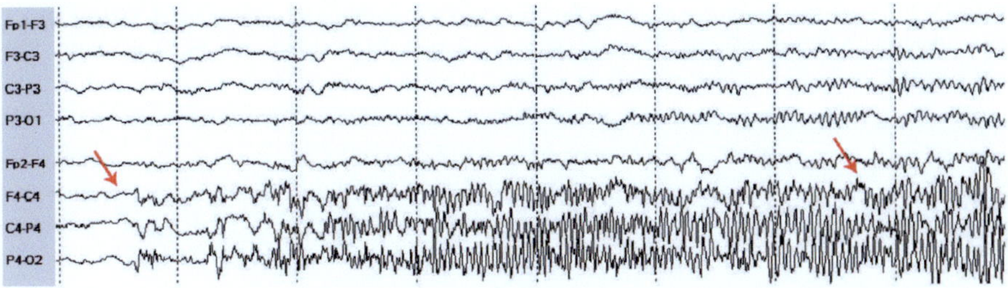

Fig. 5.11 Evolving right hemispheric seizure in the beta frequency range

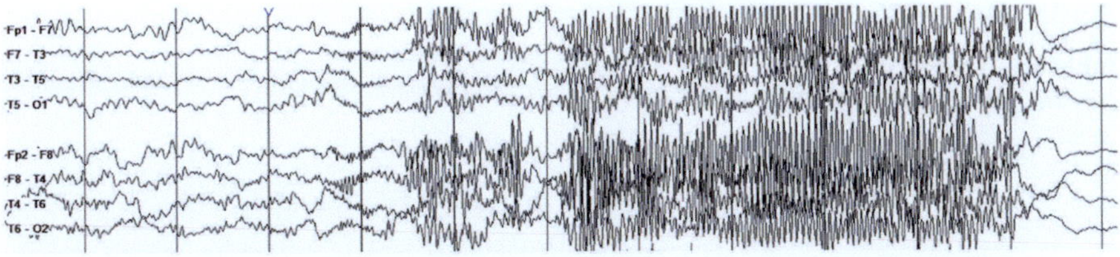

Fig. 5.12 Generalized paroxysmal fast activity (GPFA)

5.4 Periodic EEG Patterns

Periodic EEG patterns are discharges typically seen in critically ill patients. The American Clinical Neurophysiology Society (ACNS) has proposed recommendations for critical care EEG which are adopted in this chapter. The ACNS acronyms are described in Tables 5.1 and 5.2 [2].

These patterns can be generalized (G), lateralized (L), bilateral independent (BI), or multifocal (Mf). Based on whether such patterns are periodic discharges (PD) with an inter-burst interval, rhythmic delta activity (RDA), or Sharp/spike wave (SW), various patterns have been proposed to be renamed as shown in Table 5.2 with additional qualifiers indicating the distribution (e.g., Frontal, F; Occipital, O).

Lateralized periodic discharges (LPDs): These are unilateral complexes or bilateral with clear and consistent (≥80%) lead-in from one side or consistently of higher amplitude (≥ two times the amplitude on the other side). They may be followed by a slow wave and recur typically at 0.5–2 Hz frequency. If LPDs occur at ≥2.5 Hz, the periodic term can be only applied if they last for less than 10 s; otherwise (≥10 seconds), they qualify as electrographic seizures (Fig. 5.13). LPDs occur in acute nonspecific brain dysfunction or unilateral brain lesions such as stroke, tumor (e.g., GBM) or infection (e.g., abscess, HSV). LPDs are typically transient and gradually subside over 2–3 weeks. Although considered interictal in general, being frequency and duration dependent, LPDs fall into the ictal-interictal continuum if they last >10 s and meet certain criteria (>1 Hz and ≤ 2.5, or ≥ 0.5 Hz with a plus modifier).

Bilateral periodic discharges (BIPDs): BIPDs describe the occurrence of (LPDs) bilaterally in an asynchronous manner (Fig. 5.14). Typically seen with acute structural lesions, the most common causes are cerebral anoxia and CNS infections. The prognosis is typically worse than with unilateral LPDs.

Table 5.1 Proposed naming convention for periodic EEG patterns (G- generalized; L- lateralized; BI- bilateral; Mf- multifocal; PD- periodic discharge; RDA- rhythmic delta activity; SW- spike wave)

Distribution	Pattern	Qualifier
G L BI Mf	PD 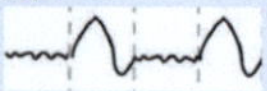 RDA 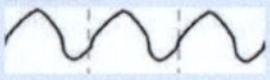 SW 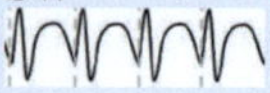	Several qualifiers including: • Prevalence (e.g., continuous, abundant, rare) • Duration (e.g., long, brief) • Frequency (e.g., 2/s) • Sharpness (e.g., spiky, sharp) • Amplitude (e.g., very low, high) • Lobe (e.g., frontal, occipital)

Table 5.2 Current and previous terminology of some common EEG patterns

Current terminology [1]	Previous terminology
GPDs	GPEDs
LPDs	PLEDs
BIPD	BIPLEDs
GRDA frontal	FIRDA
GRDA occipital	OIRDA
GPD triphasic morphology	Triphasic waves

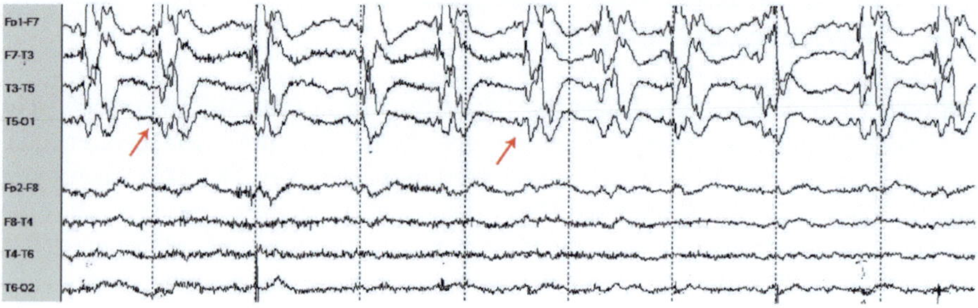

Fig. 5.13 LPDs showing unilateral discharges

Generalized periodic discharges (GPDs): These are defined as symmetric, diffuse and synchronous high amplitude discharges over bilateral hemispheres (Figure 5.15). Etiologies include cerebral anoxia and subclinical status epilepticus. GPDs seen in CJD and SSPE are described below. GPDs may be associated with seizures in up to 50% of cases.

Stimulus-induced Rhythmic, Periodic or Ictal Discharges (SIRPIDs): The term "SIRPIDs" refers to stimulus-induced (or stimulus-exacerbated) rhythmic, periodic, or ictal-appearing discharges and is a term that includes all SI- patterns together (SI-RDA, SI-PDs, SI-SW, SI-IIC, SI-BIRDs, or SI-seizures). Now, it is recommended to be more specific, referring to "SI-" followed by the exact pattern (ex. SI-LRDA; SI-LPD, etc.). These were first described in 2004. These patterns are periodic, rhythmic, or frankly ictal and are consistently reproducible, elicited or exacerbated by stimulation (auditory,

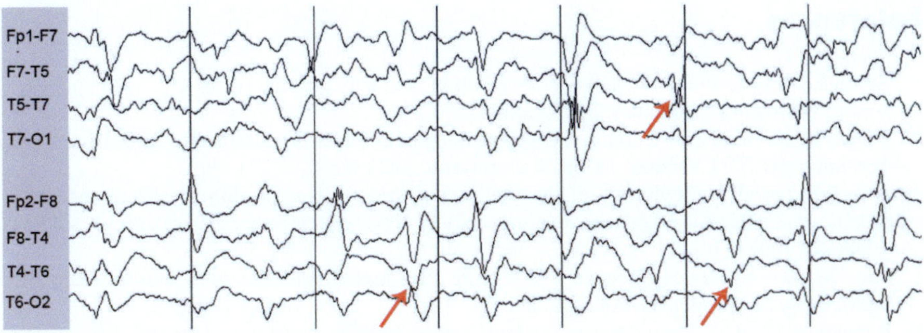

Fig. 5.14 Bilateral periodic discharges (BiPDs). Arrows point to the asynchronous discharges

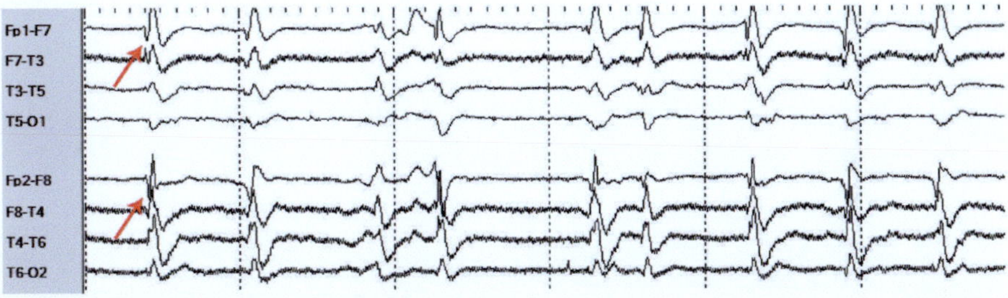

Fig. 5.15 Generalized periodic discharges (GPDs): arrows point to the synchronous discharges

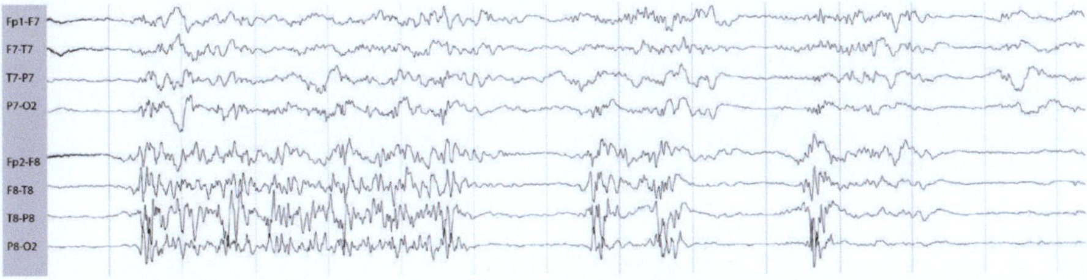

Fig. 5.16 Brief potentially ictal rhythmic discharges (BIRDs)

sternal rub, examination, suctioning etc.) of the critically ill patient. Their clinical significance is still unclear.

Brief Potentially Ictal Rhythmic Discharges (BIRDs): BIRDs include lateralized, bilateral independent, multifocal, or generalized rhythmic activity of ≥4 Hz (at least six waves at a regular rate) lasting ≥0.5 s but less than 10 s. BIRDs pattern has been seen with similar morphology and location as seizures in the same patients, and thus, the term brief (<10 s) and "potentially" ictal. If the pattern lasts more than 10 s, it qualifies for electrographic seizure (Fig. 5.16).

References

1. Smith SJM. EEG in the diagnosis, classification, and management of patients with epilepsy. J Neurol Neurosurg Psychiatry. 2005;76:ii2–7.
2. Hirsch et al. American Clinical Neurophysiology Society's Standardized Critical Care EEG Terminology: 2021 Version. J Clin Neurophysiol. 2021 Jan 1;38(1):1–29.
3. King MA, et al. Epileptology of the first-seizure presentation: a clinical, electroencephalographic, and magnetic resonance imaging study of 300 consecutive patients. Lancet Lond Engl. 1998;352:1007–11.
4. Tao JX, Ray A, Hawes-Ebersole S, Ebersole JS. Intracranial EEG substrates of scalp EEG interictal spikes. Epilepsia. 2005;46:669–76.
5. Ramantani G, et al. Simultaneous subdural and scalp EEG correlates of frontal lobe epileptic sources. Epilepsia. 2014;55:278–88.
6. Glennon JM, et al. Interictal epileptiform discharges have an independent association with cognitive impairment in children with lesional epilepsy. Epilepsia. 2016;57:1436–42.

6 Source Localization

Gamaleldin Osman and Jay R. Gavvala

EEG analysis has primarily relied on straightforward visual analysis of EEG tracings to identify epileptogenic patterns for seizure localization. However, for patients with focal epilepsy who are potential surgical candidates, a more precise prediction of the actual irritative cortex and ictal onset area is necessary to guide invasive explorations and resection. Advanced neurophysiologic techniques have allowed transformation of EEG and magnetoencephalography (MEG) tracings into models of the data (EEG and MEG source imaging). Source imaging utilizing EEG and MEG waveforms will be discussed in the following sections.

6.1 Electroencephalography (EEG)

Electrical potential is generated by a large number of geometrically aligned neurons (pyramidal cells) that simultaneously activate. The orientation of this voltage field is orthogonal to the surface of the generating cortex [1]. This is typically negative when recorded from electrodes on or above the cortical surface (Fig. 6.1). On the opposite side of this active cortical layer, either in underlying white matter or even the opposing scalp, there is a positive potential. Thus, EEG fields have a dipolar configuration with maximal negativity and positivity.

G. Osman
Division of Child Neurology, Department of Pediatrics, University of Texas Health Science Center at Houston, Houston, TX, USA

J. R. Gavvala (✉)
Neurology, Texas Comprehensive Epilepsy Program, University of Texas Health Science Center at Houston, Houston, TX, USA
e-mail: jay.r.gavvala@uth.tmc.edu

Z. Haneef (ed.), *Epilepsy Fundamentals*,
https://doi.org/10.1007/978-3-031-77741-7_6

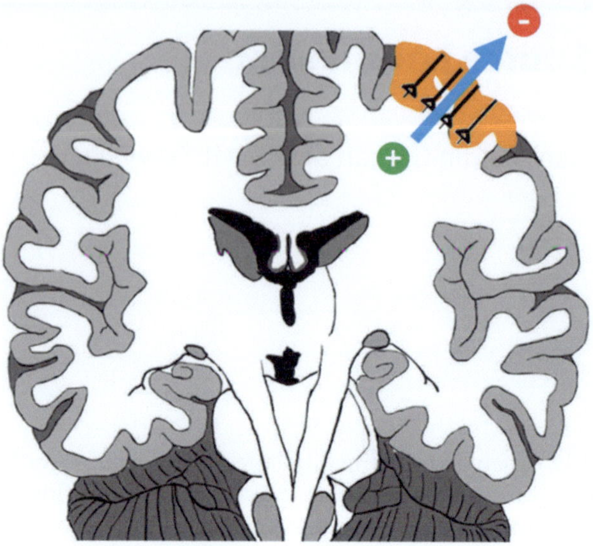

Fig. 6.1 Schematic showing geometrically aligned pyramidal cells (in black) of an electrically active cortex (in orange). These pyramidal cells produce an extracellular voltage field orthogonal to the surface of the generating cortex (blue arrow) with negativity directed towards the cortical surface

6.2 Magnetoencephalography (MEG)

MEG measures the weak magnetic fields generated by the human brain. Electric current is generated in post-synaptic (intracellular) pyramidal neurons and a corresponding magnetic field "loops" around the current source [2]. The magnetic current generated by the brain produces two sets of concentric circles representing the magnetic flux exiting the head and the re-entering flux. The two points where the recorded flux has the highest value are called extrema. The source is halfway between the extrema, and the source is at a depth proportional to the distance between the extrema (i.e., extrema close together indicate a source close to the surface of the brain with a source deeper in the brain produces extrema that are further apart) [3].

In contrast to EEG, purely radial sources do not produce a detectable magnetic field. MEG can only detect sources that have a tangential component (Fig. 6.2). Furthermore, the activity must be unopposed or it will cancel each other out (i.e., symmetric activation of a sulci will not be visible on MEG). The majority of MEG dipoles are found in the depths of sulci. MEG is of particular interest due to the fact that the skull and scalp are "magnetically transparent" and do not interfere with cortical signals, as seen with EEG. Its spatial sampling is higher than that of EEG (typical systems have 200–300 sensors) [4].

6.2.1 Technical Details

Magnetic fields are recorded by special sensors labeled "superconductive quantum interference device (SQUID)" sensors. These sensors consist of magnetic flux transformers which amplify the low-magnitude magnetic fields. Flux transformers

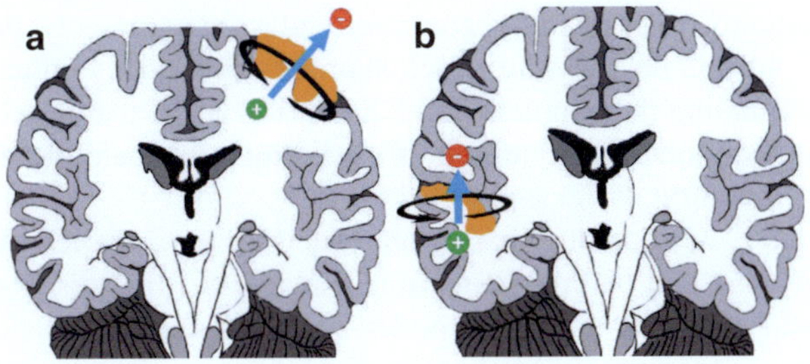

Fig. 6.2 Schematic showing an electrically active cortex (in orange) with an electrical voltage field (blue arrow) that is radial (left) or tangential (right) to the overlying cortex. The electrical field produces magnetic currents that loop around (black arrow). Note in radial sources the magnetic currents are nonlocalizable but are readily detected in tangential sources

consist of 1 or more pickup coils that sense the magnetic field of interest and an input coil that connects the transformers to SQUIDs. Magnetic flux transformer types include magnetometers, which are simple coils directly measuring magnetic fields perpendicular to the sensors and gradiometers analogous to conventional bipolar EEG and measure magnetic field gradients across two or more pickup coils. These include "planar gradiometers" consisting of two eight-shaped coils located in the same plane and wound up in opposing fashion and "axial gradiometers" consisting of two or more coils placed along the same vertical axis. Planar gradiometers preferentially pick up magnetic fields closest to the sensors and are therefore most sensitive to superficial sources and the least susceptible to external noise. Magnetometers and axial gradiometers are more sensitive to deep sources while magnetometers are the most susceptible to artifacts. The recording device consists of a helmet worn by the patient containing an array of 100+ magnetometers, gradiometers or a combination of both [3, 5–7].

6.2.2 Recording Magnetic Activity

The magnetic fields that reach the head surface are extremely small, approximately one million times weaker than the ambient magnetic field of the earth. As ambient magnetism from the earth's magnetic field (microteslas) can significantly interfere with recording, a magnetically shielded room is necessary for MEG, contributing significantly to the expense of the test [8]. Because the magnetic fields are extremely small, magnetometers/gradiometers must be superconductive (have extremely low resistance). Resistance in wires is lowered when wires are cooled to extremely low temperatures. The magnetometer/gradiometer wires are housed in a thermally insulated drum filled with liquid helium. The liquid helium keeps the wires at a temperature of about 4 degrees Kelvin, allowing them to be superconductive [8].

MEG can be successfully recorded in the majority of patient populations, including infants, patients with vagus nerve and brain stimulators, pacemakers and drug pumps. There are two absolute contraindications for MEG. These are the presence of cochlear implants or a special type of ventricular shunt, namely, the Medtronic® Strata™ NSC adjustable pressure valve shunt. Both devices contain permanent magnets contaminating the magnetic field, hindering good-quality recording. Brain and vagus nerve stimulators are turned off and drug pumps are placed in "standby mode" prior to initiation of the study [9–11].

Simultaneous EEG recording is recommended by the American Clinical MEG Society (ACMEGS) as it complements time analysis of MEG discharges, provides unique information on radial dipoles missed by MEG, supplements MEG analysis of discharges with low signal-to-noise ratio and helps elucidate the benign nature of MEG-unique variants [12]. Simultaneous EKG recording for identification of EKG artifacts is also recommended. We recommend referring to the ACMEGS and International Federation of Clinical Neurophysiology (IFCN) guideline statements for further recommendations on technical aspects [9, 10].

6.3 Principles of Source Localization

Source localization or source imaging is a model-based imaging technique that integrates temporal and spatial EEG or MEG data components to identify the generating source of electrical potentials recorded on the scalp [13]. There are two main problems that must be overcome to appropriately perform source localization known as the inverse and forward problems [2, 8].

Inverse problem refers to the problem of identifying the cortical generator of a given surface field potential. Simplistically, we would like to assume that a single source within the brain produces a discharge (Scalp EEG sharp/spike wave). However, multiple sources may contribute to scalp-recorded field potentials. Two commonly used models to overcome this are the equivalent current dipole (ECD) model and a distributed source model (Fig. 6.3) [8]

The basic assumption of an **ECD model** is that at any given instant, the detected potential represents activity from a single, infinitely small area of active cortex. This does not provide any information on the extent of the source. In contrast to the single dipole model, a **distributive source model** is based on the assumption that multiple sources can be simultaneously active across many locations at any given time. The distributed model reconstructs cerebral activity at each point on a 3D grid. Commonly utilized techniques include minimum norm estimates (MNE), dynamic statistical parametric mapping (dSPM), standard low-resolution brain electromagnetic tomography (sLORETA), EPIFOCUS, sLORETA weighted accurate minimum norm (SWARM) in addition to multiresolution focal undetermined system solution (MR-FOCUSS) [8, 14]. An alternative approach named "beamformers" utilizes spatial filters to maximize the signal of interest relative to background noise at predefined points, followed by an estimate of the activity of interest at each points [15]. Functional and effective connectivity analyses of the resting and interictal states are emerging as promising tools for the evaluation of epilepsy networks [16–20].

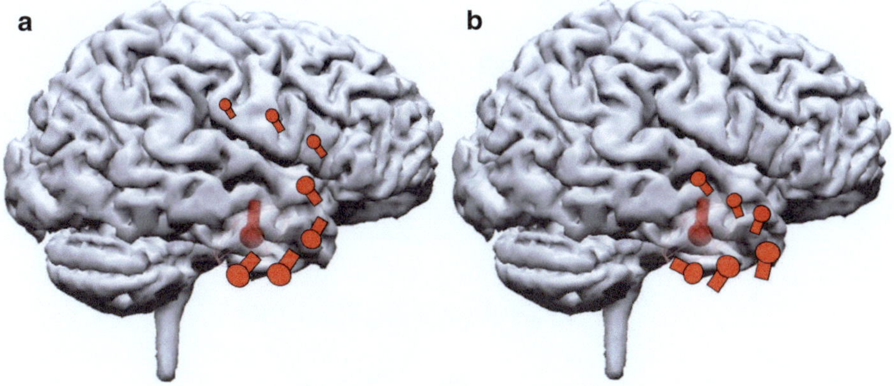

Fig. 6.3 Propagating left temporal spike, modeled as an equivalent current dipole using a three spherical model (**a**) and a boundary element model (**b**). Spherical head modeling errors (in **a**) are most exemplified in the regions of the brain that most deviate from a sphere (basal frontal, temporal and occipital sources)

Forward problem refers to the problem of identifying the source of a given surface field potential. Stated differently, within the conductor (the head), where is the given source of a field potential (EEG/MEG spike)? The earliest models use spherical head models; however, the accuracy of modeling, particularly for EEG dipoles, is decreased in the regions of the brain that conformed less to a sphere with larger confidence volumes, most notably the basal temporal regions of the brain [21]. Therefore, models based on the patient's MRI are now preferred. These include boundary element models (BEM) illustrated in Fig. 6.3 and finite element models (FEM) [3, 22, 23].

6.4 Electrical Source Imaging (ESI)

Electrical source imaging refers to the process of analyzing EEG data and using mathematical models to estimate the location of the cortical generator. The standard 10–20 EEG coverage is notable for its under-sampling of the inferior temporal lobe and basal temporo-occipital regions (Fig. 6.4). Additional subtemporal electrodes have been commonly utilized in ESI to better localize the area of maximal negativity on a sublobar level. Higher density arrays of EEG electrodes including 64, 128, and even 256 electrodes have been applied in ESI with increased sensitivity and specificity in identifying the epileptogenic focus with higher numbers of electrodes as well as results that are closer to the focus when compared to intracranial EEG data [24, 25]. A recent study attempted to validate various ESI methods by combining high-density array source analysis with intracranial EEG recording during single pulse electrical stimulation and found that localization accuracy of 1 cm can be achieved with the most accurate source analysis methods. Dipolar methods fared better than distributed methods [26]. Another recent study demonstrated substantial

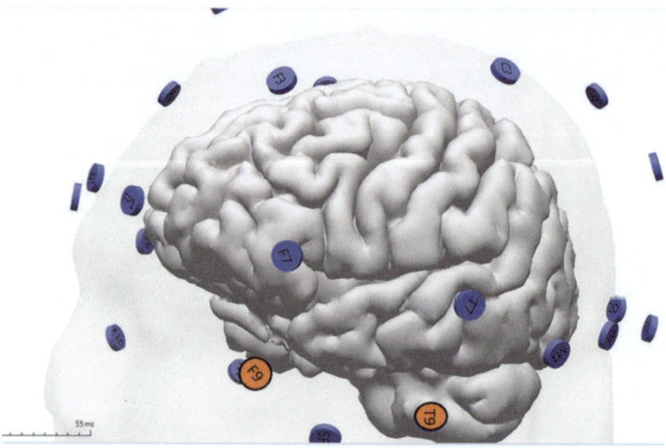

Fig. 6.4 Brain model with superimposed location of EEG electrodes. Note the placement of electrodes T7 and F7, while representing the most inferior set of 10–20 electrodes only offers coverage of the superior temporal gyrus with limited middle and inferior temporal gyrus coverage. The placement of additional subtemporal electrodes such as F9 and T9 (in orange) offers a more complete representation of the cortical activity (Image created using CURRY 7 software, Compumedics Neuroscan)

inter-analyzer agreement for interictal and ictal ESI among six experts in the field [27]. A recent study evaluated a semi-automated method for interictal ESI using long-term VEEG recordings and demonstrated excellent localizing accuracy and increased chances for seizure freedom when sources computed via this methodology were included in the surgical resection [28].

6.5 Magnetic Source Imaging (MSI)

Magnetic source imaging utilizes magnetoencephalographic data and source estimating models to project the location of the cortical generator (Figs. 6.5 and 6.6). Early prospective studies have demonstrated that MSI changed the approach of epilepsy surgery either by adding additional intracranial electrodes or changing the surgical approach in about 20–30% of all patients [29, 30]. A more recent retrospective study from Europe involving 1000 patients undergoing MEG as part of the presurgical evaluation showed that MEG provided additional information to the presurgical hypothesis in 32% of patients [31]. A recent prospective study including 141 patients showed that localization accuracy of electromagnetic source imaging (EMSI) ranged between 44–57% which was comparable to that of MRI and PET studies. EMSI additionally provided better localization accuracy than either modality alone [32].

MEG dipoles with uniform localization are commonly referred to as "clusters." A minimum of 5 dipoles are recommended per ACMEGS to define a cluster [9]. Clusters are further classified based on proximity to each other into "tight," "loose," and "scattered" dipole clusters (Fig. 6.7). While there are no universally accepted

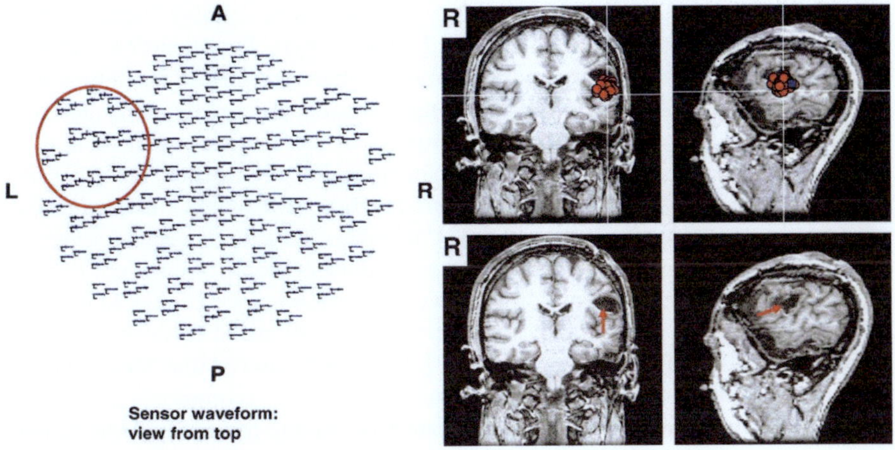

Fig. 6.5 MEG (left) showing epileptiform spikes from antero-lateral left fronto-temporal region and MSI (right) showing MEG spikes (top row) and dipole direction (bottom row) in the anterior portion of a left frontal resection cavity

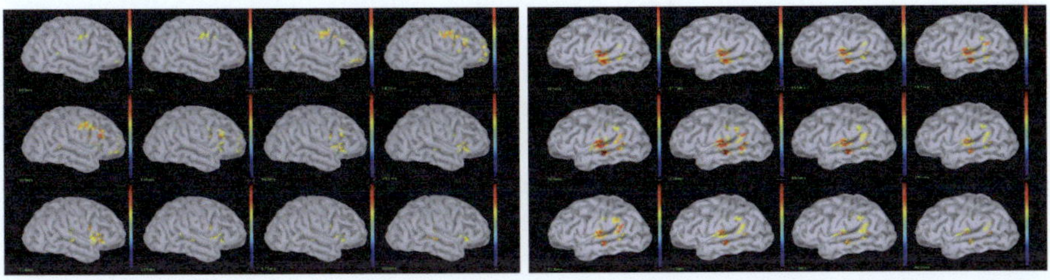

Fig. 6.6 MEG statistical parametric maps (SPM) obtained during an auditory language task provide information about the statistical reliability of the estimated signal at each location. There is only minimal activation over the right hemisphere (left panel), with prominent activation in the language areas over the left hemisphere (right panel)

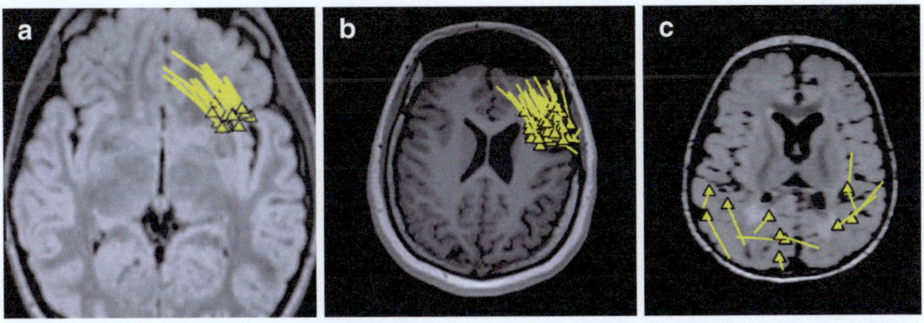

Fig. 6.7 Classification of MEG clusters: (**a**) left anterior insular tight cluster. (**b**) Left frontal loose cluster. (**c**) Scattered dipoles over the bilateral posterior head regions

definitions for these terms, Murakami et al. [33] suggested defining "tight clusters" as clusters localized to a single sulcus and 2 adjacent gyri, "loose clusters" as those occurring within the same sub-lobar region and "scattered clusters" as those within more than 1 sub-lobar region. Monofocal tight clusters are usually associated with focal pathology and inclusion of MEG clusters in the surgical resection is more likely to be associated with Engel class I outcomes. Seizure freedom is more likely in patients with a single tight cluster compared to those with a tight cluster plus scattered discharges or loose clusters [31, 33]. Dipole orientation is also an important classifying feature and can be considered as variable or stable orientation. Patients harboring tight clusters with uniform "stable" orientation perpendicular to the closest major sulcus are more likely to be seizure free compared to those harboring clusters with other orientation patterns [33]. One study showed that MEG detected about 56% of all interictal electrocorticography spikes, most commonly in the interhemispheric, orbitofrontal, peri-rolandic, superior frontal, parietal and lateral temporal neocortex [34]. However, MEG was not performed simultaneously with intracranial EEG in that study and the effect of antiseizure medications on suppressing interictal discharges in some patients cannot be ruled out. A more recent study involving 14 patients with interictal spikes on simultaneous MEG and SEEG recordings showed that MEG and SEEG were equally sensitive in the detection of superficial cortical interictal discharges, while SEEG was significantly more sensitive in the detection of deep interictal discharges [35].

Three patterns of MEG dipole orientation can be seen in patients with temporal lobe epilepsy. These include anterior temporal horizontal, anterior temporal vertical and mid-posterior vertical dipoles (Fig. 6.8). Anterior temporal horizontal and vertical dipoles are frequently associated with mesial temporal epilepsy, while mid-posterior vertical dipoles are commonly associated with lateral neocortical or pseudo-temporal sources [36, 37]. However, posterior temporal dipoles have also been reported in cases with mesial temporal epilepsy with excellent surgical outcomes after anterior temporal lobectomy despite the persistence of postoperative posterior temporal MEG discharges [38, 39]. Interictal discharges confined to mesial temporal structures are commonly missed on MEG and scalp EEG and anterior temporal horizontal dipoles are associated with propagation to a temporal pole while anterior temporal vertical dipoles suggest basal temporal lobe activation [37].

Dipole orientation also provides unique lateralizing information in patients with interhemispheric MEG discharges as dipoles point towards the epileptogenic hemisphere in nearly all cases with interhemispheric discharges (Fig. 6.9). Dipole orientation provides important localizing information for opercular discharges as superior orientation suggests fronto-parietal opercular source while inferior orientation suggests temporal opercular source [40]. Furthermore, consistent dipole orientation pointing anterior or posterior to a sulcus strongly implicates activation of the anterior or posterior bank of the gyrus, respectively. MEG can be particularly helpful in patients with operculo-insular epilepsy, including those with no apparent MRI

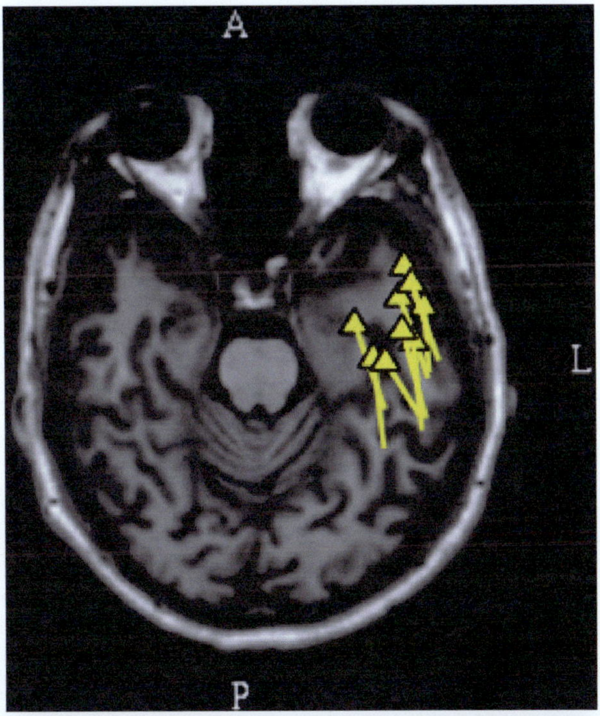

Fig. 6.8 Left anterior temporal horizontal MEG dipoles consistent with left mesial temporal epilepsy

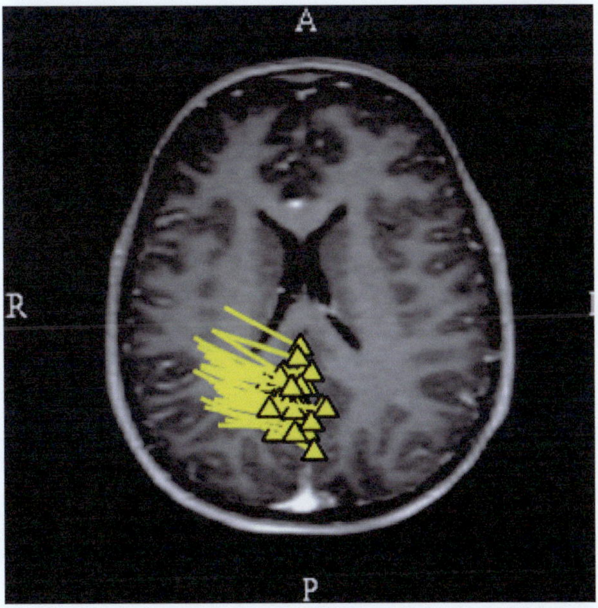

Fig. 6.9 A loose cluster of MEG discharges within the right mesial occipito-parietal region with uniform orientation pointing towards the right hemisphere consistent with right hemispheric lateralization

lesion and negative interictal epileptiform discharges, and is more likely to provide additional localizing information than either PET or SPECT in these patients [40–42].

MEG can also provide unique localizing information guiding surgical resective or ablative procedures in patients with multifocal brain pathologies such as the Tuberous sclerosis complex. One study comparing high-density ESI and MSI results showed that MSI clusters were located closer to presumed epileptogenic tubers than ESI clusters in a majority of patients [43]. Another study showed that MEG was superior to interictal SPECT in localizing epileptogenic tubers and correlated with ictal SPECT results [44]. A more recent study demonstrated that resection of MSI clusters was associated with favorable outcomes. Ictal MEG provided additional localizing information in that cohort [45]. Figure 6.10 shows MEG clusters adjacent to 2 presumed epileptogenic tubers. MEG can be similarly helpful in outlining the extent of resection in patients with porencephalic cysts and areas of encephalomalacia where MEG is superior to EEG in the detection of epileptiform discharges [46].

The systematic reporting of MEG findings is variable among institutions. While simultaneous EEG is recorded alongside MEG at clinical MEG centers, EEG signal analysis may not always be performed. An averaging of similar MEG discharges can highlight low amplitude EEG discharges missed otherwise and highlight the temporal relationship between EEG and MEG. Some centers utilize statistical thresholds to select model-worthy dipoles. Criteria for statistically significant sources at our institution include reduced Chi-square ≥1 and ≤2, confidence volume <1000 mm [3], goodness of fit ≥80% and dipole strength of 100–500 nAm [12].

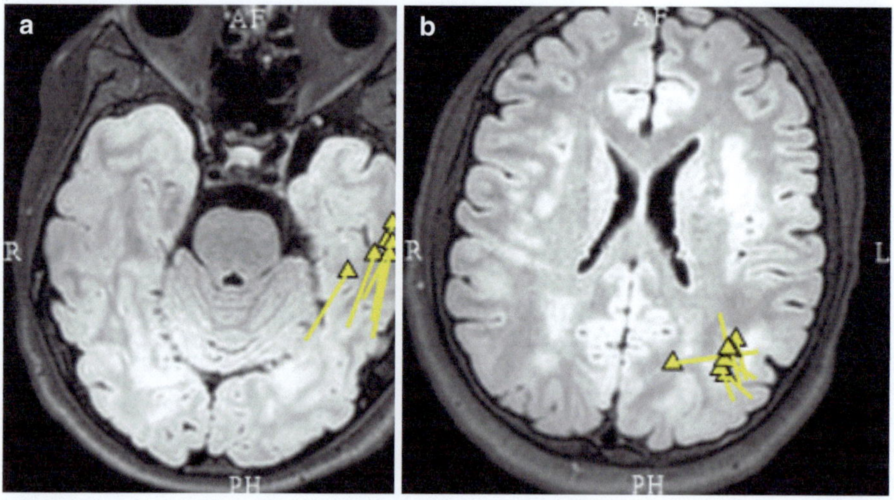

Fig. 6.10 Left mid-temporal (**a**) and posterior parietal (**b**) MEG clusters adjacent to cortical tubers in a patient with TSC with multifocal tubers

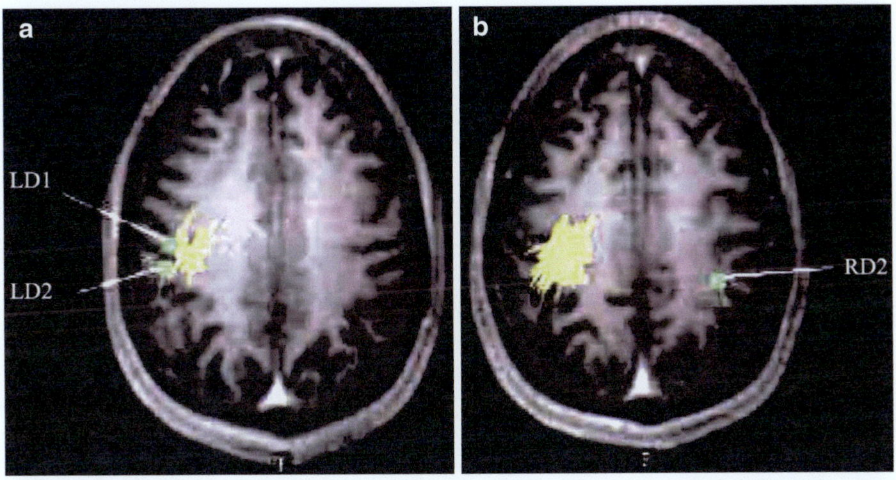

Fig. 6.11 Spontaneous MEG dipoles shown in yellow in a patient with suspected right frontal epilepsy. Motor-evoked potentials from stimulation of the left (LD2, **a**) and right (RD2, **b**) index fingers result in the MEG dipoles shown in green

Despite the limited duration of the MEG studies, seizures are recorded in 7–24% of patients undergoing MEG for presurgical evaluation. Post-processing and movement compensation algorithms allow for accurate localization of seizure onset despite relatively low SNR compared to interictal discharges. Dipole modeling, distributed source analysis and beamformer methods have been employed for seizure onset analysis [47–53].

MEG is also useful in brain mapping as it allows real-time tracking of brain activity to study somatosensory, motor, language, visual, auditory, and cognitive functions (Figs. 6.6 and 6.11). Early data seems to suggest a similar ability to lateralize language compared to Wada testing and to localize sensorimotor cortices as compared to fMRI [54–57].

With the multitude of available testing at a clinician's disposal, it can be difficult to best determine which testing would be most useful in a given scenario. MEG has been felt to offer the most utility in the following scenarios [2, 58]:

(1) No clear hypothesis regarding seizure onset
(2) Insular onset suspected
(3) Interhemispheric (especially frontal) onset suspected
(4) MRI negative suspected mesial temporal lobe epilepsy
(5) Equivocal apparent bilateral or generalized EEG patterns
(6) Intra-sylvian onset suspected
(7) Multiple or very large epileptogenic lesions on MRI
(8) EEG-negative cases
(9) Prior craniotomy or epilepsy surgery re-operation

Pearls: Source Localization

- Electrical activity detected by EEG has a dipolar configuration with negativity typically projected outside and orthogonal to the surface of the generating cortex.
- MEG has superior spatial sampling than EEG and magnetic waves are not distorted by the skull like electrical currents.
- MEG has superior localization accuracy compared to EEG. Whole-head MEG systems have 3–4 mm localization accuracy for dipoles with good SNR while the localization accuracy of HD-EEG does not exceed 7–8 mm [45].
- At least 5 MEG discharges within the same region are required for accurate localization of discharges [44].
- Use of a boundary element model to model epileptiform data is more accurate than a spherical model particularly in the basal regions of the brain [22, 23] for EEG. The finite element model provides more accurate ESI localization in patients with skull defects or brain edema [3].
- MEG dipolar distribution creates two focal points with maximal signal magnitude called "extrema." Superficial sources create proximate extrema while deep sources create distant extrema [3].
- Resection of a tightly clustered MSI source is associated with a higher likelihood of seizure freedom. Loose clusters with variable dipole orientation and scattered dipoles suggest a regional or broad network rather than a confined focal onset.
- Averaging of discharges can help improve SNR and accentuate low-amplitude MEG or EEG discharges. However, it can mask location differences between discharges, leading to false localization and should therefore be limited to discharges with identical source and field distribution [44].
- MEG has been shown to be most valuable in patients with non-localizing video EEG studies and in interhemispheric and operculo-insular epilepsies, including those with negative EEG. Dipole orientation is usually indicative of the region of onset in opercular discharges where upward/anterior orientation is associated with frontal opercular sources, upward/posterior orientation is associated with parietal opercular sources and downward orientation is associated with temporal opecular sources [44–47].
- Anterior temporal horizontal dipoles suggest temporal tip activation while anterior temporal vertical dipoles suggest mesio-basal temporal source. Both are consistent with mesial temporal lobe epilepsy.
- The occurrence of posterior temporal vertical dipoles raises concern for neo-cortical temporal or pseudo-temporal epilepsy [37, 59, 60].
- MEG dipoles are noted within 3 cm of epileptic lesions in a majority of lesional FLE patients with monofocal MEG clusters. The occurrence of monofocal MEG clusters is associated with a favorable prognosis. Incomplete resection of MEG clusters is associated with poor prognosis even with complete lesionectomy [61].

(continued)

- Dipole orientation determines the epileptogenic hemisphere in patients with inter-hemispheric epilepsy. MEG dipoles are usually oriented towards the epileptogenic side. This can help resolve EEG false lateralization occasionally encountered in these cases [43–45, 62].
- MEG can help with the identification of epileptogenic tubers for potential resection in patients with tuberous sclerosis complex (TSC). MEG provides more accurate localization than EEG in those patients, and MEG clusters are usually closer to epileptogenic tubers than HD-EEG clusters [43–45, 62].
- MEG can assist with delineating the extent of resection in patients with porencephalic cysts and encephalomalacia, where MEG is more sensitive than EEG in the detection of epileptiform discharges [46, 63].

References

1. Ebersole JS. Noninvasive localization of epileptogenic foci by EEG source modeling. Epilepsia. 2000;41(Suppl 3):S24–33.
2. Kharkar S, Knowlton R. Magnetoencephalography in the presurgical evaluation of epilepsy. Epilepsy Behav. 2015;46:19–26.
3. Papanicolaou AC. Clinical magnetoencephalography and magnetic source imaging. Cambridge: Cambridge University Press; 2009.
4. Ebersole JS, Ebersole SM. Combining MEG and EEG source modeling in epilepsy evaluations. J Clin Neurophysiol. 2010;27:360–71.
5. Lee Y-H, Kim K. Instrumentation for measuring MEG signals. In: Supek S, Aine CJ, editors. Magnetoencephalography: from signals to dynamic cortical networks. Berlin: Springer; 2014. p. 3–33. https://doi.org/10.1007/978-3-642-33045-2_1.
6. Vrba J, Robinson SE. SQUID sensor array configurations for magnetoencephalography applications. Supercond Sci Technol. 2002;15:R51–89.
7. Hari R, Puce AMEG. EEG primer. Oxford: Oxford University Press; 2023. https://doi.org/10.1093/med/9780197542187.001.0001.
8. Anderson CT, Carlson CE, Li Z, Raghavan M. Magnetoencephalography in the preoperative evaluation for epilepsy surgery. Curr Neurol Neurosci Rep. 2014;14:446.
9. Bagić AI, Knowlton RC, Rose DF, Ebersole JS, ACMEGS Clinical Practice Guideline (CPG) Committee. American Clinical Magnetoencephalography Society clinical practice guideline 1: recording and analysis of spontaneous cerebral activity. J Clin Neurophysiol. 2011;28:348–54.
10. Hari R, et al. IFCN-endorsed practical guidelines for clinical magnetoencephalography (MEG). Clin Neurophysiol. 2018;129:1720–47.
11. Mosher JC, Funke ME. Towards best practices in clinical magnetoencephalography: patient preparation and data acquisition. J Clin Neurophysiol. 2020;37:498–507.
12. Laohathai C, et al. Practical fundamentals of clinical MEG interpretation in epilepsy. Front Neurol. 2021;12:722986.
13. Kaiboriboon K, Luders HO, Hamaneh M, Turnbull J, Lhatoo SD. EEG source imaging in epilepsy–practicalities and pitfalls. Nat Rev Neurol. 2012;8:498–507.
14. Tenney JR, Fujiwara H, Rose DF. The value of source localization for clinical magnetoencephalography: beyond the equivalent current dipole. J Clin Neurophysiol. 2020;37:537–44.
15. Westner BU, et al. A unified view on beamformers for M/EEG source reconstruction. NeuroImage. 2022;246:118789.

16. van Mierlo P, Höller Y, Focke NK, Vulliemoz S. Network perspectives on epilepsy using EEG/MEG source connectivity. Front Neurol. 2019;10:721.
17. Evaluation of brain connectivity: the role of magnetoencephalography - Burgess - 2011 - Epilepsia - Wiley Online Library. https://doi.org/10.1111/j.1528-1167.2011.03148.x.
18. Brookes MJ, et al. Measuring functional connectivity using MEG: methodology and comparison with fcMRI. NeuroImage. 2011;56:1082–104.
19. Jin S-H, Chung CK. Towards brain connectivity in epilepsy using MEG. In: Supek S, Aine CJ, editors. Magnetoencephalography: from signals to dynamic cortical networks. Berlin: Springer; 2014. p. 843–8. https://doi.org/10.1007/978-3-642-33045-2_40.
20. Xu N, Shan W, Qi J, Wu J, Wang Q. Presurgical evaluation of epilepsy using resting-state MEG functional connectivity. Front Hum Neurosci. 2021;15:649074.
21. Ebersole JS, Hawes-Ebersole S. Clinical application of dipole models in the localization of epileptiform activity. J Clin Neurophysiol. 2007;24:120–9.
22. Fuchs M, Wagner M, Kastner J. Boundary element method volume conductor models for EEG source reconstruction. Clin Neurophysiol. 2001;112:1400–7.
23. Scheler G, et al. Spatial relationship of source localizations in patients with focal epilepsy: comparison of MEG and EEG with a three spherical shells and a boundary element volume conductor model. Hum Brain Mapp. 2006;28:315–22.
24. Brodbeck V, et al. Electroencephalographic source imaging: a prospective study of 152 operated epileptic patients. Brain. 2011;134:2887–97.
25. Mégevand P, et al. Electric source imaging of interictal activity accurately localises the seizure onset zone. J Neurol Neurosurg Psychiatry. 2014;85:38–43.
26. Pascarella A, et al. An in–vivo validation of ESI methods with focal sources. NeuroImage. 2023;277:120219.
27. Mattioli P, et al. Electric source imaging in presurgical evaluation of epilepsy: an inter-analyser agreement study. Diagnostics. 2022;12:2303.
28. Spinelli L, et al. Semiautomatic interictal electric source localization based on long-term electroencephalographic monitoring: a prospective study. Epilepsia. 2023;64:951–61.
29. Sutherling WW, et al. Influence of magnetic source imaging for planning intracranial EEG in epilepsy. Neurology. 2008;71:990–6.
30. Knowlton RC, et al. Effect of epilepsy magnetic source imaging on intracranial electrode placement. Ann Neurol. 2009;65:716–23.
31. Rampp S, et al. Magnetoencephalography for epileptic focus localization in a series of 1000 cases. Brain J Neurol. 2019;142:3059–71.
32. Duez L, et al. Electromagnetic source imaging in presurgical workup of patients with epilepsy. Neurology. 2019;92:e576–86.
33. Murakami H, et al. Correlating magnetoencephalography to stereo-electroencephalography in patients undergoing epilepsy surgery. Brain. 2016;139:2935–47.
34. Agirre-Arrizubieta Z, Huiskamp GJM, Ferrier CH, Van Huffelen AC, Leijten FSS. Interictal magnetoencephalography and the irritative zone in the electrocorticogram. Brain. 2009;132:3060–71.
35. Vivekananda U, et al. The use of simultaneous stereo-electroencephalography and magnetoencephalography in localizing the epileptogenic focus in refractory focal epilepsy. Brain Commun. 2021;3:72.
36. Ebersole JS, Smith JR. MEG spike modeling differentiates baso-mesial from lateral cortical temporal epilepsy. Electroencephalogr Clin Neurophysiol. 1995;95:20.
37. Baumgartner C, Pataraia E, Lindinger G, Deecke L. Neuromagnetic recordings in temporal lobe epilepsy. J Clin Neurophysiol. 2000;17:177–89.
38. Iwasaki M, et al. Surgical implications of neuromagnetic spike localization in temporal lobe epilepsy. Epilepsia. 2002;43:415–24.
39. Assaf BA, et al. Magnetoencephalography source localization and surgical outcome in temporal lobe epilepsy. Clin Neurophysiol. 2004;115:2066–76.
40. Kakisaka Y, et al. Magnetoencephalography in fronto-parietal opercular epilepsy. Epilepsy Res. 2012;102:71–7.

41. Mohamed IS, et al. The utility of magnetoencephalography in the presurgical evaluation of refractory insular epilepsy. Epilepsia. 2013;54:1950–9.
42. Chourasia N, Quach M, Gavvala J. Insular magnetoencephalography dipole clusters in patients with refractory focal epilepsy. J Clin Neurophysiol. 2021;38:542–6.
43. Jansen FE, et al. Identification of the epileptogenic tuber in patients with tuberous sclerosis: a comparison of high-resolution EEG and MEG. Epilepsia. 2006;47:108–14.
44. Kamimura T, et al. Magnetoencephalography in patients with tuberous sclerosis and localization-related epilepsy. Epilepsia. 2006;47:991–7.
45. Koptelova A, et al. Ictal and interictal MEG in pediatric patients with tuberous sclerosis and drug resistant epilepsy. Epilepsy Res. 2018;140:162–5.
46. Kakisaka Y, et al. MEG may reveal hidden population of spikes in epilepsy with porencephalic cyst/encephalomalacia. J Clin Neurophysiol. 2017;34:546–9.
47. Fujiwara H, et al. Ictal MEG onset source localization compared to intracranial EEG and outcome: improved epilepsy presurgical evaluation in pediatrics. Epilepsy Res. 2012;99:214–24.
48. Plummer C, et al. Interictal and ictal source localization for epilepsy surgery using high-density EEG with MEG: a prospective long-term study. Brain. 2019;142:932–51.
49. Alkawadri R, Burgess RC, Kakisaka Y, Mosher JC, Alexopoulos AV. Assessment of the utility of ictal magnetoencephalography in the localization of the epileptic seizure onset zone. JAMA Neurol. 2018;75:1264–72.
50. Assaf BA, et al. Ictal magnetoencephalography in temporal and extratemporal lobe epilepsy. Epilepsia. 2003;44:1320–7.
51. Stefan H, Rampp S. Interictal and ictal MEG in presurgical evaluation for epilepsy surgery. Acta Epileptol. 2020;2:11.
52. Ramanujam B, et al. Can ictal-MEG obviate the need for phase II monitoring in people with drug-refractory epilepsy? A prospective observational study. Seizure. 2017;45:17–23.
53. Garcia Dominguez L, Tarazi A, Valiante T, Wennberg R. Beamforming seizures from the temporal lobe using magnetoencephalography. Can J Neurol Sci J Can Sci Neurol. 2023;50:201–13.
54. Papanicolaou AC, et al. Magnetocephalography: a noninvasive alternative to the Wada procedure. J Neurosurg. 2004;100:867–76.
55. Bowyer SM, Zillgitt A, Greenwald M, Lajiness-O'Neill R. Language mapping with magnetoencephalography: an update on the current state of clinical research and practice with considerations for clinical practice guidelines. J Clin Neurophysiol. 2020;37:554–63.
56. Kreidenhuber R, De Tiège X, Rampp S. Presurgical functional cortical mapping using electromagnetic source imaging. Front Neurol. 2019;10:628.
57. De Tiège X, Bourguignon M, Piitulainen H, Jousmäki V. Sensorimotor mapping with MEG: an update on the current state of clinical research and practice with considerations for clinical practice guidelines. J Clin Neurophysiol. 2020;37(6):564–73.
58. Bagić AI, et al. The 10 common evidence-supported indications for MEG in epilepsy surgery: an illustrated compendium. J Clin Neurophysiol. 2020;37:483–97.
59. Ebersole JS. Magnetoencephalography/magnetic source imaging in the assessment of patients with epilepsy. Epilepsia. 1997;38(Suppl 4):S1–5.
60. Pataraia E, Lindinger G, Deecke L, Mayer D, Baumgartner C. Combined MEG/EEG analysis of the interictal spike complex in mesial temporal lobe epilepsy. NeuroImage. 2005;24:607–14.
61. Stefan H, et al. MEG in frontal lobe epilepsies: localization and postoperative outcome. Epilepsia. 2011;52:2233–8.
62. Wu JY, et al. Magnetic source imaging localizes epileptogenic zone in children with tuberous sclerosis complex. Neurology. 2006;66:1270–2.
63. Bennett-Back O, et al. Magnetoencephalography helps delineate the extent of the epileptogenic zone for surgical planning in children with intractable epilepsy due to porencephalic cyst/encephalomalacia. J Neurosurg Pediatr. 2014;14:271–8.

Neonatal and Pediatric EEG

7

Joyce H. Matsumoto and Jason T. Lerner

The EEG of the neonate is one of the most challenging topics in neurophysiology. During the neonatal time period, the background patterns of the EEG change every few weeks with the growth and development of the brain. Development is rapid during the prenatal time period before birth.

7.1 Maturation of the EEG from the Neonate to Childhood

A pattern that is normal at one time may be abnormal a month later (Fig. 7.1). For example, the background is 100% synchronous in a 28-week-old premature baby, 70% synchronous in a 34-week-old and 100% again in a 38-week-old (Table 7.1, Figs. 7.2, 7.3, 7.4, 7.5, 7.6, 7.7, 7.8, 7.9, and 7.10). Additionally, this is all a continuum, and there can be a range in the age at which the changes occur (Table 7.2).

Positive sharp wave transients are sharp waves seen in the Rolandic head regions (C3, C4 and the vertex) that may be associated with underlying white matter injury, including periventricular leukomalacia. Excessive sharp transients may be indicative of focal pathology (white matter injury or hemorrhage). Positive sharp discharges may be seen in healthy preterm infants at 3/h and in term infants at 1.5/h.

Multifocal independent sharp discharges (MISD) are defined as excessive negative sharp-wave transients that are seen in at least 3 non-adjacent electrodes (Fig. 7.1). Physiologic sharp transients may be seen mostly in central and temporal locations, generally occurring less than once per page (<11/h in normal preterm neonates and

J. H. Matsumoto (✉)
Division of Pediatric Neurology, David Geffen School of Medicine at UCLA,
UCLA Mattel Children's Hospital, Los Angeles, CA, USA
e-mail: JMatsumoto@mednet.ucla.edu

J. T. Lerner
Biohaven Pharma, New Haven, CT, USA

Z. Haneef (ed.), *Epilepsy Fundamentals*,
https://doi.org/10.1007/978-3-031-77741-7_7

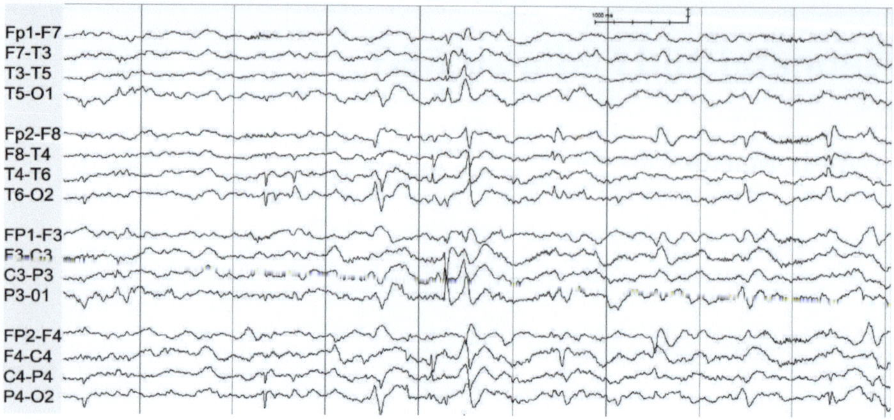

Fig. 7.1 Multifocal independent sharp discharges

Table 7.1 Development of the EEG in neonates

Age (CA) (weeks)	Continuity	Synchrony (%)	Burst composition	Reactivity	Voltage (μV)
24–29	Tracé discontinu	100	Monorhythmic delta intermixed with temporal theta	None	<2
30–34	Periods of continuity increase in wake and active sleep	70–80	Increase in theta, delta brushes in active sleep > quiet sleep	Some change (diffuse attenuation) with stimulation	<25
35–36	Tracé alternant	85	Mix of all frequencies, delta brushes in quiet sleep	Stimulation in quiet sleep provokes diffuse attenuation	≥25
37–40	Continuous in wake and active sleep	100	Encoches frontales appear, delta brushes in quiet sleep	Consistent reaction to stimulation	>25
40–44	Continuous in wake and active sleep, tracé alternant in quiet sleep	100	Delta brushes in quiet sleep	Consistent reaction to stimulation	75–100
44–48	Continuous background	100	Delta brushes disappear, spindles start in the midline	Consistent reaction to stimulation	75–100

<13/h in normal term neonates). Excessive negative sharp transients may be concentrated in one or more regions or are more frequent than expected for physiologic sharps.

Brief rhythmic discharges (BRDs) are rhythmic waveforms that appear similar to seizures; however, they are brief (<10 s) and are frequently seen in the background of an abnormal EEG. The significance of BRDs is unclear.

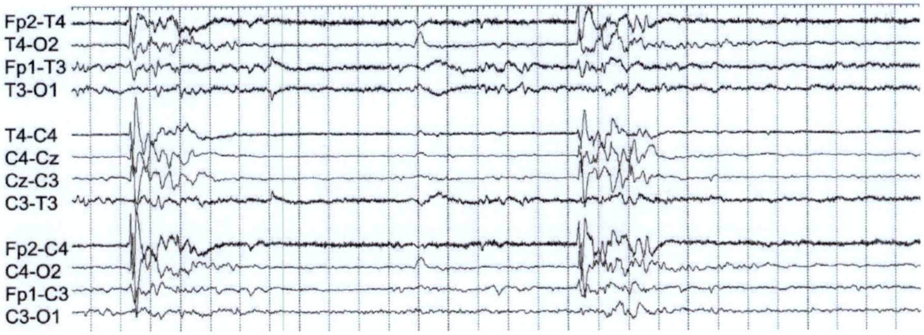

Fig. 7.2 24–26 weeks of gestation: synchronous, burst suppression-like pattern: Wakefulness and sleep are electrographically similar, characterized by burst suppression pattern with interburst intervals <5 μV

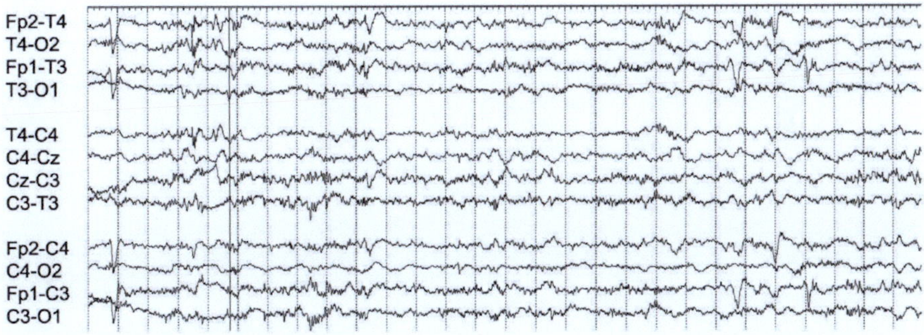

Fig. 7.3 30–40 weeks of gestation: Activité Moyenne pattern: This occurs in healthy term neonates in the awake state with eyes open. The EEG shows continuous, low to medium voltage (25–50 uV) mixed frequency activity with a predominance of theta and delta and overriding beta. The term activité moyenne roughly means "average or medium" (EEG activity)

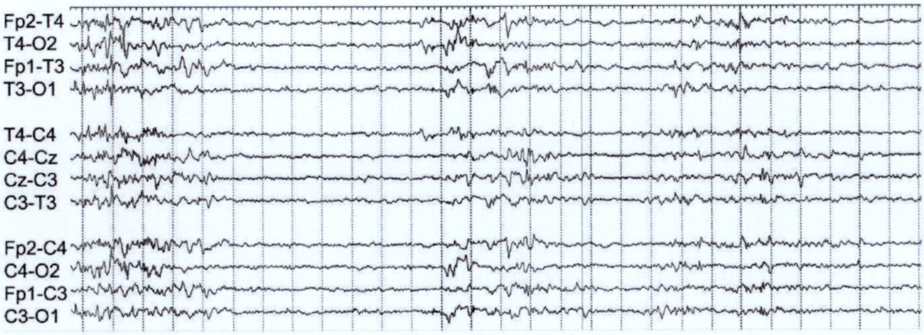

Fig. 7.4 28–32 weeks of gestation: Tracé discontinu pattern of quiet sleep: In contrast to the continuous activité moyenne pattern of wakefulness and active sleep, quiet sleep is composed of a discontinuous tracé discontinu pattern in which interburst intervals remain <25 μV

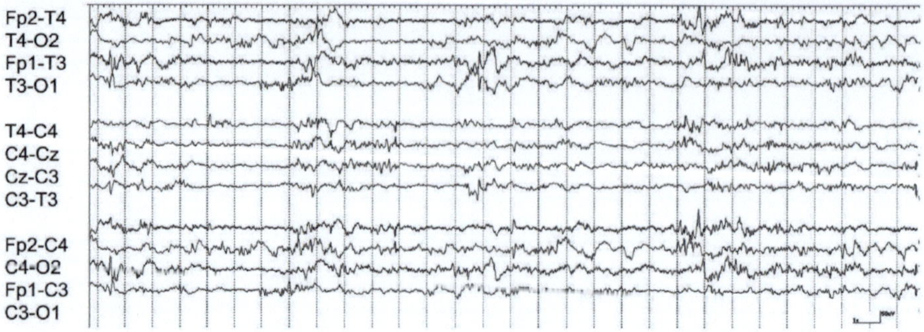

Fig. 7.5 35–37 weeks of gestation: Tracé alternant pattern of quiet sleep: Activité moyenne continues to characterize wakefulness and active sleep. Interburst intervals gradually shorten, and interburst voltage increases to >25 μV)

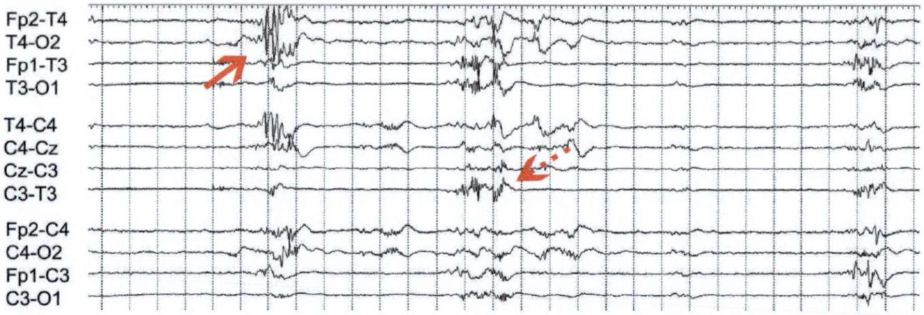

Fig. 7.6 Temporal theta bursts (solid arrow) and central delta brushes (dashed arrow) at 30 weeks

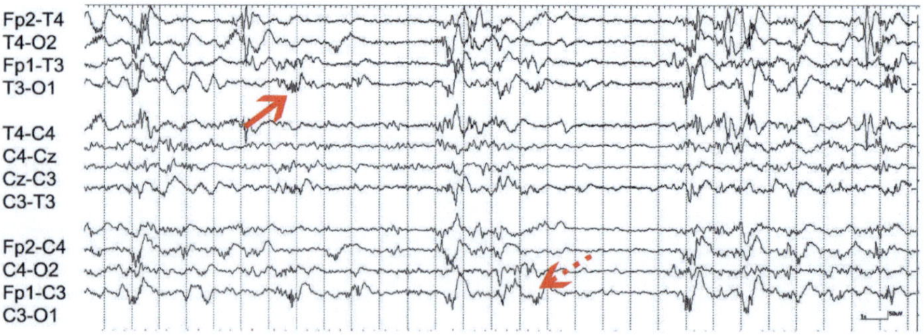

Fig. 7.7 Posterior delta brushes at 34 weeks (solid and dashed arrows)

Sleep spindles are bursts of 10–12 Hz oscillations with distinct spindle morphology which appear in the background during drowsiness and early sleep between 2 and 3 months of age (Table 7.2). They are seen initially asynchronously over the central head region, with a duration of up to 10 s (Figs. 7.9 and 7.10). Over time, spindles increase in frequency (12–14 Hz) and are mostly synchronous by 2 years

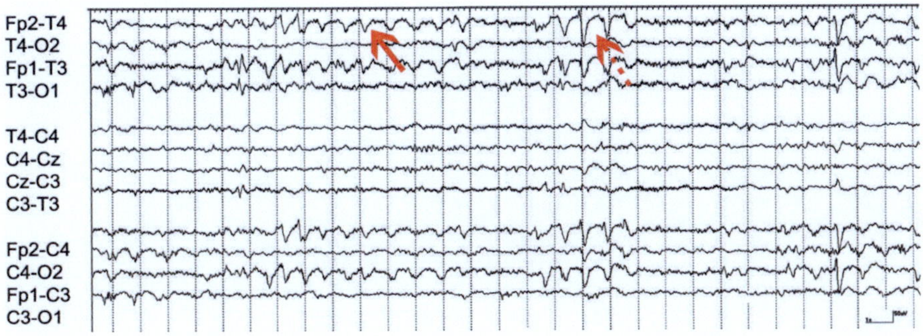

Fig. 7.8 Anterior dysrhythmia (solid arrow) and encoches frontales (dashed arrow) at 40 weeks

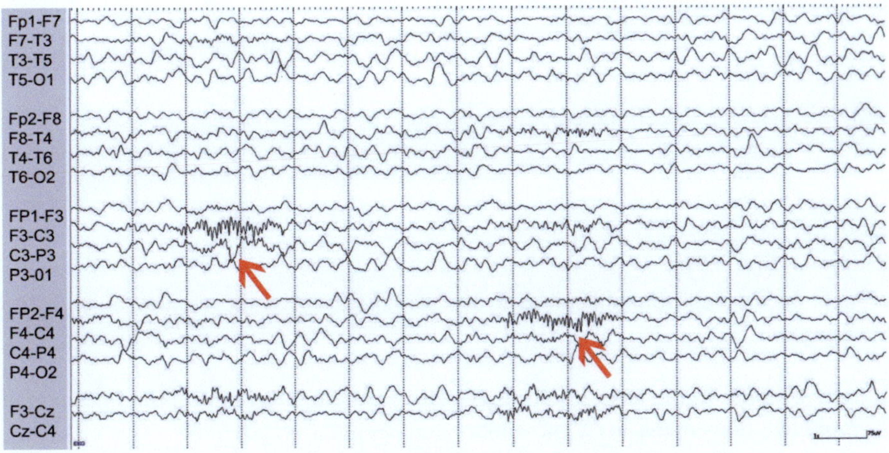

Fig. 7.9 Asynchronous sleep spindles at 4 months of age (solid and dashed arrows)

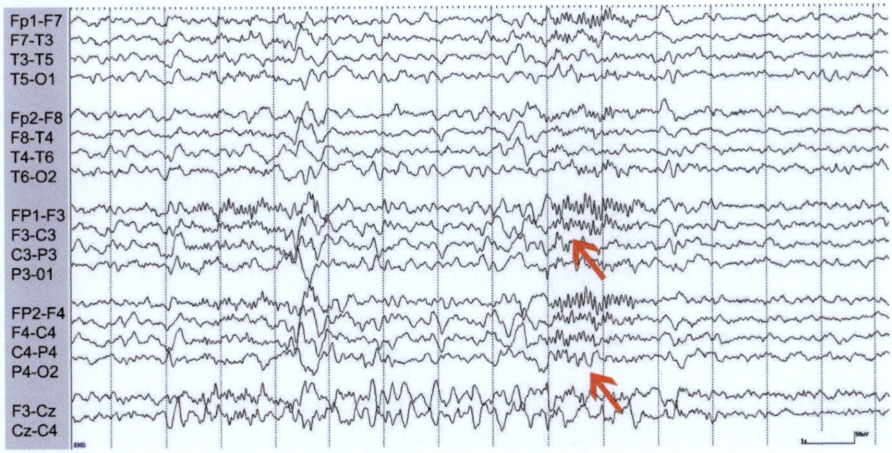

Fig. 7.10 Synchronous sleep spindles (solid and dashed arrows) and vertex waves in childhood

Table 7.2 Development of the EEG in children

Age	Awake background	Posterior dominant rhythm	Sleep background	Activation
2 months to 1 year	Delta frequencies with a steady increase in theta	3 Hz at 3 months 6 Hz at 1 year	Spindles seen at 3 months; vertex waves seen at 4 months; hypnagogic hypersynchrony seen	Photic stimulation produces driving at lower frequencies
1–3 years	Steady increase in theta activity	8 Hz at 3 years	Spindles are mostly synchronous by 2 years Hypnagogic and hypnopompic hypersynchrony seen	Photic stimulation produces driving at lower frequencies
3–6 years	Posterior slow waves of youth and rolandic mu seen	8–10 Hz	POSTS seen	Photic stimulation produces driving at <8 Hz, hyperventilation produces prominent diffuse delta
6–12 years	Posterior slow waves of youth and rolandic mu	10 Hz	Hypnagogic hypersynchrony less frequent after 6 years	Photic stimulation produces driving at 6–16 Hz, hyperventilation produces high-amplitude delta

of age. In late childhood, spindles are prominent over the frontal and central head regions.

- *Active sleep*: In the term infant active sleep is defined as intermittent periods of rapid eye movement, body movement and irregular respirations
- *Continuity*: The degree to which the background activity is sustained. The background is discontinuous when periods of electrical activity (bursts) are interrupted by periods of attenuation (interburst intervals)
- *Delta brushes*: High amplitude slow waves with superimposed fast activity
- *Encoches frontales*: Frontal sharp discharges with a small initial negative deflection followed by a larger positive deflection
- *Quiet sleep*: In the term infant quiet sleep is defined as the absence of rapid eye movement, minimal movements and closed eyes
- *Reactivity*: Changes in the EEG background in response to external stimulation
- *Synchrony*: Bursts of activity simultaneous in the right and left hemispheres
- *Tracé discontinu*: Absence of sustained continuity
- *Tracé alternant*: Transitions from complete discontinuity to a continuous background
- *Voltage*: Amplitude of the background

Posterior slow waves of youth are a normal variant consisting of delta waves seen in the occipital head region intermixed with alpha activity. These are seen when the eyes are closed and blocked with eye-opening. Posterior slow waves of

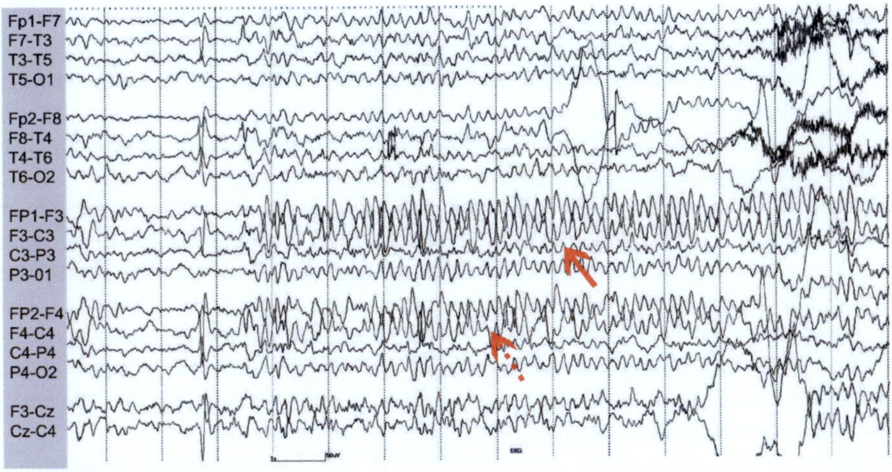

Fig. 7.11 Hypnopompic hypersynchrony in a 3-year-old child, most prominent in bifrontal regions (solid and dashed arrows)

youth appear around 3 years and persist throughout childhood but disappear in young adulthood (Table 7.2).

Positive occipital sharp transients of sleep (POSTS) are synchronous sharp waves in the posterior head region that are surface positive, occurring in singlets or brief runs during sleep. These are most common in young adults. POSTS are considered a normal variant, which occurs in approximately 10% of studies and may be mistaken for sharp discharges. ***Hypersynchrony*** is a high-voltage monomorphic slow rhythmic activity seen diffusely but most prominently in the bilateral frontal head regions (Fig. 7.11). This typically occurs during drowsiness (hypnagogic) and upon waking (hypnopompic). A similar pattern is also seen during hyperventilation.

7.2 Abnormal EEG Findings

There are a number of neonatal and childhood epilepsy syndromes that must be considered with pediatric EEGs. These syndromes have characteristic ages, seizure types, EEG findings, and prognostic implications.

7.2.1 Epileptic Encephalopathies

It is important to recognize epileptic encephalopathy syndromes because these are associated with permanent cognitive deterioration. In some cases, early recognition and prompt definitive treatment of these seizure disorders can halt the regression and enable restoration of the child's developmental trajectory. There are several age-dependent epileptic encephalopathies, and some affected children will transition from one syndrome to the next as they grow older. A number of underlying etiologies are possible, including structural brain abnormalities, perinatal insults, genetic anomalies, and metabolic disorders. In many cases, the underlying etiology may remain unknown.

Early-infantile developmental and epileptic encephalopathy (EIDEE), also commonly known as ***Ohtahara syndrome***, begins in the first 3 months of life and is characterized by frequent, often clustered, tonic seizures which can be referred to as "tonic spasms." The characteristic interictal EEG background consists of a "suppression-burst" pattern, in which high voltage 1–3 s bursts with intermixed spikes alternate regularly with nearly isoelectric attenuations lasting 2–5 s [1]. The prognosis of EIDEE is generally poor, with severe global developmental impairment and death frequently occurring during the childhood years. In some cases, prognosis may be improved if treatment of the underlying etiology is possible (such as with respective epilepsy surgery for large cortical malformations).

Infantile epileptic spasms syndrome (IESS), also known as West syndrome is comprised of the triad of infantile spasms (IS), hypsarrhythmia and cognitive impairment. IS is a specific seizure type with the peak age of onset at 4–6 months of age (generally <2 years of age), consisting of clusters of brief stereotyped flexion movements of the trunk and neck, with simultaneous elevation of the arms. The ictal signature of each spasm movement is known as an electrodecremental response, which consists of a generalized high-amplitude slow wave followed by a brief 1–2 s diffuse attenuation (Fig. 7.12) [2]. In some cases, IS consists of subtler stereotyped head drops, eye-rolling, or shoulder shrug movements. Hypsarrhythmia is a disorganized, high voltage (>200 μV), chaotic interictal background pattern with multifocal spike-wave discharges (Fig. 7.13).

Lennox Gastaut syndrome (LGS) typically develops after 2 years of age and is a clinical syndrome characterized by (1) multiple seizure types, typically including

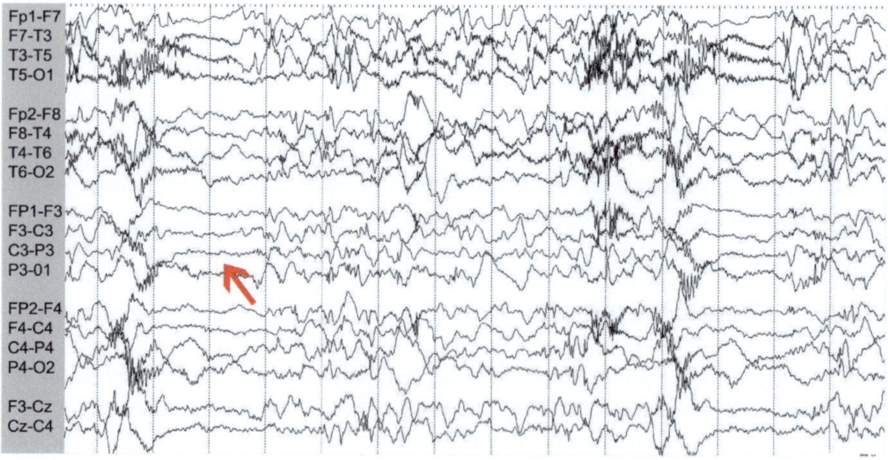

Fig. 7.12 Infantile spasms as seen in West syndrome (solid arrow: electrodecrement)

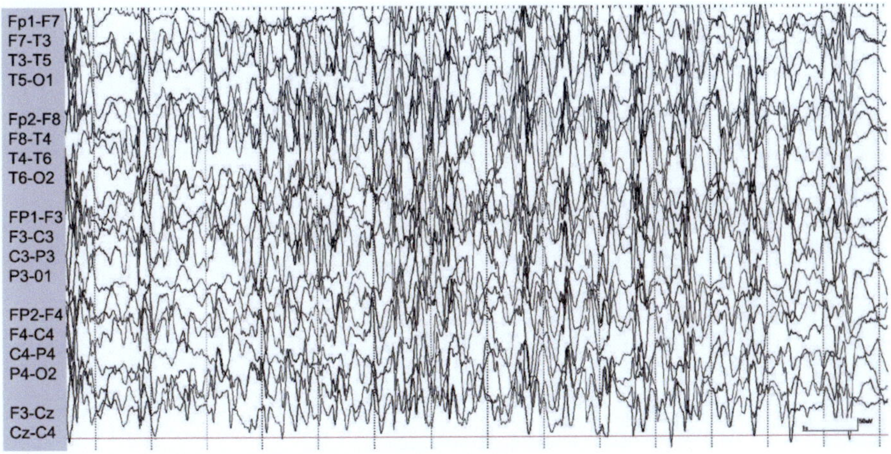

Fig. 7.13 Hypsarrhythmia

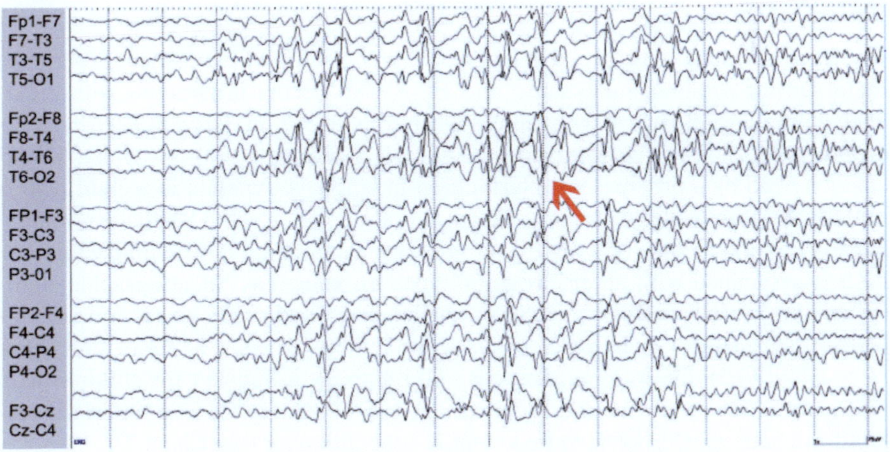

Fig. 7.14 Slow-spike wave as seen in Lennox-Gastaut syndrome (solid arrow)

nocturnal tonic seizures, (2) generalized 2–2.5 Hz spike-wave discharges, often alternatively described as a "slow spike-wave" pattern on interictal EEG (Fig. 7.14) and (3) cognitive impairment. There is no consistent underlying etiology, but structural and genetic defects should be investigated. In some cases, children with infantile spasms will clinically and electrographically transition to an LGS phenotype.

Developmental and epileptic encephalopathy (DEE) and epileptic encephalopathy (EE) with spike-and-wave activation in sleep (SWAS) are two related conditions in which developmental regression is associated with spike-wave discharges are activated by sleep and occupy an abundance of the sleep recording. DEE-SWAS

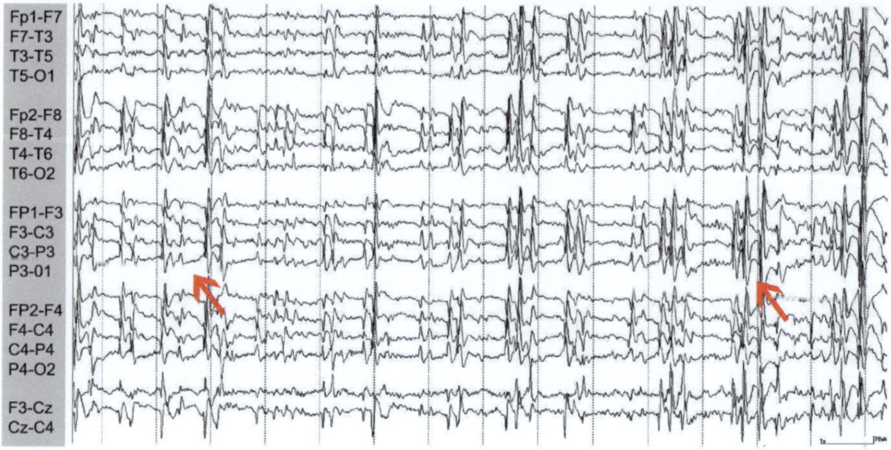

Fig. 7.15 Continuous spike-wave discharges (solid arrows) as seen in electrographic status epilepticus of sleep (ESES)

describes children with pre-existing developmental delay who then experience regression of acquired milestones, and EE-SWAS occurs in children with previously normal development who regress due to this condition. ***Landau Kleffner syndrome (LKS)***, also known as ***Acquired Epileptic Aphasia***, is a subtype of EE-SWAS characterized by a progressive deterioration in language function in a child who is initially developmentally normal. Affected children will develop language normally in early childhood, then will regress, at times even culminating in complete mutism. LKS is associated with an electrographic pattern known as "electrographic status epilepticus in slow wave sleep (ESES)," which is defined as sleep activation of epileptiform discharges such that the spike-wave activity occupies >85% of slow wave sleep (Fig. 7.15). Of note, seizures may be relatively infrequent for children with LKS, and ESES is not true status epilepticus but rather very frequent interictal epileptiform activity. Response to treatment is generally poor (see Chap. 2).

7.3 Childhood Epilepsy Syndromes

Absence epilepsy. An absence seizure consists of a brief period of behavioral arrest, sometimes accompanied by eye fluttering and chewing automatisms, which lasts for several seconds and typically occurs multiple times per day. Unlike focal dyscognitive seizures, there is no postictal state, though these events may be associated with a short period of anterograde or retrograde amnesia. The interictal EEG in the absence epilepsy is associated with frontally dominant generalized 3 Hz spike-wave (Fig. 7.17). A prolonged run of generalized spike-wave discharges lasting >10 s is considered an electrographic seizure. Ictal spike-wave bursts may be shorter in duration if they are associated with a behavioral arrest. Clinical absence seizures are often provoked by hyperventilation and are associated with a burst of generalized 3 Hz spike-wave, which is generally indistinguishable from interictal findings.

Self-limited epilepsy with centrotemporal spikes (SeLECTS), also known as ***Benign Rolandic epilepsy (BRE)***, is the most common childhood focal epilepsy syndrome. Rolandic seizures often occur shortly after sleep onset, characterized by clonic jerking of the face, often associated with a "prickling" or "freezing" sensation inside the cheek. If the child awakens due to the seizure, awareness is generally preserved unless progression to secondary convulsion occurs. Interictal rolandic discharges often demonstrate simultaneous phase reversal in both central and temporal regions, and are typically sleep-activated and seen independently in bilateral hemispheres (Fig. 7.16). Because these discharges are generally oriented tangentially to the cortical surface, in a referential montage arrangement, a "horizontal dipole" is often seen, characterized by a reversal of polarity with frontal electropositivity and temporal electronegativity (Fig. 7.17).

Childhood occipital epilepsies are less common than BECTS and can be divided into two different types. ***Self-limited epilepsy with autonomic seizures (SeLEAS)***, also known as ***Panayiotopoulos syndrome (PS)***, is the more common of the subtypes and typically affects children in early childhood. Seizures in PS are infrequent, characterized by prominent ictal vomiting, which may be followed by tonic unilateral eye deviation and secondary convulsion. The second subtype, ***Childhood Occipital Visual Epilepsy (COVE), also known as Childhood Occipital Epilepsy of Gastaut (Gastaut type-COE),*** is rare and typically begins in the late childhood years. COVE is associated with frequent daily visual

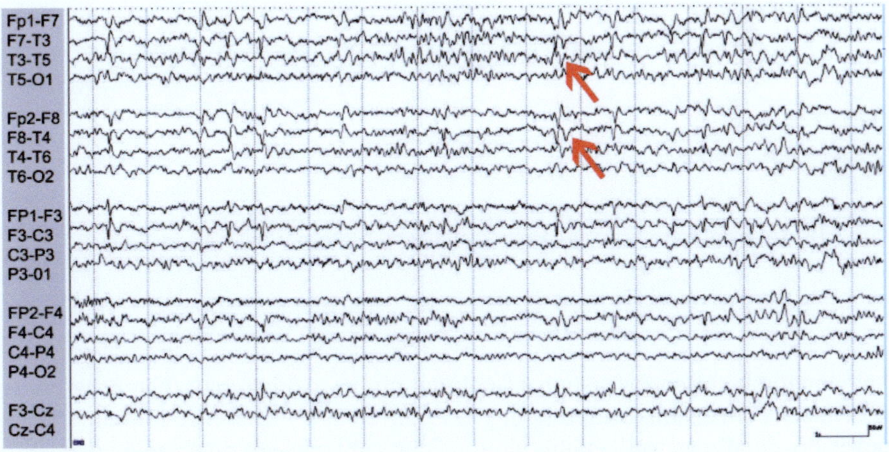

Fig. 7.16 Centrotemporal spike-wave discharges (solid arrows) as seen in SeLECTS

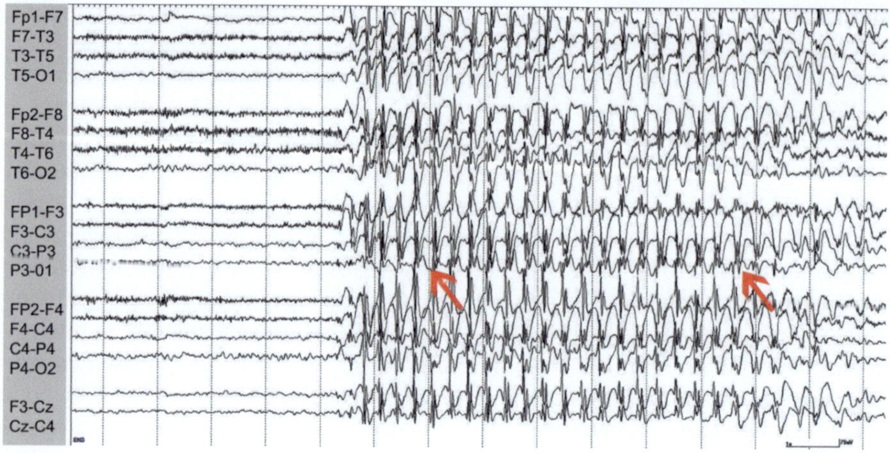

Fig. 7.17 3-Hz generalized spike-wave discharges (solid arrows) as seen in childhood absence epilepsy (CAE)

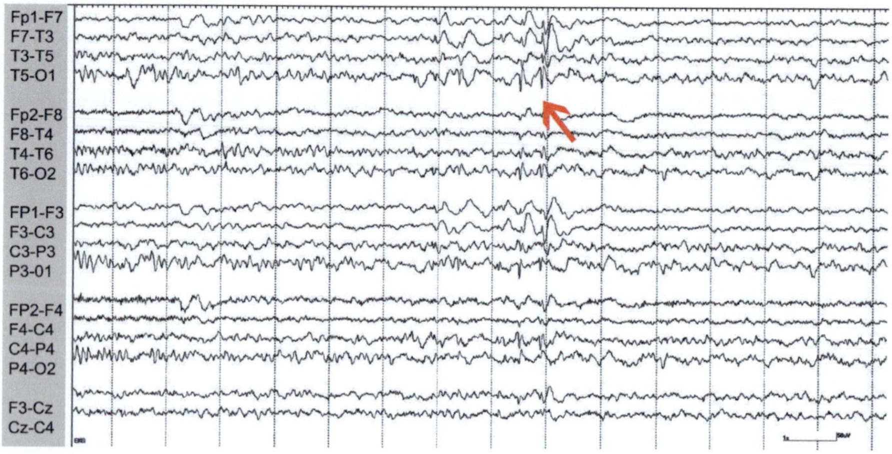

Fig. 7.18 Occipital spike-wave discharges (solid arrow) as seen in SeLEAS and COVE

hallucinations, often consisting of colored circles, which may then progress to secondary convulsion. Sufferers of COVE often also experience migraine headaches postictally as well as interictally. Both subtypes are associated with interictal bioccipital spike-wave discharges (Fig. 7.18). Syndromic features are described in greater detail in Chap. 2.

Pearls: Neonatal EEG

- **Neonatal electrode placement**: Due to space limitations on the preterm infants' scalp, only half of the electrodes are typically applied in neonatal recordings. Some laboratories use alternative locations for the frontopolar electrodes, employing an Fp3 electrode halfway between the expected placement of Fp1 and F3, and Fp4 halfway between Fp2 and F4.
- Neonatal EEG display settings: Because slower frequencies (theta and delta) predominate in neonatal brains, review settings are modified to highlight these slower patterns.
  - Paper speed: 20 s/page, or 15 mm/s (compared to 10 s/page or 30 mm/s for standard review)
  - Low-frequency filter: 0.5 Hz (1 Hz for standard review)
- Due to incomplete myelination, neonatal seizures may be confined to one electrode, with 0.5–1 Hz sharps evolving primarily in frequency and amplitude rather than distribution (Fig. 7.19).
- Knowledge of the infant's conceptional age (Gestational age at birth + chronologic age) is critical. Interpretation is simplified if one reviews the expected background findings for the infant's conceptional age prior to waveform review; understanding what is normal facilitates recognition of the abnormal.

Pearls: Pediatric EEG

- Some children may exhibit interictal findings characteristic of more than one epilepsy syndrome (e.g., generalized spike-wave discharges and centrotemporal spikes). In these cases, clinical history determines the primary diagnosis, though treatment decisions should take into consideration the potential for other seizure types.
- Because interictal epileptiform discharges occur with greater frequency at younger ages, epilepsy syndrome classification is often clarified by routine outpatient EEG findings. Detection of epileptiform findings can be enhanced by certain provocative maneuvers:
  - Sleep deprivation: Epileptiform abnormalities in certain childhood epilepsy syndromes may be activated by drowsiness and sleep. Up to 30% of children with BRE may exhibit abnormalities exclusively in sleep.
  - Hyperventilation: Generalized spike-wave discharges and absence seizures are often provoked by hyperventilation. Hyperventilation is contraindicated in patients with recent strokes, or others for whom hypocapnia-induced vasoconstriction may be detrimental.
  - Photic stimulation: Epileptic myoclonus (such as in JME) may be triggered by certain frequencies of photic stimulation.

(continued)

- Children with photosensitive epilepsy may have epileptiform discharges or seizures triggered by certain frequencies of photic stimulation (i.e., a **photoparoxysmal** or **photoconvulsive response**).
  - Photic stimulation is performed with eyes closed in order to filter out the higher-frequency colors. Longer wavelengths (red lights) are the most likely to trigger abnormalities.
  - Traditionally, some EEG laboratories have employed colored lenses to better evaluate patients exhibiting a photoparoxysmal response. Occasionally, photosensitive seizures may be improved by wearing blue polarized sunglasses.

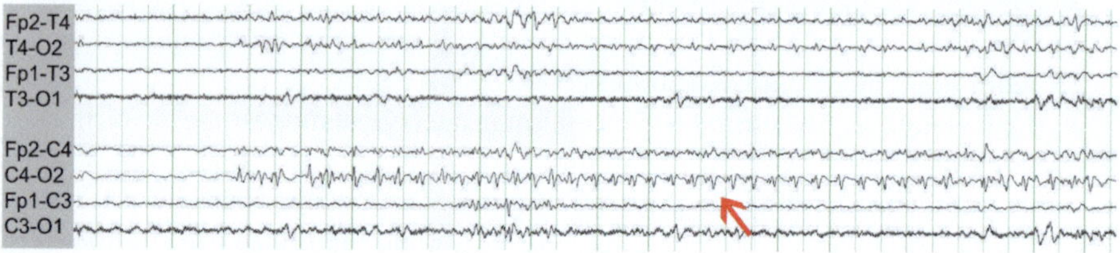

Fig. 7.19 Evolution of a neonatal seizure (solid arrow)

Acknowledgements Adam L. Numis, MD for contributions to a prior version of this chapter.

References

1. Yamatogi Y, Ohtahara S. Early-infantile epileptic encephalopathy with suppression-bursts, Ohtahara syndrome; its overview referring to our 16 cases. Brain Dev. 2002;24(1):13–23.
2. Pavone P, Striano P, Falsaperla R, Pavone L, Ruggieri M. Infantile spasms syndrome, West syndrome and related phenotypes: what we know in 2013. Brain Dev. 2014;36(9):739–51.

Pathophysiology

8

Brian Hanrahan, Spencer Hall, and Arun Raj Antony

Seizures are abnormal paroxysms of rhythmic synchronized discharges due to the imbalance between excitatory and inhibitory signaling in the brain. Most of the excitatory activity in the brain is controlled by three types of glutamate-activated receptors—kainate, AMPA and NMDA. Inhibitory neurotransmission in the brain is controlled by GABA, with two receptor subtypes ($GABA_A$ and $GABA_B$) acting through chloride and potassium channels. Epileptogenesis defines the process of alteration in the cellular, molecular, and network architecture of a normal brain to produce seizures by exposure to multiple predisposing conditions (two-hit hypothesis). Many animal models are available to study various types of epilepsy- for example, the pilocarpine and kainate models for prolonged seizures and the pentylenetetrazole model for short seizures.

8.1 Membrane Potential

The equilibrium potential of an ion is the potential at which the electrical gradient matches the concentration gradient, and there is no net movement of the ion across the membrane [1].

The equilibrium potential is calculated using the Nernst equation and does not depend on the membrane permeability. An increase in membrane permeability to a specific ion moves the membrane potential towards the equilibrium potential of that ion (Table 8.1).

B. Hanrahan (✉) · S. Hall
St. Luke's University Health Network, Bethlehem, PA, USA

A. R. Antony
Hackensack School of Medicine, Nutley, NJ, USA

Division of Epilepsy, Jersey Shore University Medical Center, Neptune, NJ, USA

Z. Haneef (ed.), *Epilepsy Fundamentals*,
https://doi.org/10.1007/978-3-031-77741-7_8

Table 8.1 Equilibrium potential of different ions (mV)

E_{K+}	−100
E_{Na+}	+40
E_{Cl^-}	−75
E_{Ca++}	+250

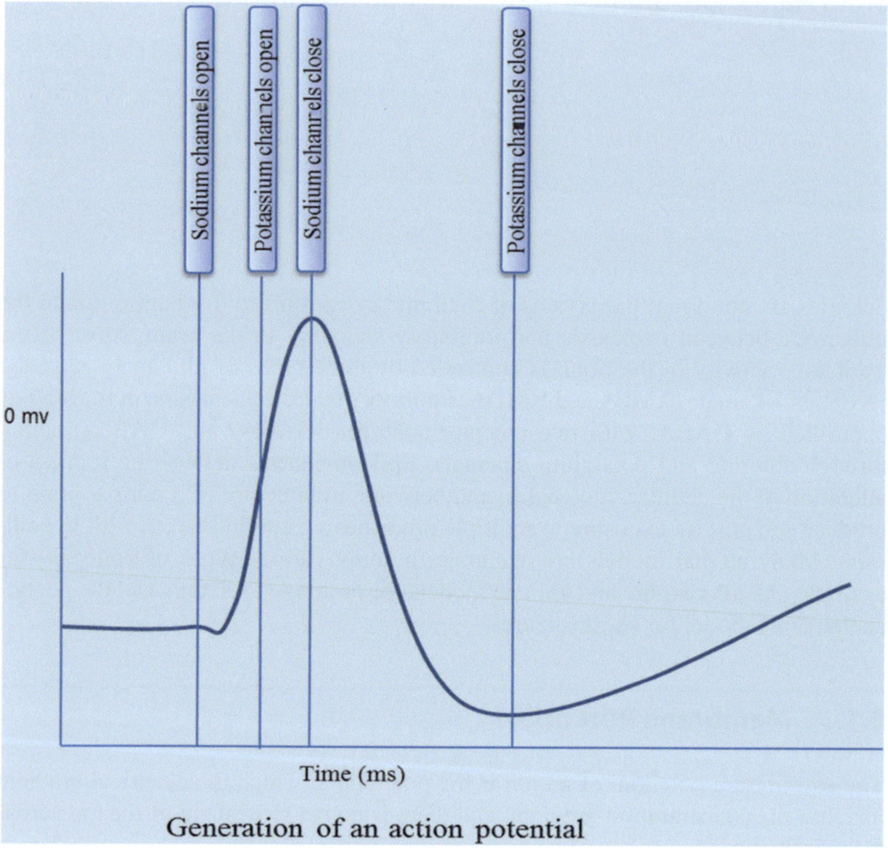

Fig. 8.1 Generation of an action potential

The Goldman-Hodgkin-Katz equation calculates the resting membrane potential from the Nernst potentials of the ions around the membrane and their permeability. The resting membrane potential of a typical human neuron is maintained at −70 mV.

When the membrane is depolarized partially (−55 to −60 mV) due to electrical, chemical or mechanical causes, the voltage-gated Na^+ channels on the surface open. This leads to a spread of voltage change to the adjacent regions, opening the nearby voltage-gated Na^+ channels, thus propagating an action potential down the axon, independent of the stimulus. The action potential is terminated by the inactivation of Na^+ channels and the delayed opening of voltage-gated K^+ channels. This is followed by a short refractory phase of hyperpolarization during which another action potential cannot be initiated in the cell (Fig. 8.1).

Action potentials and post-synaptic potentials are not usually seen in glial cells, except when the extracellular potassium concentration increases due to repetitive neuronal action potentials.

8.2 Cortical Basis of EEG

EEG waveforms are field potentials produced by summated excitatory and inhibitory post-synaptic potentials (EPSPs and IPSPs) in the dendrites of cortical pyramidal neurons. Action potentials (AP) do not have a prominent contribution to surface EEG due to the shorter duration of the waveforms, thus preventing synchronization [2]. Electronegativity on the scalp EEG suggests a superficial EPSP or a deep IPSP. When an excitatory synapse stimulates a vertical neuron, an influx of cations (Na^+ and Ca^{++}) and depolarization occurs, leading to current flow in the intracellular and extracellular space. An extracellular electrode near the synapse records "negativity" due to the flow of these cations away from the electrode (relative influx of negative ions).

Surface spikes are mostly "negative" because excitatory neurons innervate the superficial layers (layers 2 and 3) of the cortex while the inhibitory synapses end in the deeper layers, both of which produce electronegativity on the scalp. We see "positive" spikes in cases where the superficial layers of the cortex are damaged due to trauma or surgery. The brain, as a whole, is electrically neutral. The appearance of a potential in one region causes a flow of charge, leading to the appearance of the opposite potential in the nearby regions. Usually, the opposite charge appears in the deep layers and is not visible on the scalp ("radial dipole"). At times, the opposite potential appears in the superficial layers, appearing as "horizontal dipoles" on the scalp EEG (Fig. 8.2). The amplitude of epileptic activity is higher due to a higher degree of synchronization.

8.3 The Thalamocortical Circuit

The thalamus acts as the brain's chief message relay center, and the connections between the thalamus and cerebral cortex are known as thalamocortical connections or circuits. Two basic subcategories of these connections are thalamocortical fibers or radiations, which take information one way from the thalamus to specific cortex locations, and corticothalamic fibers, which send information back from the cortex to the thalamus [3].

These connections play a role in generalized seizures, especially absence epilepsy. The thalamocortical cells provide excitatory input to the thalamic reticular cells, which in turn provide feedback inhibition to the thalamocortical cells causing hyperpolarization (GABA) (Fig. 8.3) [4]. This hyperpolarization de-inactivates low threshold (T-type) Ca^{++} channels, producing bursts of action potentials and the classic 3 Hz spike and wave pattern that can be appreciated on EEG. This is the same network responsible for the formation of sleep spindles seen during N2 sleep.

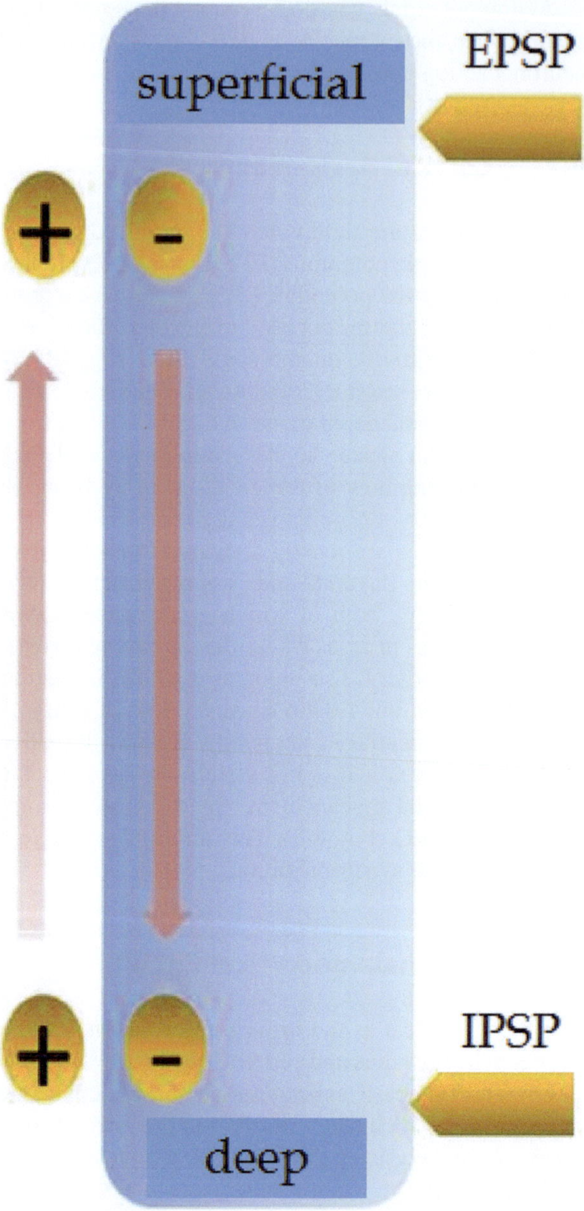

Fig. 8.2 When an excitatory synapse stimulates a vertical neuron, an influx of cations and depolarization leads to current flow in the intracellular and extracellular space

Recent research has also begun supporting the role of thalamocortical circuits in focal onset seizures [5]. Neuromodulation of the thalamus with deep brain stimulation, particularly in the anterior nucleus, can disrupt seizure-prone thalamocortical circuits in patients with focal epilepsy, leading to a reduction in seizure frequency.

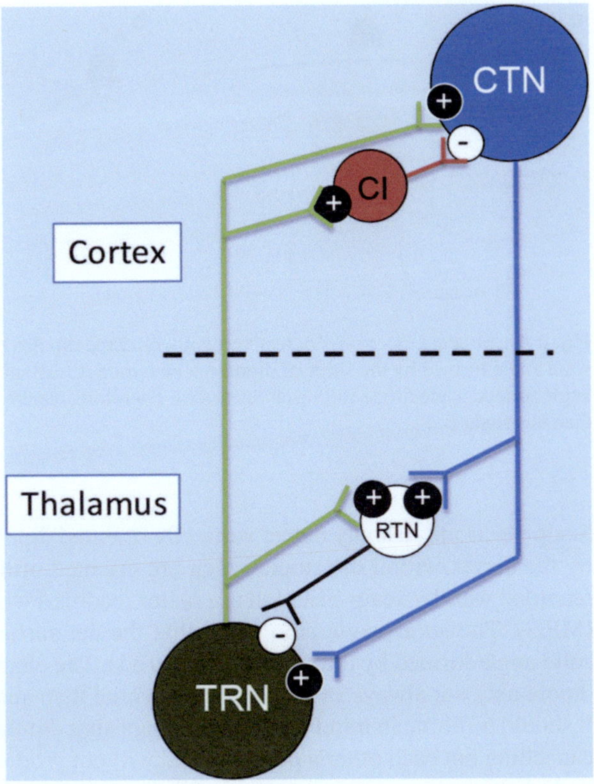

Fig. 8.3 Thalamocortical pathway. *CTN* corticothalamic neuron, *RTN* reticular thalamic neuron, *TRN* thalamic relay neuron, *CI* cortical interneuron

Similar research is actively being conducted with responsive neurostimulation (RNS) targeting both the anterior and centromedian thalamic nuclei with similar results [6, 7].

8.4 Volume Conduction

Volume conduction is the rapid, passive spread of electrical potential through the brain parenchyma between the positive and negative end of a dipole, which demonstrates a rapid fall off with distance. A near-field potential (e.g., EMG) is recorded when the recording electrode is close to the generator and a far-field potential (e.g., most evoked potentials) is obtained when the potential is recorded far from the generator (Fig. 8.4) [1].

At least 6 cm^2 of the cortex needs to be activated to record activity on the scalp EEG [8], although 10 cm^2 has also been quoted [9]. Simultaneous activation of a large cortical region is produced by a wave of cortical excitation called the paroxysmal depolarizing shift (PDS) along with overriding bursts of action potentials.

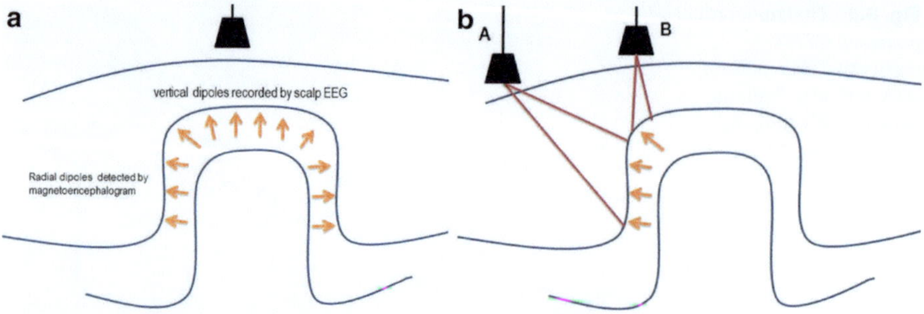

Fig. 8.4 (**a**) Scalp electrodes detect vertically oriented dipoles better than radial dipoles. (**b**) The solid angle formed by the sheet of dipoles at electrode A (distant electrode) is larger than the solid angle formed at electrode B (closer electrode). Therefore, electrode A will record a larger potential than electrode B

Scalp electrodes usually detect vertically oriented dipoles. Radial dipoles produced by the cortex within the sulci, which are oriented orthogonal to the scalp, are not recorded well by scalp EEG but are better recorded with a magnetoencephalogram (MEG). The *solid angle rule* states that the net surface potential depends on the solid angle formed by the recording electrode. The electrode directly over a sheet of dipole may not always show a higher potential than another electrode at a distance. It should be borne in mind that adjacent opposing dipoles will produce a closed field cancelling out each other's potential and will not produce significant changes in the scalp EEG.

Pearls: Pathophysiology

- The immediate precursor to GABA is glutamate, and it is ultimately metabolized back to glutamate.

Acknowledgments Atul Maheshwari—for authoring previous versions of this chapter.

References

1. Ebersole JS, Nordli DR, Husain AM. Current practice of clinical electroencephalography. Philadelphia: Lippincott Williams & Wilkins; 2014.
2. Blum AS, Rutkove SB. The clinical neurophysiology primer. Totowa: Humana Press; 2007.
3. Shine JM, Lewis LD, Garrett DD, Hwang K. The impact of the human thalamus on brain-wide information processing. Nat Rev Neurosci. 2023;24:416–30.
4. George K, Das M, J. Neuroanatomy, thalamocortical radiations. In: StatPearls. Treasure Island: StatPearls Publishing; 2023.
5. He X, et al. Disrupted basal ganglia-thalamocortical loops in focal to bilateral tonic-clonic seizures. Brain J Neurol. 2020;143:175–90.

6. Fisher R, et al. Electrical stimulation of the anterior nucleus of thalamus for treatment of refractory epilepsy. Epilepsia. 2010;51:899–908.
7. Roa JA, et al. Responsive neurostimulation of the thalamus for the treatment of refractory epilepsy. Front Hum Neurosci. 2022;16:926337.
8. Cooper R, Winter AL, Crow HJ, Walter WG. Comparison of subcortical, cortical and scalp activity using chronically indwelling electrodes in man. Electroencephalogr Clin Neurophysiol. 1965;18:217–28.
9. Tao JX, Ray A, Hawes-Ebersole S, Ebersole JS. Intracranial EEG substrates of scalp EEG interictal spikes. Epilepsia. 2005;46:669–76.

Genetics of Epilepsy

9

Stuti Joshi and Dennis Lal

Although the etiology of epilepsy is heterogeneous, the significance of genetic factors in its expression and severity has been recognized for a long time. Over the past decade, advancements in molecular techniques, particularly the emergence and widespread accessibility of next-generation sequencing (NGS) technology, have significantly enhanced the identification of genes associated with epilepsy. These developments have not only elucidated the genetic underpinnings of the condition but have also broadened our comprehension of the phenotypic spectrum associated with known genetic epilepsy syndromes [1]. A number of epilepsies previously deemed "cryptogenic" are now reclassified as genetic.

Identifying a pathogenic genetic variant is important for several reasons. It not only ensures diagnostic certainty but also aids in the identification and monitoring of associated comorbidities. Having a genetic diagnosis may ultimately lead to individualized treatment, which may include the selection of specific anti-seizure medications (ASMs) and, as we enter the era of precision medicine, the implementation of gene therapies. Finally, for genes that have been thoroughly researched, the identification of pathogenic variants facilitates the prognosis of recurrence risk estimation and guides decisions in reproductive planning.

9.1 Introduction

Early evidence from familial aggregation and twin studies indicates that epilepsy is highly heritable, and generalized epilepsies overall are more heritable than focal epilepsies, with 82% compared to 36% concordance rates in twin studies, respectively [2]. In generalized epilepsy, the risk in first-degree

S. Joshi (✉) · D. Lal
University of Texas Health Science Center at Houston, McGovern Medical School, Houston, TX, USA
e-mail: Stuti.H.Joshi@uth.tmc.edu

Z. Haneef (ed.), *Epilepsy Fundamentals*,
https://doi.org/10.1007/978-3-031-77741-7_9

relatives with epilepsy is 5–10 times greater than that in the general population as compared to 2–3 times for focal epilepsy [3, 4]. In the absence of a family history or evidence of a genetic syndrome, the risk of epilepsy in the general population is approximately 1%. For a child born to a mother with epilepsy, the risk ranges from 2.8% to 8.7% and for a child of a father with epilepsy, it falls between 1% and 3.6%. The risk escalates if the parent develops epilepsy before the age of 20 years [5].

The genetic underpinnings of epilepsy exhibit considerable complexity, underscored by the distinction between monogenic and polygenic forms of the disorder. While severe forms of epilepsy with intellectual disability are often traced back to single de novo variants, elucidating a clear genetic etiology, families with multiple affected individuals typically present a polygenic inheritance pattern, challenging detection through conventional clinical genetic testing. This discrepancy highlights the limitations of current genetic diagnostics, which excel at identifying potent single-gene mutations but falter in capturing the nuanced contributions of multiple genes. Such intricacies emphasize the need for advanced genetic analyses to unravel the multifaceted genetic architecture of epilepsy, promising to refine diagnostic, prognostic, and therapeutic strategies in the genomic era of medicine.

Most monogenic epilepsies are observed in developmental and epileptic encephalopathies (DEEs). The best-known example of this is Dravet syndrome in which more than 80% of patients have a pathogenic variant of *SCN1A*. A significant proportion of monogenic epilepsies are due to de novo variants that arise sporadically during gametogenesis, either in the oocyte or spermatocyte. There is no relevant family history in these cases. More recently, the importance of post-zygotic mosaic variants (mutations that occur *after* fertilization in some cells) has been recognized in individuals with epilepsy. Mosaic mutations have been implicated in the pathogenesis of focal brain lesions such as malformations of cortical development, which are important to recognize as they may be amenable to surgical resection. They may be difficult to test for as genetic testing can be negative in one tissue and positive in another (Fig. 9.1).

Variants in genes implicated in inherited monogenic epilepsies most commonly follow an autosomal dominant inheritance pattern, with autosomal recessive and X-linked inheritance occurring less commonly. In most common forms of genetic epilepsies, the inheritance pattern appears to be more intricate, following a polygenic or complex pattern incorporating the influence of multiple potentially interacting genes and/or epigenetic factors. This is particularly seen in genetic generalized epilepsies and some cases of sporadic focal epilepsies. In research studies, the common polygenic variant burden for epilepsy can be measured and is different amongst patients with epilepsy and those without, and between those with generalized versus focal epilepsy [6]. However, these approaches do not yet have enough accuracy for clinical application.

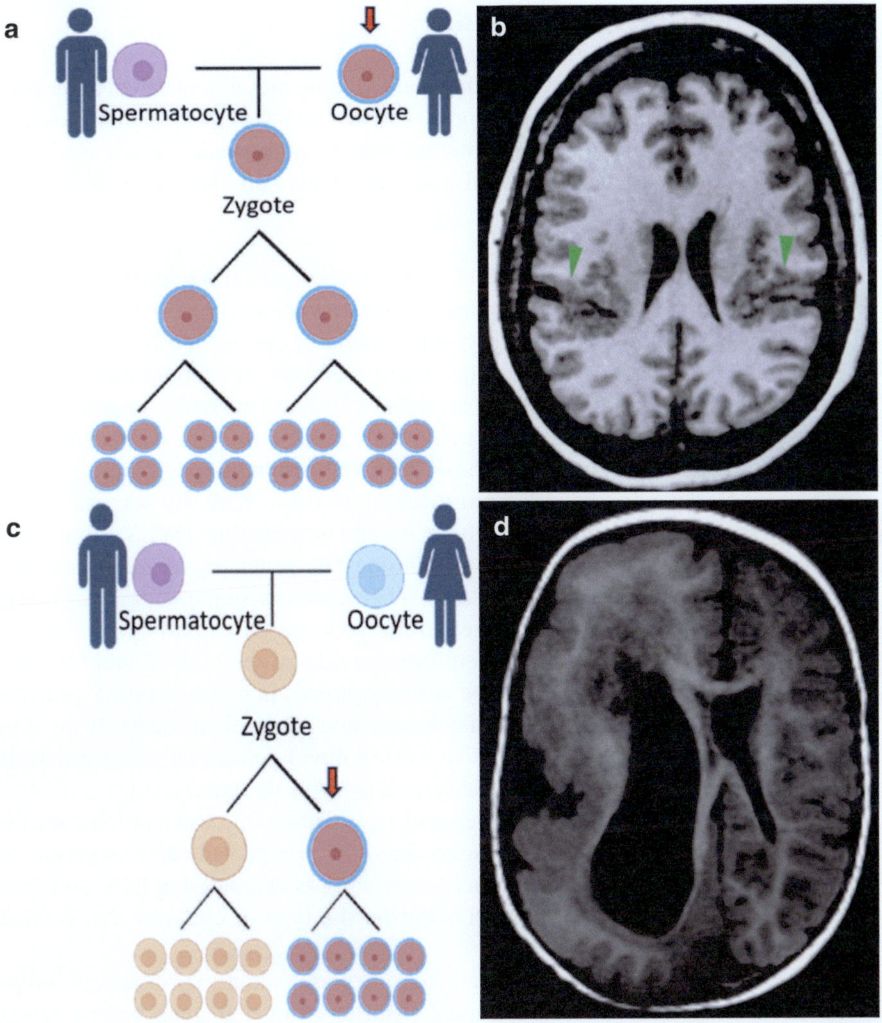

Fig. 9.1 De novo germline and somatic mutations causing disease. (**a**) A de novo mutation arises during oogenesis (red arrow) and is passed on to all subsequent cells (blue halo indicates mutated cells). This pattern is typical of autosomal dominant conditions. (**b**) An axial T1-weighted image from an MRI brain of a patient with bilateral peri-sylvian polymicrogyria (green arrowheads) due to a de novo variant in *NEDD4L*. (**c**) A post-zygotic somatic mutation arises late (red arrow) and is present only in certain tissues, in this case, in half the brain. (**d**) Axial T1-weighted MRI image shows right-sided hemimegalencephaly with enlargement of the right hemisphere, thickened and abnormal grey matter and an enlarged ventricle (Figure created with BioRender.com)

9.2 Genetic Testing Methods

The testing methods commonly used in genetic diagnostics aim to detect pathogenic single nucleotide variants (SNV) or copy number variants (CNV) such as deletions and duplications. Other genetic variants that may be pathogenic include nucleotide repeat expansions, complex structural rearrangements and aberrations in chromosomal number.

1. Next-Generation Sequencing (NGS): NGS techniques allow for the simultaneous sequencing of many genes and genomic regions, which is particularly valuable for studying the genetic heterogeneity of epilepsy. NGS has replaced traditional Sanger single-gene sequencing methods and should be utilized as the first-line investigation where available [7]. Until recently, targeted multigene panels, which involve sequencing a specific set of genes associated with a particular group of conditions, have been utilized in clinical practice as they are cost-effective and have a rapid turnaround time. With the decreasing cost and increasing availability of whole exome sequencing (WES) and whole genome sequencing (WGS), these techniques are replacing targeted gene panels as the first line in many centers. WES involves sequencing the protein-coding regions (exons) of all genes in the genome. Most pathogenic variants lie within the exome, which is therefore considered the highest yield for testing. WGS involves sequencing the entire genome, including coding and non-coding regions, which provides an unbiased and comprehensive analysis of the entire genome and is useful for discovering novel genes and structural variations. Both comprehensive genetic tests also offer the potential for reanalysis in the future as knowledge evolves and new pathogenic associations are discovered. WGS has the highest diagnostic yield as expected (48%) whereas the yield of WES and multi-gene panel sequencing is estimated at 24% and 19%, respectively [7]. The yield is dependent on the type of epilepsy that is being sequenced.

   Nevertheless, suspected genetic syndromes, mitochondrial disorders, uniparental disomy, and nucleotide repeat disorders targeted gene panels remain appropriate for well-defined genetic conditions. WES and WGS are valuable in cases where a genetic cause is suspected without a clear candidate gene or in cases where the etiology is complex or poorly defined (Fig. 9.2).
2. Chromosomal Microarray (CMA): CMA has largely replaced conventional karyotyping and is used to detect medium-large scale CNVs or conditions where chromosomal imbalances are suspected, which may be missed by sequencing-based tests. In such cases, it may be used as a screening tool to identify a significant CNV and is of highest yield in patients with suspected DEE or if there is unexplained epilepsy and comorbid intellectual disability, autism, dysmorphism or schizophrenia where the yield may increase up to 16% [8]. Of note, CMAs are unable to detect gross structural rearrangements such as inversions, translocations and ring chromosomes in which case classical karyotyping would need to be requested if previous testing methods were

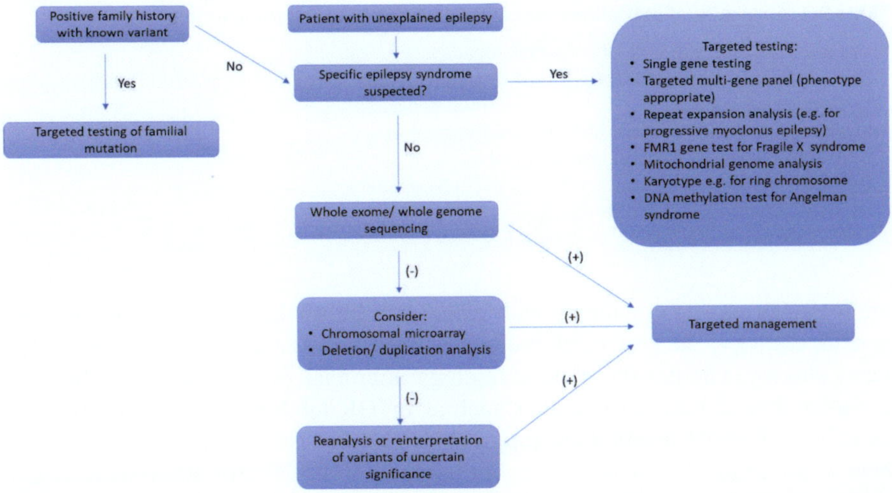

Fig. 9.2 Genetic testing strategy for patients with unexplained epilepsy. (+) indicates a positive test result and (−) indicates a negative test result

negative and a structural variant-associated phenotype was observed in the patient. CMA is not sensitive to the detection of SNVs or sequence variants where NGS techniques are required.

Genetic testing can now be performed using a buccal swab for saliva, which may be particularly useful in patients with intellectual disabilities, for whom a blood draw may be difficult. Pre- and post-test genetic counseling should be provided to patients and their families. In the pretest counseling phase, it is important to discuss the purpose and form of testing, the costs involved, anticipated results, the effect on treatment and the likelihood of incidental findings. Once the results are available, the discussion revolves around explaining the findings, implications for treatment, and implications for other family members, including prenatal testing where relevant. In patients with a 100% penetrant syndrome, prenatal diagnosis and genetic counseling for a specific epilepsy syndrome can be straightforward. However, in situations of incomplete penetrance and de novo mutations, the process can be complicated and should be performed with the assistance of a genetic counselor. Many genetic testing laboratories offer genetic counseling as part of their services at no additional cost to the patient. However, all epilepsy centers should have access to medical genetics consultations as required.

9.2.1 When Should We Pursue Genetic Testing?

The threshold for requesting genetic testing in epilepsy is falling, and where possible, all patients with unexplained epilepsy should be offered genetic testing regardless of age [9]. Often, a patient's epilepsy may be classified into more than one

Table 9.1 Examples of indications currently recommended for genetic testing

• Developmental and epileptic encephalopathies
• Epilepsy and comorbid dysmorphism, intellectual disability, autism or developmental delay
• Neonatal or infantile onset seizures of unknown cause
• Phenotype suggestive of a specific genetic syndrome
• Malformations of cortical development
• Inherited metabolic or mitochondrial encephalopathies
• Medically intractable epilepsy of uncertain etiology
• Presurgical workup of nonlesional focal epilepsy

etiology. For example, patients with tuberous sclerosis have both a structural and genetic etiology; the structural etiology is important for the consideration of epilepsy surgery whereas the genetic etiology is crucial for genetic counseling and consideration of targeted therapies such as mTOR inhibitors. Therefore the presence of a structural lesion does not preclude genetic testing. The highest yield has been reported in those with neonatal or infantile-onset seizures, those with DEE or epilepsy with comorbid neurodevelopmental disorders [7]. With increasing age, the likelihood of identifying a genetic etiology decreases; however, the age at testing should not influence the decision to test. Adults who developed epilepsy at an early age may have experienced the onset of their condition in a period when genetic testing was not readily accessible and should be offered testing now, especially in the case of drug-resistant epilepsy without a clear cause, malformations of cortical development and during the presurgical workup of nonlesional focal epilepsy. Table 9.1 outlines the groups in whom genetic testing is likely to be of the highest yield.

9.3 Specific Clinical Scenarios

Developmental and Epileptic Encephalopathies DEE refers to a group of severe neurodevelopmental disorders characterized by the co-occurrence of early-onset seizures and significant cognitive and developmental impairments. Epilepsy itself often contributes to cognitive and developmental regression. The seizures are typically resistant to treatment. Genetic testing will yield the highest in this group, and a monogenic pathogenic variant will be identified in up to 40% of patients [10]. In addition to screening pathogenic single nucleotide variants, all patients should have a copy number variant study (ideally WES or WGS) to identify abnormal duplications or deletions that may be wholly causative or partially contributory to the etiology. Several of the most common DEEs are the result of channelopathies, where there is either gain or loss of function of voltage-gated or ligand-gated ion channels (Table 9.2). The severity of the condition correlates with the extent of impairment of the channel involved. Additionally, gain or loss of function of the same ion channel can result in epilepsy but with different phenotypes, e.g., in *GRIN2A*, *SCN2A*, *SCN1A* and *SCN8A*-related epilepsies.

Table 9.2 Common monogenic mutations associated with developmental and epileptic encephalopathies [10, 11]

Gene	Protein	Phenotype
SCN1A	Sodium channel	Loss of function variants Dravet syndrome (refractory early-onset epilepsy with severe developmental delay): typically presents with prolonged, hemi-clonic febrile seizures and later development of afebrile seizures including GTCS, focal seizures with impaired awareness, atypical absences and myoclonic seizures Gain of function mutations Genetic epilepsy with febrile seizures plus (GEFS+)–Multiple febrile seizures including GTCS, absences, myoclonic seizures and even focal seizures, with a family pedigree of seizures with heterogenous semiology and variable developmental delay Early onset epileptic encephalopathy, familial hemiplegic migraine, arthrogryposis
SCN2A	Sodium channel	Broad phenotypic spectrum ranging from benign epilepsy in neonates and infancy to severe epileptic encephalopathy. Can also present with febrile seizures, West syndrome, Ohtahara syndrome or epilepsy of infancy with migrating focal seizures. MRI is often abnormal and patients have variable degrees of developmental delay
KCNQ2	Potassium channel	Heterogenous phenotypic spectrum comprising severe developmental and epileptic encephalopathy, self-limited familial neonatal and infantile epilepsies or Ohtahara syndrome
CDKL5	Cell adhesion protein	X-linked condition that has a female predominance and is typically lethal in male infants. Causes a severe, early onset refractory epilepsy with onset within the first 2 months of life, and severe neurodevelopmental impairment. Symptoms can overlap with those of Rett syndrome. Clusters of infantile spasms and tonic seizures can predominate early in life, with evolution to multifocal and myoclonic epilepsy with tonic seizures, myoclonus, absences or multifocal seizures later on
PCDH19	Cell adhesion protein	PCDH19 clustering epilepsy is characterized by focal and/or generalized seizures, commonly fever-induced and in clusters, as well as behavioral and psychiatric comorbidity and varying degrees of intellectual disability. Nearly all cases inherited in an X-linked pattern (de novo or inherited). Females are mainly affected with unaffected male carriers
SLC2A1	Glucose transporter type 1	Epilepsy, developmental delay, microcephaly and complex movement disorders. Multiple seizure types (myoclonic, atonic, absence, generalized tonic-clonic and focal seizures), seizure clusters and status epilepticus which begin before age 3. Early-onset absence seizures (onset before age 4) should also prompt testing for this condition. Seizures typically occur during fasting, e.g., early morning before breakfast
STXBP1	Synaptic vesicle fusion	Early infantile epileptic encephalopathy and neurodevelopmental delay within the first few months of life, often with multiple seizure types. Comorbid movement disorders are common. Variable phenotype that may resemble West syndrome, Ohtahara syndrome or Dravet syndrome

(continued)

Table 9.2 (continued)

Gene	Protein	Phenotype
CACNA1A	Calcium channel	Developmental and epileptic encephalopathy, episodic ataxia, familial hemiplegic migraine
GRIN2A	Glutamate receptor	Epilepsy and intellectual disability, epileptic encephalopathy, epilepsy with centro-temporal spikes

Generalized Epilepsies Clinical genetic testing for genetic generalized epilepsies (GGE) is usually negative for large impact variants as the majority of cases are thought to occur through a complex inheritance pattern. A pathogenic CNV is identified in approximately 3% of patients with GGE syndrome. The presence of 15q13.3 microdeletion is a major risk factor for the development of GGE and is present at over 50 times higher rates in GGE patients than in the general population [12]. Genetic testing should also be considered in patients with an atypical phenotype. For example, 10% of patients with early-onset absence epilepsy (<4 years of age) are found to have pathogenic variants in *SLC2A1*, the gene encoding glucose transporter type 1 (GLUT1) [13]. This is particularly important to recognize as epilepsy, in these cases, is exquisitely responsive to the ketogenic diet. With early recognition and treatment, there is an improvement in cognitive abilities. Most cases of juvenile myoclonic epilepsy (JME) are caused by a polygenic inheritance pattern, and although there are rare forms that are caused by single gene mutations, e.g., in *GABRA1*, most clinicians would not perform genetic testing unless there were atypical features. Similarly, genetic testing is usually negative in most patients who have epilepsy with eyelid myoclonia (Jeavons syndrome), although de novo pathogenic variants have been reported in *CHD2, SYNGAP1* and *KCNB1* in patients with a similar phenotype often in association with cognitive impairment [14].

Focal Epilepsies More than 50% of focal epilepsies show no structural abnormalities on imaging, and although a single gene disorder is rarely found, genetic factors are thought to contribute to the etiology of these. The nonlesional focal epilepsies comprise a broad spectrum of epilepsy syndromes, spanning self-limited neonatal, infantile and childhood epilepsies to well-defined genetic syndromes with distinct phenotypes that persist into adulthood [15]. Table 9.3 highlights the notable nonlesional focal epilepsies inherited in an autosomal dominant pattern and may persist into adulthood. Family history is often sparse, even in those that follow an autosomal dominant inheritance pattern, because the penetrance can be variable even amongst individuals with the same mutation. Nevertheless, it is important to conduct a detailed family history and create a pedigree accordingly.

Malformations of Cortical Development (MCD) The MCDs encompass structural abnormalities in the cerebral cortex and are frequently associated with intractable focal epilepsy. They can occur due to abnormal neuronal or glial proliferation or apoptosis (e.g., focal cortical dysplasia), abnormal neuronal migration (e.g., lissencephaly) or abnormal cortical organization (e.g., polymicrogyria). The genetics of MCDs are diverse, involving both sporadic and inherited mutations (Table 9.4).

Table 9.3 Selected nonlesional focal epilepsy syndromes [15, 16]

Syndrome	Onset age	Seizure characteristics	Genes
Sleep-related hypermotor epilepsy (formerly autosomal dominant nocturnal frontal lobe epilepsy)	Typically in childhood or adolescence	Clusters of brief nocturnal focal seizures characterized by hypermotor, tonic or dystonic motor features. There is significant variability in the severity even within family members carrying the same gene	Neuronal nicotinic acetylcholine receptor mutations are found in 20% of patients (*CHRNA4, CHRNA2, CHRNB2*). A more severe form with intellectual disability is associated with *KCNT1* mutation. Others include *DEPDC5, NPRL2, NPRL3, PRIMA1*. Penetrance is around 70%
Epilepsy with auditory features (EAF)	Adolescence or early adulthood	The most prominent seizure types are focal aware seizures with a distinct auditory aura or receptive aphasia. The auditory auras usually comprise elementary sounds such as humming or buzzing whereas complex auditory hallucinations are uncommon	*LGI1* (50% of cases), *RELN, MICAL1* Penetrance can range from 55–78%
Familial mesial temporal lobe epilepsy (FMTLE)	Adolescence or early adulthood	Typically present with focal aware seizures with profound déjà vu, and less commonly epigastric auras or anxiety. Focal seizures with impaired awareness and generalized tonic-clonic seizures are rare	Monogenic cases associated with *DEPDC5, NPRL2, NPRL* 3. Penetrance is highly variable, ranging from 25 to 100% and recent data raises the possibility of a complex polygenic inheritance [16]
Familial focal epilepsy with variable foci (FFEVF)	Typically first or second decade	Focal seizures depending on cortical area involved. There is marked variability of seizure semiology amongst family members and long periods of seizure freedom may occur	*DEPDC5, NPRL2, NPRL3* (all part of GATOR1 complex that acts as an inhibitor of the mTOR pathway). *TSC1, TSC2*

Sporadic cases often result from de novo mutations affecting genes crucial for neuronal migration, differentiation, and cortical organization.

The mammalian target of rapamycin (mTOR) signaling pathway is important in regulating cell growth, migration, proliferation and protein synthesis. Mutations in regulatory genes within the mTOR pathway (such as *DEPDC5, MTOR, NPRL2, NPRL3*) can be associated with nonlesional focal epilepsies as described above but can also give rise to a spectrum of MCDs, e.g., focal cortical dysplasia and can contribute to multi-system disorders such as tuberous sclerosis complex (*TSC1* and *TSC2*) [17]. These can occur as germline mutations that may be detected on a

Table 9.4 Selected malformations of cortical development that may cause epilepsy [17, 18]

Phenotype	Characteristics	Commonly affected genes
Focal cortical dysplasia	Gyral and sulcal irregularities, increased cortical thickness, blurring of the grey-white matter junction and white matter T2-hyperintensity. Cortical dyslamination ± abnormal cell types (dysmorphic neurons and balloon cells) on histopathology	mTOR pathway genes including *TSC1, TSC2, DEPDC5, NPRL2* and *NPRL3*
Hemimegalencephaly	Enlarged and dysplastic cerebral hemisphere	*TSC1, TSC2, MTOR, DEPDC5* and *NPRL3*. Somatic mutations in *AKT3* implicated in some cases
Polymicrogyria	Excessive number of abnormally small gyri with cortical overfolding and an irregular pebbled cortical surface and a stippled grey-white matter boundary. One of the most frequent MCDs and considered highly epileptogenic	*PTEN, PIK3R2, TUBA1A, TUBB2A, COL4A1/2, WDR62* and *NEDD4L*
Periventricular nodular heterotopia	Nodular masses of grey matter along the ventricular walls. May be unilateral or bilateral and often occur with other malformations	More than 50% of patients with bilateral PVNH have *FLNA* mutations which are inherited in an X-Linked dominant manner which affects females and are typically lethal in male infants. Others include *ARGEF2, MAP1B, NEDD4L* and *22q11.2* deletion
Lissencephaly spectrum	"Smooth brain"—Abnormal gyral pattern which includes agyria (absent gyri), pachygyria (broad gyri) and subcortical band heterotopia	*DCX, LIS1, RELN, VLDLR, TUBA1A, TUBB2B, NDE1, DYNC1H1, ARX, CEP85L* and Miller-Dieker deletion syndrome
Subcortical band heterotopia	Part of the lissencephaly spectrum. Band of grey matter separated from the cortex and lateral ventricles by zones of grey matter	*LIS1, DCX*
Schizencephaly	Full thickness cerebral cleft lined with grey matter which extends from the ventricular surface to the meningeal surface	*EMX2, GPR56*

targeted gene panel and as somatic mosaic mutations in the brain, which may only be discovered when the affected brain tissue is sampled. Knudson's two-hit hypothesis suggests that patients with germline variants can go on to develop focal lesions in the presence of a somatic second hit, e.g., during brain development. The double germline and somatic mutations are most commonly seen in the setting of focal cortical dysplasia type II and TSC [18]. Of note, somatic mutations are less likely to produce bilateral, symmetric MCDs, which are more classically seen with germline variants.

The genetic diagnosis is important as there is increasing evidence to suggest that patients with refractory epilepsy due to pathogenic variants in the mTOR pathway have a higher chance of achieving seizure freedom after resective surgery. For example, individuals with TSC stand a 50–60% chance of achieving long-term seizure freedom following epilepsy surgery, including those with infantile spasms [10]. On the other hand, outcomes are less favorable in patients with mutations causing synaptopathies or channelopathies [18]. Genetic testing is increasingly used in the presurgical evaluation of both lesional and nonlesional focal epilepsies to help facilitate a more accurate etiologic classification and therefore predictions regarding postoperative outcomes. WES is recommended for nonlesional cases whereas single gene analysis or targeted gene panels can be utilized in those with structural abnormalities. In individuals with focal epilepsy in whom a mutation in an mTOR gene is discovered, a high-resolution MRI brain should be repeated to look for subtle cortical dysplasia that may have been missed on prior imaging.

9.3.1 Interpretation of Results

Genetic variants are classified into five different categories based on their clinical significance—benign, likely benign, variant of unknown significance (VUS), likely pathogenic and pathogenic—as per the American College of Medical Genetics (ACMG) guidelines [19]. The categorizations are made based on the variant type, whether it has previously been reported in individuals or families with the disease, the frequency in the population, predicted impact on protein function and phenotype match. The guidelines provide the level of evidence for this classification (e.g., very strong, strong, moderate, supporting etc.) which is derived from multiple evidence types such as functional data, segregation analysis, allelic data, computational and predictive data and case reports. For certain categories, the ACMG guidelines recommend quantitative criteria for variant classification, such as allele frequency cutoffs in affected versus control populations. This is particularly relevant for epilepsy, where the distinction between rare benign variants and those likely contributing to disease phenotypes can be subtle and necessitates careful consideration of population data. It is important to note that variant classification can evolve as new evidence becomes available, and periodic re-evaluation of variants may be necessary to ensure accurate and up-to-date clinical interpretations.

The ACMG guidelines also provide a framework for evaluating the gene-disease association. The pathogenicity of variants in certain genes may be well-established for specific epilepsy syndromes, whereas for others, the association may be less clear. Understanding the strength of the association is crucial when interpreting the variants in the context of epilepsy.

The guidelines provide recommendations for standardized reporting of variants, including how to convey uncertainty, especially for VUS, and the importance of transparently reporting the evidence basis for each classification to facilitate clinical interpretation and research. Finally, they also address ethical considerations, such as the importance of patient consent for genetic testing, privacy issues, and the potential psychosocial impact of genetic findings. These considerations are vital in epilepsy care, where genetic findings can have significant implications for prognosis, treatment choices, and family planning.

9.3.2 Targeted Management Modified by Genetic Diagnosis

Precision medicine, which aims to provide personalized treatment tailored to the patient's specific needs, requires an etiological diagnosis. In patients with epilepsy, making a genetic diagnosis is associated with improved outcomes. In a cross-sectional study among 418 patients with epilepsy who received a genetic diagnosis, nearly 50% had changes to their clinical management [20]. Out of those, nearly 74% experienced an improvement in outcomes, which was a reduction or elimination of seizures in most cases.

Anti-seizure Medication Selection The identification of a genetic etiology can lead to specific choices of ASM in a small but growing number of genetic epilepsies:

- Sodium channel blockers (e.g., phenytoin, carbamazepine and lamotrigine) can exacerbate seizures in DEEs caused by "loss of function" mutations (e.g., Dravet syndrome caused by *SCN1A* loss of function) and improve seizure control in those with "gain of function" mutations e.g. in *SCN1A, SCN2A* and *SCN8A* related DEEs [21].
- Dravet syndrome caused by a loss of function *SCN1A* gene mutation responds well to medications such as clobazam, valproic acid, stiripentol, cannabidiol and fenfluramine. On the other hand, the condition can worsen with sodium channel blockers, prolonged use of which is associated with negative effects on cognition [22]. Cognition may improve after stopping these medications, even amongst older patients.
- In *KCNQ2*-related developmental and epileptic encephalopathy, treatment with sodium channel blockers such as carbamazepine has been shown to lead to seizure freedom in over 60% of patients [23].
- Ganaxolone was recently approved by the U.S. Food and Drug Administration (FDA) for the treatment of seizures associated with *CDKL5* deficiency disorder (CDD), a rare X-linked developmental and epileptic encephalopathy, after a randomized controlled trial showed a 31% median reduction in major motor seizure frequency [24]. Ganaxolone represents the first disease-specific treatment for CDD. It has also shown promising results in a proof-of-concept trial for the treatment of *PCDH19*-related epilepsy [25].
- In patients with tuberous sclerosis complex, vigabatrin is effective for epileptic spasms. There is evidence to suggest that treatment with vigabatrin before the onset of clinical seizures in patients with TSC can lead to a lower incidence and later time to onset of infantile spasms although the overall incidence of epilepsy is unchanged [26].

- Valproic acid is contraindicated in patients with mitochondrial disorders such as *POLG1* due to a high risk of causing increased oxidative stress and hepatic failure.

HLA Polymorphisms and Pharmacogenetics in Epilepsy The *HLA-B*1502* allele confers a high risk of severe hypersensitivity reactions to carbamazepine. This allele is particularly prevalent in the Han Chinese population, and the carriers are almost 90 times more likely to develop carbamazepine-induced Stevens Johnson Syndrome and Toxic Epidermal Necrolysis (SJS/TEN), which are potentially fatal neurocutaneous syndromes [27]. The predisposition to adverse skin reactions extends beyond carbamazepine. *HLA*B-1502* allele carriers are vulnerable to develop SJS/TEN upon exposure to oxcarbazepine, phenytoin and lamotrigine. This allele-drug reaction association has also been shown to be common in the Thai, Malaysian and Indian populations, so patients of these backgrounds should have HLA testing prior to initiating these medications [28].

Dietary Modification In epilepsies caused by mutations affecting metabolic pathways, correction or replacement of the metabolic deficit can diminish the pathophysiologic dysfunction.

- GLUT1 deficiency syndrome is due to mutations in *SLC2A1,* which encodes the glucose type 1 transporter. This results in impaired transport of glucose across the blood-brain barrier. The ketogenic diet is effective because it provides an alternative fuel source (ketones) to the brain and bypasses the neuroglycopenia caused by the glucose transport deficit. Early diagnosis and initiation of the ketogenic are crucial, and an improvement in seizure control was noted within weeks of commencing the ketogenic diet [29].
- Pyridoxine-dependent epilepsy is an autosomal recessive condition caused by mutations in the *ALDH7A1* gene, which encodes for the enzyme antiquitin. Antiquitin is involved in the catabolism of lysine in the brain. Deficiency leads to the accumulation of alpha-aminoadipic semialdehyde, which is neurotoxic and impairs the metabolism of pyridoxine. Supplemental pyridoxine in high doses can improve seizure control, and early treatment with adjunctive lysine reduction therapies (LRT) has been shown to mitigate the neurodevelopmental impairment with improved fine motor function, cognition, behavioral and psychiatric manifestations [30].

9.3.3 Other Precision Therapies

- The mTOR inhibitor rapamycin (also known as sirolimus) can prevent the occurrence of seizures in polyhydramnios, megalencephaly, and symptomatic epilepsy syndrome (PMSE), an autosomal-recessive disorder associated with mutations in the *STRADA* gene. In TSC, treatment with everolimus (a synthetic mTOR inhibitor) before seizure onset can prevent the development of epilepsy.

Everolimus is also showing promising results in other GATOR1-related epilepsies [31, 32].

- The FDA has recently approved the use of intracerebroventricular enzyme replacement therapy (ERT) for Neuronal ceroid lipofuscinosis (NCL) caused by *CLN2* gene mutations, which encodes for the enzyme TPP1. *CLN2* disease manifests between 2 and 4 years of age with speech delay and developmental regression followed by the development of refractory epilepsy with multiple seizure types. Recombinant human cerliponase alfa is administered into the cerebrospinal fluid via the lateral ventricles every 2 weeks for the rest of their lives to slow the progression of motor and language deficits. *CLN2* disease was fatal within the first 10 years of life until the advent of enzyme replacement therapy, which represents the first disease-modifying therapy for any NCL [33].

Gene Therapies Gene therapy for epilepsies represents a promising and rapidly evolving field offering hope for individuals with genetic forms of epilepsy that are often resistant to traditional treatments. This can be difficult for epilepsies because of the vast heterogeneity and difficulty in penetrating the blood-brain barrier. For the channelopathies, the complexity arises from the fact that epilepsy can result from both gain-of-function and loss-of-function mutations in the same gene, and each type of mutation may require specific therapies to restore normal function without over-correcting the deficit. The CRISPR-Cas9 system is a revolutionary gene-editing tool that allows for precise modification of DNA in almost any cell type and organism. Huge advances have been made in this field of gene editing over the past 5 years, and several companies are now using CRISPR/Cas9 to develop new gene therapies (Fig. 9.3).

Glossary of commonly used genetic terminology	
Single nucleotide variations (SNV)	Single nucleotide changes that occur in a gene either in the coding (exons) or in the noncoding (intron) sections. This includes SNP (Single nucleotide polymorphisms), single nucleotide insertions/ deletions (indels), and point mutations. SNVs can be synonymous (not changing the encoded amino acid) or non-synonymous (changing the encoded amino acid). They are often clinically silent but can rarely cause disease.
Copy number variants (CNV)	Segmental deletions or duplications of DNA that, depending on their extent and location, may predispose to disease. The segment may range from one kilobase to several megabases in size. CNVs are often detected using molecular genetic techniques, such as chromosomal microarray analysis (CMA), comparative genomic hybridization (CGH), or next-generation sequencing (NGS). These methods can identify variations in copy number by comparing an individual's DNA to a reference genome.
Chromosomal microarray analysis (CMA)	Genomic analysis technique that provides high resolution, genome-wide information about chromosomal abnormalities and copy number variants. Includes both comparative genomic hybridization (CGH) and SNP arrays.
Comparative genome hybridization (CGH)	CGH is a molecular-cytogenetic method for the analysis of CNVs or copy number changes (gains or losses) in the DNA. The DNA test sample is competitively hybridized with a reference sample of DNA of known sequence to a DNA microarray, used to detect copy number changes in the test sample.

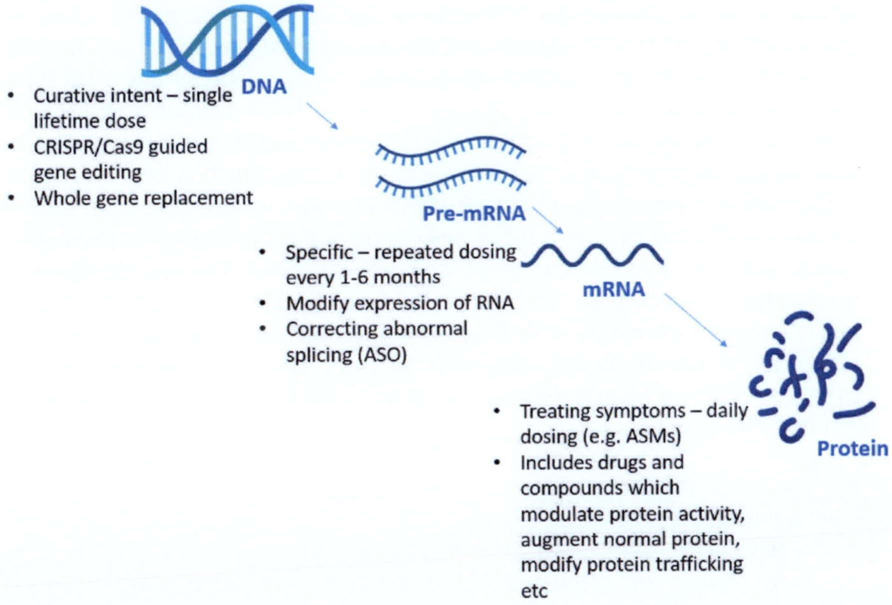

Fig. 9.3 Therapeutic targets in epilepsy (Figure created with BioRender.com)

Glossary of commonly used genetic terminology	
Next generation sequencing	DNA sequencing technology that permits rapid sequencing of large portions of the genome. Includes whole exome and whole genome sequencing.
Exome	The protein-coding portions of the genome.
Intron	The noncoding segments of DNA. These are transcribed into RNA but ultimately removed from the transcript by splicing together the exons to produce messenger RNA (mRNA).
Karyotype	A karyotype is an individual's complete set of chromosomes. The term also refers to a laboratory-produced image of a person's chromosomes isolated from an individual cell and arranged in numerical order. A karyotype may be used to look for abnormalities in chromosome number or structure.
Mitochondrial DNA	DNA found specifically in the mitochondria. Mitochondrial DNA is a single, circular chromosome. In mammals, this DNA is transmitted primarily from mother to offspring via maternal egg cells.

The two major techniques for delivering gene therapy are non-viral and viral-mediated [34]. Among the non-viral gene therapies, antisense oligonucleotides are the most promising. Viral-mediated gene therapy is particularly versatile because viruses can express full-length genes and can therefore be used for gene replacement.

- Antisense oligonucleotide therapies (ASO): ASOs are short nucleotide sequences that are chemically modified to enter cells without degradation. They are disease-specific and are designed to target and degrade mutant RNA or to promote alternative splicing. There is a multicenter trial underway led by Stoke therapeutics to investigate the efficacy of an ASO administered via an intrathecal injection every 4 months. The therapy is designed to upregulate the expression of the unaffected copy of *SCN1A*.

- Adeno-associated virus (AAV) therapies: This technology uses specific AAV serotypes as vectors to deliver therapeutic genetic material to the target cells. The therapeutic gene is inserted into the AAV vector, which is then introduced to the patient, typically via an intracerebroventricular injection for neurological diseases. AAVs can cross cell membranes and enter the nucleus, where the therapeutic gene becomes integrated into the host cells' DNA. Encoded therapeutics is investigating a potential one-time gene therapy for Dravet's syndrome, which involves administering an AAV-based transcription factor that upregulates the expression of the unaffected copy of *SCN1A*. An AAV-based gene therapy targeting *MECP2* in mouse models for Rett syndrome has shown promising results [35].

Clinical Pearls

- All individuals with unexplained epilepsy, regardless of age, should be offered genetic testing although the yield is highest in patients with comorbid neurodevelopmental delay or with a developmental and epileptic encephalopathy.
- Comprehensive, multi-gene testing such as whole exome sequencing or whole genome sequencing should be performed as a first line in most cases.
- Genetic testing should be considered during the presurgical evaluation of patients with drug-resistant focal epilepsy.
- "Precision medicine" focuses on the identification of an underlying genetic etiology, allowing personalized therapeutic choices.
- As knowledge of gene-disease associations evolves, reanalysis of whole exome or whole genome data from previously unsolved cases may be helpful.

References

1. Ruggiero S, Xian J, Helbig I. The current landscape of epilepsy genetics: where are we, and where are we going? Curr Opin Neurol. 2023;36(2):86–94.
2. Berkovic SF, Howell RA, Hay DA, Hopper JL. Epilepsies in twins: genetics of the major epilepsy syndromes. Ann Neurol. 1998;43(4):435–45.
3. Peljto AL, Barker-Cummings C, Vasoli VM, Leibson CL, Hauser WA, Buchhalter JR, Ottman R. Familial risk of epilepsy: a population-based study. Brain. 2014;137(3):795–805.
4. Helbig I, Scheffer IE, Mulley JC, Berkovic SF. Navigating the channels and beyond: unravelling the genetics of the epilepsies. Lancet Neurol. 2008;7(3):231–45.
5. Winawer MR, Shinnar S. Genetic epidemiology of epilepsy or what do we tell families? Epilepsia. 2005;46(suppl 10):24–30.
6. Leu C, Stevelink R, Smith AW, Goleva SB, Kanai M, Ferguson L, Campbell C, Kamatani Y, Okada Y, Sisodiya SM, Cavalleri GL, Koeleman BP, Lerche H, Jehi L, Davis LK, Najm IM,

Palotie A, Daly MJ, Busch RM, Lal D. Polygenic burden in focal and generalized epilepsies. Brain. 2019;142(11):3473–81.

7. Sheidley BR, Malinowski J, Bergner AL, Bier L, Gloss DS, Mu W, Mulhern MM, Partack EJ, Poduri A. Genetic testing for the epilepsies: a systematic review. Epilepsia. 2022;63(2):375–87.
8. Borlot F, Regan BM, Bassett AS, Stavropoulos DJ, Andrade DM. Prevalence of pathogenic copy number variation in adults with pediatric-onset epilepsy and intellectual disability. JAMA Neurol. 2017;74(11):1301.
9. Smith L, Malinowski J, Ceulemans S, Peck K, Walton N, Sheidley BR, Lippa N. Genetic testing and counseling for the unexplained epilepsies: an evidence-based practice guideline of the National Society of Genetic Counselors. J Genet Couns. 2023;32(2):266–80.
10. Guerrini R, Balestrini S, Wirrell EC, Walker MC. Monogenic epilepsies: disease mechanisms, clinical phenotypes, and targeted therapies. Neurology. 2021;97(17):817–31.
11. Symonds JD, Zuberi SM, Stewart K, McLellan A, O'Regan M, MacLeod S, Jollands A, Joss S, Kirkpatrick M, Brunklaus A, Pilz DT, Shetty J, Dorris L, Abu-Arafeh I, Andrew J, Brink P, Callaghan M, Cruden J, Diver LA, Findlay C, Gardiner S, Grattan R, Lang B, MacDonnell J, McKnight J, Morrison CA, Nairn L, Slean MM, Stephen E, Webb A, Vincent A, Wilson M. Incidence and phenotypes of childhood-onset genetic epilepsies: a prospective population-based national cohort. Brain. 2019;142(8):2303–18.
12. Helbig I, Mefford HC, Sharp AJ, Guipponi M, Fichera M, Franke A, Muhle H, de Kovel C, Baker C, von Spiczak S, Kron KL, Steinich I, Kleefuss-Lie AA, Leu C, Gaus V, Schmitz B, Klein KM, Reif PS, Rosenow F, Weber Y, Lerche H, Zimprich F, Urak L, Fuchs K, Feucht M, Genton P, Thomas P, Visscher F, de Haan GJ, Møller RS, Hjalgrim H, Luciano D, Wittig M, Nothnagel M, Elger CE, Nürnberg P, Romano C, Malafosse A, Koeleman BP, Lindhout D, Stephani U, Schreiber S, Eichler EE, Sander T. 15q13.3 microdeletions increase risk of idiopathic generalized epilepsy. Nat Genet. 2009;41(2):160–2.
13. Suls A, Mullen SA, Weber YG, Verhaert K, Ceulemans B, Guerrini R, Wuttke TV, Salvo-Vargas A, Deprez L, Claes LR, Jordanova A, Berkovic SF, Lerche H, De Jonghe P, Scheffer IE. Early-onset absence epilepsy caused by mutations in the glucose transporter GLUT1. Ann Neurol. 2009;66(3):415–9.
14. Smith KM, Youssef PE, Wirrell EC, Nickels KC, Payne ET, Britton JW, Shin C, Cascino GD, Patterson MC, Wong-Kisiel LC. Jeavons syndrome: clinical features and response to treatment. Pediatr Neurol. 2018;86:46–51.
15. Karge R, Knopp C, Weber Y, et al. Genetics of nonlesional focal epilepsy in adults and surgical implications. Clin Epileptol. 2023;36:91–7.
16. Harris RV, Oliver KL, Perucca P, Striano P, Labate A, Riva A, Grinton BE, Reid J, Hutton J, Todaro M, O'Brien TJ, Kwan P, Sadleir LG, Mullen SA, Dazzo E, Crompton DE, Scheffer IE, Bahlo M, Nobile C, Gambardella A, Berkovic SF. Familial mesial temporal lobe epilepsy: clinical spectrum and genetic evidence for a polygenic architecture. Ann Neurol. 2023;94(5):825–35.
17. Møller RS, Weckhuysen S, Chipaux M, Marsan E, Taly V, Bebin EM, Hiatt SM, Prokop JW, Bowling KM, Mei D, Conti V, De La Grange P, Ferrand-Sorbets S, Dorfmüller G, Lambrecq V, Larsen LHG, Leguern E, Guerrini R, Rubboli G, Cooper GM, Baulac S. Germline and somatic mutations in the MTOR gene in focal cortical dysplasia and epilepsy. Neurol Genet. 2016;2:e118.
18. Moloney PB, Dugan P, Widdess-Walsh P, Devinsky O, Delanty N. Genomics in the presurgical epilepsy evaluation. Epilepsy Res. 2022;184:106951.
19. Richards S, Aziz N, Bale S, Bick D, Das S, Gastier-Foster J, Grody WW, Hegde M, Lyon E, Spector E, Voelkerding K, Rehm HL, ACMG Laboratory Quality Assurance Committee. Standards and guidelines for the interpretation of sequence variants: a joint consensus recommendation of the American College of Medical Genetics and Genomics and the Association for Molecular Pathology. Genet Med. 2015;17(5):405–24.
20. McKnight D, Morales A, Hatchell KE, et al. Genetic testing to inform epilepsy treatment management from an international study of clinical practice. JAMA Neurol. 2022;79(12):1267–76.

21. Brunklaus A, Feng T, Brünger T, Perez-Palma E, Heyne H, Matthews E, Semsarian C, Symonds JD, Zuberi SM, Lal D, Schorge S. Gene variant effects across sodium channelopathies predict function and guide precision therapy. Brain. 2022;145(12):4275–86.
22. de Lange IM, Gunning B, Sonsma ACM, van Gemert L, van Kempen M, Verbeek NE, Nicolai J, Knoers NVAM, Koeleman BPC, Brilstra EH. Influence of contraindicated medication use on cognitive outcome in Dravet syndrome and age at first afebrile seizure as a clinical predictor in SCN1A-related seizure phenotypes. Epilepsia. 2018;59(6):1154–65.
23. Kuersten M, Tacke M, Gerstl L, Hoelz H, Stülpnagel CV, Borggraefe I. Antiepileptic therapy approaches in KCNQ2 related epilepsy: a systematic review. Eur J Med Genet. 2020;63(1):103628.
24. Knight EMP, Amin S, Bahi-Buisson N, Benke TA, Cross JH, Demarest ST, Olson HE, Specchio N, Fleming TR, Aimetti AA, Gasior M, Devinsky O, Marigold Trial Group. Safety and efficacy of ganaxolone in patients with CDKL5 deficiency disorder: results from the double-blind phase of a randomised, placebo-controlled, phase 3 trial. Lancet Neurol. 2022;21(5):417–27.
25. Sullivan J, Gunning B, Zafar M, Guerrini R, Gecz J, Kolc KL, Zhao Y, Gasior M, Aimetti AA, Samanta D. Phase 2, placebo-controlled clinical study of oral ganaxolone in PCDH19-clustering epilepsy. Epilepsy Res. 2023;191:107112.
26. Bebin EM, Peters JM, Porter BE, McPherson TO, O'Kelley S, Sahin M, Taub KS, Rajaraman R, Randle SC, McClintock WM, Koenig MK, Frost MD, Northrup HA, Werner K, Nolan DA, Wong M, Krefting JL, Biasini F, Peri K, Cutter G, Krueger DA; PREVeNT Study Group. Early treatment with vigabatrin does not decrease focal seizures or improve cognition in tuberous sclerosis complex: the PREVeNT trial. Ann Neurol. 2023. https://doi.org/10.1002/ana.26778.
27. Ferrell PB Jr, McLeod HL. Carbamazepine, HLA-B*1502 and risk of Stevens-Johnson syndrome and toxic epidermal necrolysis: US FDA recommendations. Pharmacogenomics. 2008;9(10):1543–6.
28. Balestrini S, Sisodiya SM. Pharmacogenomics in epilepsy. Neurosci Lett. 2018;667:27–39.
29. Schwantje M, Verhagen LM, van Hasselt PM, Fuchs SA. Glucose transporter type 1 deficiency syndrome and the ketogenic diet. J Inherit Metab Dis. 2020;43(2):216–22.
30. Tseng LA, Abdenur JE, Andrews A, Aziz VG, Bok LA, Boyer M, Buhas D, Hartmann H, Footitt EJ, Grønborg S, Janssen MCH, Longo N, Lunsing RJ, MacKenzie AE, Wijburg FA, Gospe SM, Coughlin CR, van Karnebeek CDM. Timing of therapy and neurodevelopmental outcomes in 18 families with pyridoxine-dependent epilepsy. Mol Genet Metab. 2022;135(4):350–6.
31. Crino PB. The mTOR signalling cascade: paving new roads to cure neurological disease. Nat Rev Neurol. 2016;12(7):379–92.
32. Moloney PB, Kearney H, Benson KA, Costello DJ, Cavalleri GL, Gorman KM, Lynch BJ, Delanty N. Everolimus precision therapy for the GATOR1-related epilepsies: a case series. Eur J Neurol. 2023;30(10):3341–6.
33. Schulz A, Ajayi T, Specchio N, de Los Reyes E, Gissen P, Ballon D, Dyke JP, Cahan H, Slasor P, Jacoby D, Kohlschütter A, CLN2 Study Group. Study of intraventricular cerliponase alfa for CLN2 disease. N Engl J Med. 2018;378(20):1898–907.
34. Goodspeed K, Bailey RM, Prasad S, Sadhu C, Cardenas JA, Holmay M, Bilder DA, Minassian BA. Gene therapy: novel approaches to targeting monogenic epilepsies. Front Neurol. 2022;13:805007.
35. Powers S, Likhite S, Gadalla KK, Miranda CJ, Huffenberger AJ, Dennys C, Foust KD, Morales P, Pierson CR, Rinaldi F, Perry S, Bolon B, Wein N, Cobb S, Kaspar BK, Meyer KC. Novel MECP2 gene therapy is effective in a multicenter study using two mouse models of Rett syndrome and is safe in non-human primates. Mol Ther. 2023;31(9):2767–82.

Pathology

10

Nithisha Thatikonda and Todd Masel

10.1 Mesial Temporal Sclerosis

Mesial Temporal Sclerosis (MTS) or hippocampal sclerosis (HS) is the most common cause of drug-resistant epilepsy (DRE)in adults, present in 36.4% of cases undergoing surgical resection [1]. The pathogenesis of MTS is still debatable, with febrile seizures being an important harbinger [2]. Hyperintensity on T2-weighted sequence of MRI and atrophy of the hippocampus and surrounding limbic network is diagnostic of MTS (Fig. 10.1) [1, 2]. Other features, such as ipsilateral temporal horn enlargement, thinning of the fornix and mamillary body, and loss of hippocampal head interdigitations are variably seen [2].

Segmental neuronal loss, reactive gliosis, and granule cell loss play a critical role in epileptogenesis in MTS [3, 4]. Mossy fiber sprouting is another characteristic feature of HS, with debatable epileptogenic potential [4]. HS is classified into three types as per the International League Against Epilepsy (ILAE) (Table 10.1) [3]. Timely diagnosis and removal of abnormal tissue results in seizure freedom in 64% of patients at 12 months follow-up [5].

N. Thatikonda · T. Masel (✉)
Department of Neurology, University of Texas Medical Branch, Galveston, TX, USA
e-mail: tsmasel@UTMB.EDU

Z. Haneef (ed.), *Epilepsy Fundamentals*,
https://doi.org/10.1007/978-3-031-77741-7_10

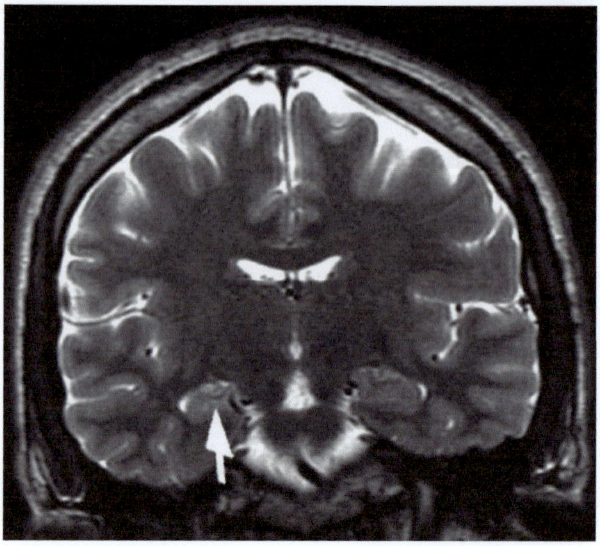

Fig. 10.1 Coronal T2-weighted imaging with the arrow showing right hippocampal hyperintensity and atrophy

Table 10.1 ILAE classification and histopathologic subtypes of hippocampal sclerosis in patients with temporal lobe sclerosis [3] (images from Jardim et al. [4] under CC-BY 4.0 license)

Type 1 Predominant **CA4** (90%) and **CA1** (>80%) loss Variable CA2 and CA3 involvement 50–60% granule cell loss	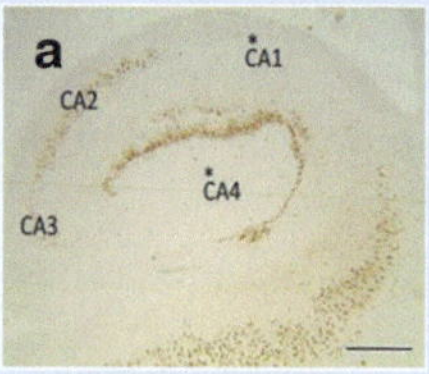	**Type 2** Predominant **CA1** loss (80%) Variable CA2, CA3, and CA4 involvement No severe granule cell loss	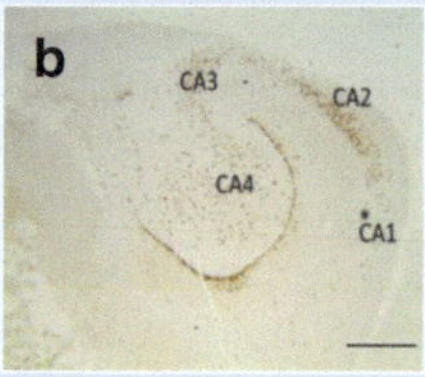
Type 3 Predominant **CA4** loss (50–60%) Variable CA1, CA2, and CA3 involvement 35% granule cell loss	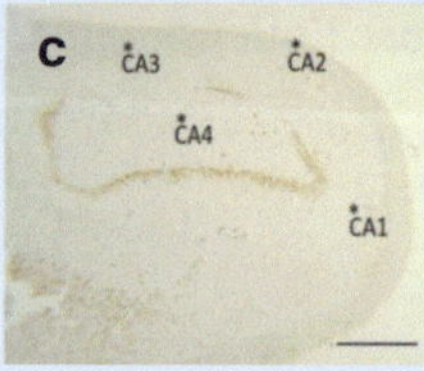	No HS, gliosis only No significant neuronal loss; only reactive gliosis No severe granule cell loss	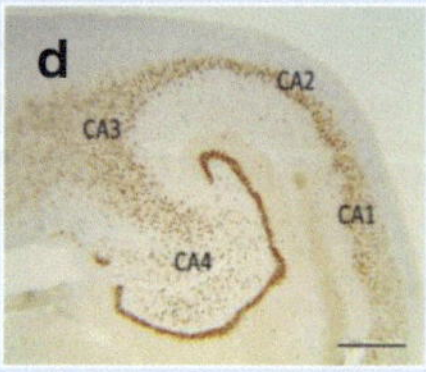

CA cornua ammonis

10.2 Malformations of Cortical Development

Malformations of cortical development (MCD) result from the disruption of cortical development and are classified into three main categories based on the developmental stage affected [1] (Table 10.2). Focal cortical dysplasia (FCD), which is the most common type of MCD is discussed separately (Table 10.3).

FCD represents 75% of all MCD cases, contributing to 19.8% of patients undergoing surgical resection for DRE [1, 3].As it is often subtle in its expression, FCD is often "MRI-negative." [2] Patients with focal epilepsy and normal MRIs are often

Table 10.2 ILAE classification of MCD along with imaging features (images from Sufiani et al. [6] (a), Jaiswal et al. [7] (b), and Oegema et al. [8] (c–f) under CC-BY 4.0 license)

1. Malformations due to *abnormal neuronal proliferation/ apoptosis* (a) Hemimegalencephaly Overgrowth of unilateral cerebral hemisphere (figure a). MRI shows gray matter thickening, polymicrogyria, marked dysplasia, pachygyria, and ventricular enlargement (figure b)	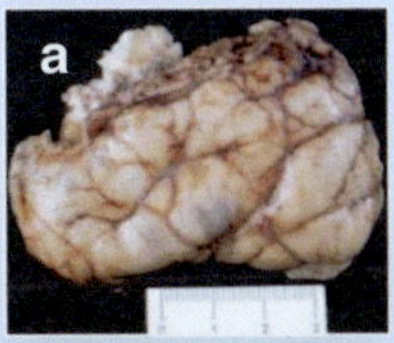 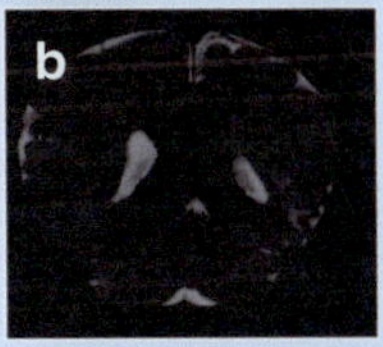
2. Malformations due to *abnormal neuronal migration* (a) Lissencephaly Loss of normal gyral patterns resulting in agyria and loss of convolution of cortex [1] MRI often shows a thick, bilateral, symmetric cerebral cortex with reduced gyration (figure c, white arrows) (b) Cobblestone complex Overmigration of neurons and cortical disorganization result in holes/defects of the pia matter, especially in frontoparietal regions [1] MRI shows characteristic jagged cortical–white matter border with frequent vertical striations [1] (c) Heterotopia Abnormal migration of normal neurons in an inappropriate location, most commonly in periventricular region [1, 9] Microscopically, these clusters of abnormal neurons form nodule-like structures (figure d: MRI shows gray matter lesions often with associated pachygyria of the left (arrow) more than right (arrowhead) occipital horns of the lateral ventricles)	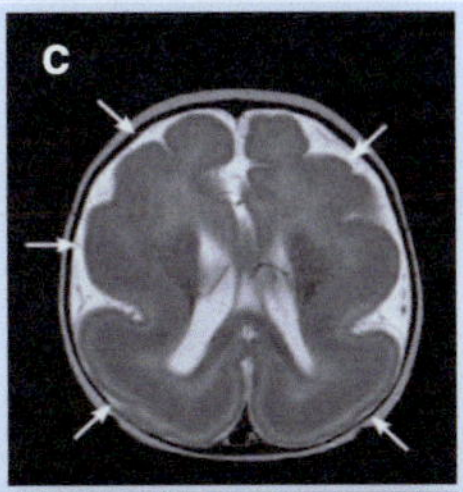 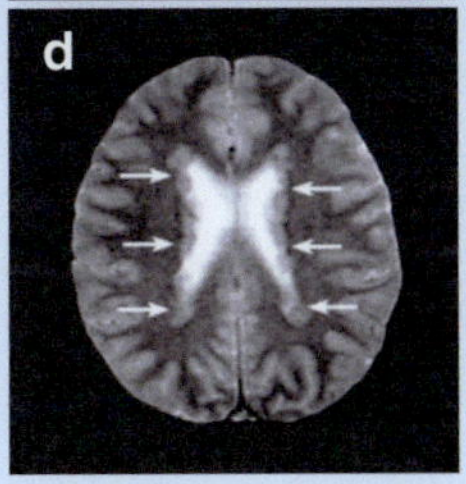

(continued)

Table 10.2 (continued)

3. Malformations due to *abnormal cortical organization* (including post-neuronal migration)
   (a) Polymicrogyria
   Cortical disorganization with excessive abnormally small gyri and shallow sulci gives rise to irregular cortical surface [9]
   MRI shows numerous small gyri with shallow sulci on T1 (figure e: arrows), most commonly in the perisylvian insular cortex, along with irregular cortical-subcortical junction
   (b) Schizencephaly (polymicrogyria with clefts)
   Characterized by abnormal slits/clefts lined by polymicrogyria [1]
   There are two types of schizencephalic clefts—closed (type 1) or open (type 2). Closed-lip clefts often present with hemiparesis/motor delay, whereas open-lip clefts present with hydrocephalus
   MRI shows clefts communicating with ventricles and lined polymicrogyria, commonly involving perisylvian cortex (figure f: MRI showing schizencephaly, characterized by a cleft (arrow) extending from the cortex to the ventricle)

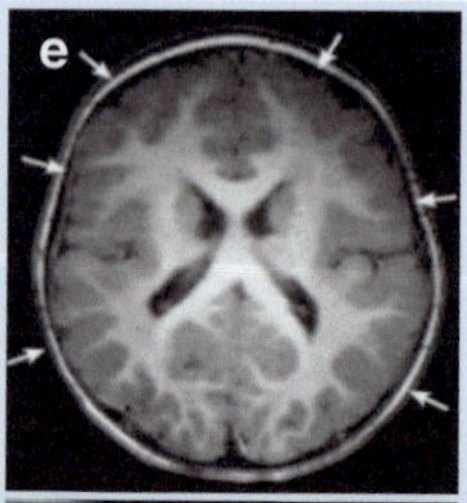

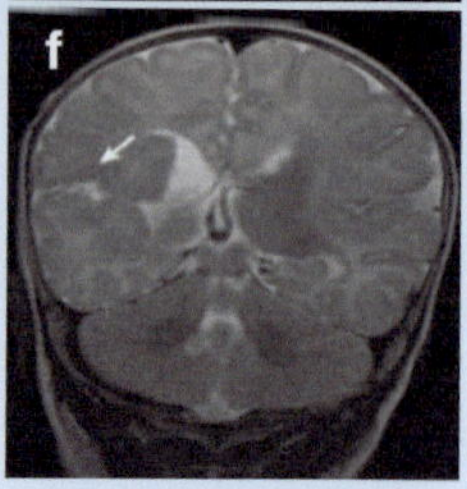

Table 10.3 ILAE classification of FCD along with histopathology and radiology features (images from Sufiani et al. [6] (a–c), Berlangieri et al. [10] (d), and Matsuo et al. [11] (e) under CC-BY 4.0 license)

Histopathology characteristics	MRI features
Ia: Abnormal *radial* cortical lamination with preservation of cortical layering (figure a) 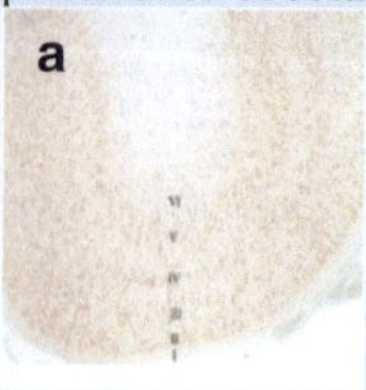 Ib: Abnormal *tangential* cortical lamination	Ia and Ib • Subtle features are often invisible • Hypoplasia or thin cortex • Blurring of the gray-white matter junction • Abnormally shaped sulci • Cannot differentiate the subtypes

(continued)

Table 10.3 (continued)

Histopathology characteristics	MRI features
IIa: *Dysmorphic* neurons (figure b) 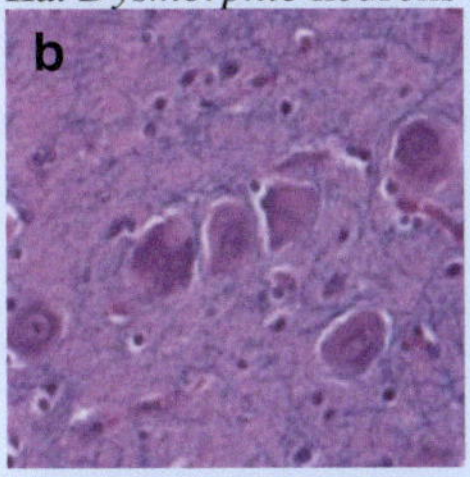 IIb: Dysmorphic neurons with *balloon cells* (figure c, arrowheads) 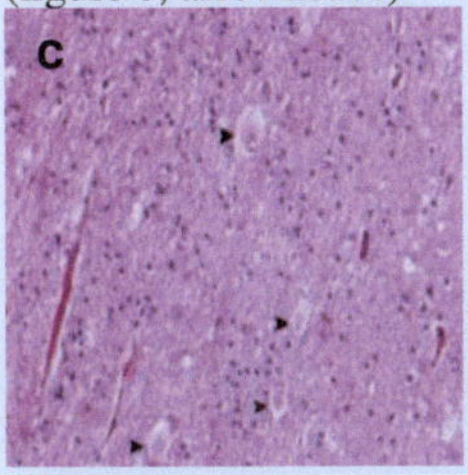	IIa and IIb • Easier to identify than FCD type I • Bottom-of-sulcus dysplasia: dysplastic features maximum at sulcal depth, highly amenable to successful surgical treatment (figure d, arrows) 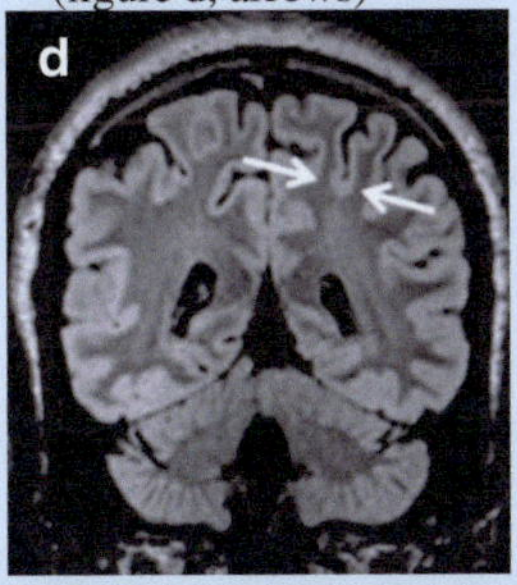 IIb: "Transmantle sign" is a long region of T2/FLAIR hyperintense signal tapering between the affected cortex and the ventricular wall (figure e, arrowhead) 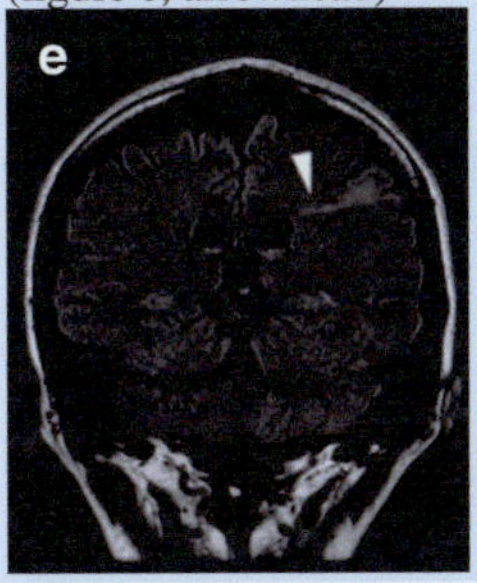
IIIa: Cortical lamination abnormalities associated with *hippocampal sclerosis* IIIb: Cortical lamination abnormalities adjacent to a *glioneuronal tumor* IIIc: Cortical lamination abnormalities adjacent to *vascular malformation* IIId: Cortical lamination abnormalities adjacent to *any other lesion* acquired during early life (e.g., trauma, ischemic injury, and encephalitis)	IIIa–d: Variable combination of MRI features of FCD described above and the associated lesion in the same region

ILAE international league association of epilepsy, *FCD* focal cortical dysplasia

presumed to have FCD. FCDs are classified into three subgroups based on ILAE classification [1, 2, 6, 12] (Table 10.3). The median age of seizure onset is 5 years in FCD type I and 7 years in FCD type II. Isolated focal epilepsy is often the only clinical manifestation in small FCD. At the same time, larger FCDs in temporal or occipital locations are associated with severe cognitive impairment [12]. While FCD type I is not associated with any specific scalp EEG findings, FCD type II is often accompanied by focal rhythmic interictal repetitive spikes and bursts of fast rhythms (brushes) [12]. Surgical outcomes of FCD type II are better than type I, with 62–64% of patients with FCD II becoming seizure-free after surgery. Due to mTOR pathway gene mutations in FCD type II, pharmacological mTOR inhibitors such as everolimus are being tested in clinical trials [12].

10.3 Neoplasms

Although any brain tumor involving the cortex can cause seizures, slow-growing tumors are more likely to be epileptogenic than rapidly growing tumors and have been categorized as long-term epilepsy-associated tumors (LEATs) [2]. LEATs account for 22–24% of cases of intractable epilepsy undergoing surgery [1, 3]. Recently, a large number of brain tumors with neuroepithelial origin were included, expanding the LEAT subtypes (Figs. 10.2 and 10.3 and Table 10.4) [1, 2, 13].

Glioneuronal Tumors Glioneuronal tumors have intrinsic epileptogenicity due to glutamatergic dysplastic neurons functionally integrated within excitatory circuits. Epilepsy results from an imbalance between glutamatergic excitation and GABAergic inhibition. Glioneuronal tumors consist of gangliogliomas (GGs) and dysembryoplastic neuroepithelial tumors (DNETs). They most commonly occur in

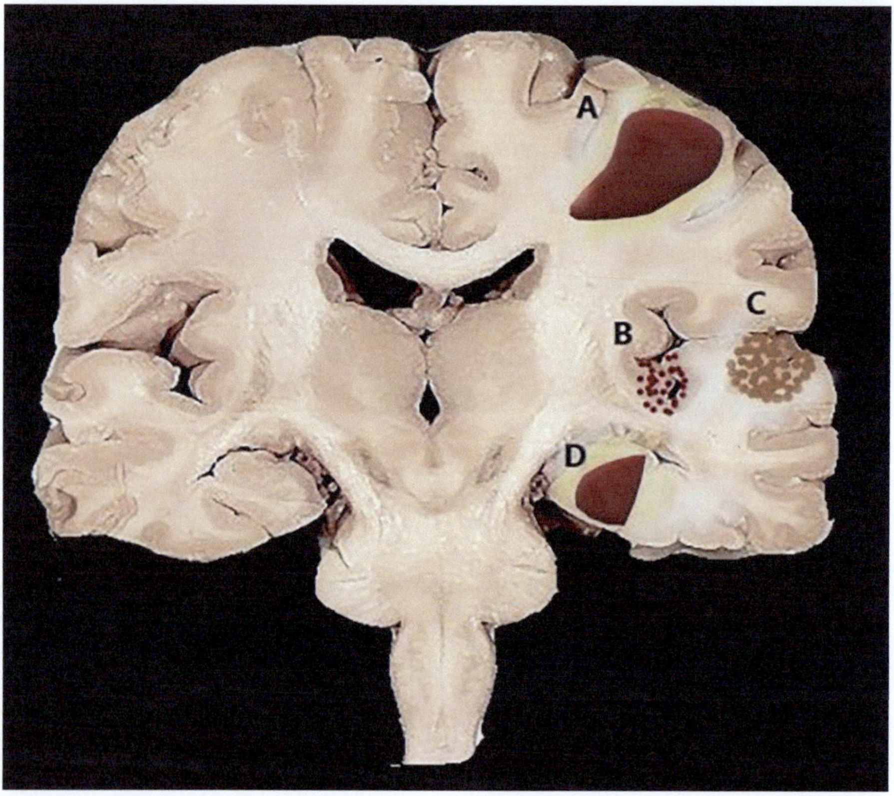

Fig. 10.2 Graphic illustration of a few common LEAT subtypes showing the typical locations of their occurrence, namely, (**a**) pleomorphic xanthoastrocytoma, (**b**) multinodular and vacuolating neuronal tumor (MVNT), (**c**) dysembryoplastic neuroepithelial tumor (DNET), and (**d**) ganglioglioma (GG). Image modified from "Beal [13]–Human Brain Frontal (Coronal) Section by John A Beal," license: Creative Commons Attribution-2.5 Generic

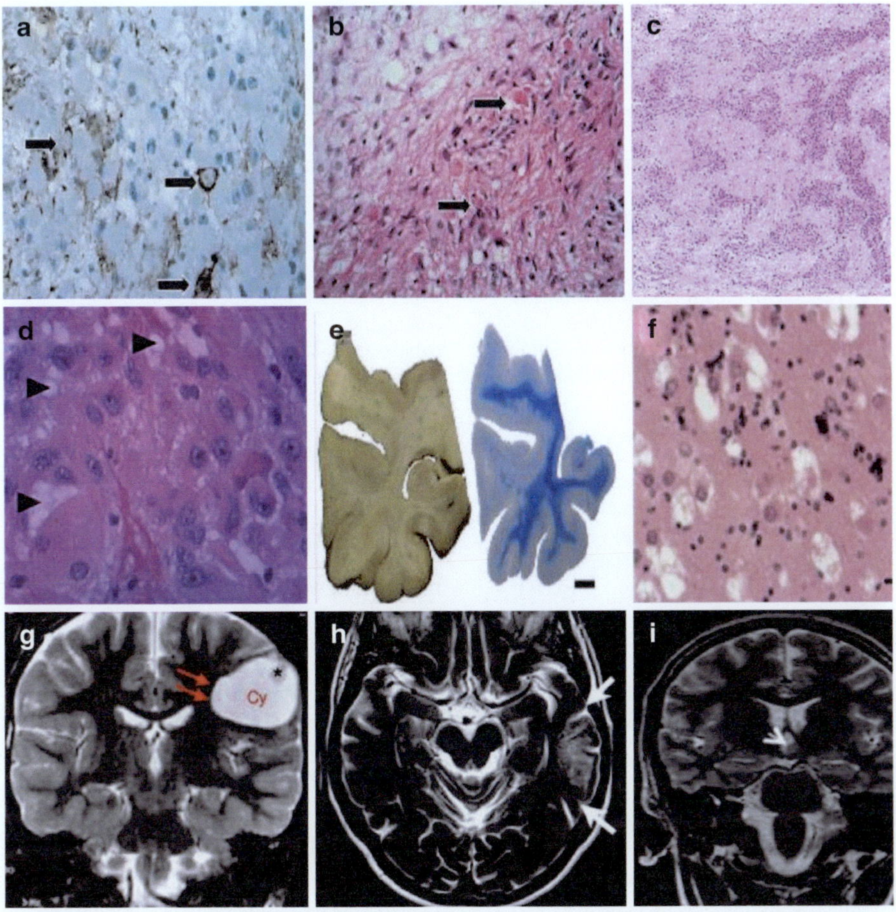

Fig. 10.3 Common types of LEATs. Ganglioglioma shows neuronal clusters of CD34+ atypical ganglion cells (**a**) and multiple eosinophilic granular bodies within glia (**b**); dysembryoplastic neuroepithelial tumor showing neuronal microcolumns with interposing myxoid matrix (**c**); pleomorphic xanthoastrocytoma (PXA) shows xanthomatous changes with an accumulation of lipid droplets (**d**), appearing as T2 hyperintense cystic lesion with mural nodule (**g**, red arrows); multinodular and vacuolating neuronal tumor (MVNT) from autopsy revealing numerous coalescing grey nodules along the junction of the cortex and grey matter, which are hypomyelinated nodules by Nissl stain (**e**), vacuolated large cells (**f**), and appearing as hyperintense and nodular in MRI (**h**, white arrows). Hypothalamic hamartoma appears as a hypointense lesion arising from the hypothalamus protruding into the third ventricle (**i**, arrowhead). Images from Resta et al. [14] (**a**, **b**), Rahim et al. [15] (**c**), Furtado et al. [16] (**d**, **g**), Bodi et al. [17] (**e**, **f**, **h**), and Han et al. [18] (**i**) under CC-BY 4.0 license

the temporal lobes (50–75%) followed by the frontal lobes (8–10%) with a mean age of onset of 16 years [20]. When they occur in association with FCDs, they are classified as FCD type IIIb. GGs are composed of nodular, compact masses of dysplastic neurons, often located in the peripheral cortex [19, 20] (Fig. 10.2). On MRI,

Table 10.4 Types of long-term epilepsy-associated tumors (LEATs)

Clinical presentation and histopathology	Imaging and outcomes
Low-grade gliomas (grade 2 astrocytoma, oligodendroglioma, oligoastrocytoma)	
• Typically, temporal lobes with focal epilepsy • Rosenthal fibers and pilocytic processes in astrocytoma and nodules of round cells arranged in rosettes are seen on microscopy [1]	• T1-hypointense lesion without contrast enhancement, often with calcifications [2] • Surgery is the mainstay of treatment, with seizure freedom at 67% and a recurrence rate of 17% [19]
Multinodular and vacuolating neuronal tumors (MVNTs)	
• Variable in location with an indolent course • Abnormal clustering and vacuolation of neuronal cells seen on microscopy [17] (Fig. 10.3e, f)	• Non-enhancing T2-hyperintense lesion containing discrete ovoid nodules giving bubbly appearance surrounding the sulcus [2] (Figs. 10.2 and 10.3h) • Resection is considered in drug-resistant epilepsy
Pleomorphic xanthoastrocytoma	
• Typically occur in the superficial cortex adjacent to leptomeninges, protruding into pia matter (Fig. 10.3b) • Xanthomatous changes (large cells with accumulation of lipid droplets) and eosinophilic inclusion bodies are characteristic (Fig. 10.3d)	• Both cystic (T1-hypointense) and solid (T2 hyper-intense) components give the appearance of a mural nodule (Fig. 10.3g). Characteristic leptomeningeal involvement with dural tail in 15–50% of cases • Surgical resection is the treatment of choice, with a 5-year survival rate of 81% [20]
Hypothalamic hamartoma	
• Originate from the ventral hypothalamus and present with gelastic seizures in one-third of patients [1] • Clusters of small neurons in a grape-like configuration	• Lesions are T1-hypointense and T2-hyperintense (Fig. 10.3i) • Surgical resection with gamma knife radiosurgery results in seizure freedom in 40% of the patients [20]
Meningioma	
• Extra-axial meningiomas located in parasagittal region/falx and sphenoid ridge are associated with high frequency of epilepsy (up to one quarter of patients) [2]	• Lesions are T1-iso to hypointense, T2-hyperintense with homogenous contrast enhancement and dural tail [1, 2] • Variable outcomes rendering up to 83% of patients are seizure free after resection [20]

they appear hypointense to isointense on T1, and hyperintense on T2/FLAIR with minimal contrast enhancement, giving the appearance of a cystic lesion with an enhancing mural nodule [1]. DNETs result from disordered neuronal migration from the subpial/subependymal germinal layers, and are arranged in microcolumns with an interposing myxoid matrix [19, 20]. MRI shows a characteristic "pseudo-cystic" or "soap bubble appearance." [1] Surgical resection leads to excellent seizure reduction (85% with Engel Class 1 outcomes) [19].

10.4 Vascular Malformations

Vascular malformations constitute 5% of epileptogenic substrates in the general population [1]. Arteriovenous malformations (AVM) may cause seizures by acute bleeding, posthemorrhagic encephalomalacia, the presence of sclerotic brain parenchyma within capillary beds, or the adjacent development of a focal cortical dysplasia type III [21]. On MRI, both T1- and T2-weighted images demonstrate serpiginous flow voids [1]. Cavernous malformations are immature endothelium channels without intervening brain tissue and are highly epileptogenic (likely due to hemosiderin deposits that accumulate around the lesion) [21, 22]. On gradient echo sequence (GRE) of MRI, these lesions have a classic "popcorn" appearance.

10.5 Neurocutaneous Disorders

While phenotypically diverse, neurocutaneous disorders are united by their origin from the primitive ectodermal tissue, which gives rise to both the skin and the nervous system. In addition, there is now emerging evidence of overlap in the cellular signaling pathways in these disorders [23]. The three main types of prototypical neurocutaneous disorders and their mechanisms of epileptogenesis are discussed in Table 10.5 and Fig. 10.4a.

Table 10.5 Common neurocutaneous disorders and their characteristics

Characteristics	Tuberous sclerosis complex (TSC)	Neurofibromatosis type 1 (NF-1)	Sturge Weber syndrome (SWS)
Gene mutations	• TSC1 and TSC2	• NF1	• GNAQ
Abnormal protein	• Hamartin/Tuberin	• Neurofibromin	• Gαq (GTPase)
Gross and histopathology	• Cortical hamartomas/tubers often at the gray-white matter junction (Fig. 10.4b), made of dysplastic neurons with characteristic balloon cells with positive silver stain (Fig. 10.4c, d) • Subependymal nodules in the periventricular region	• Astrogliosis • Focal cortical dysplasia • Heterotopias • Polymicrogyria	• Leptomeningeal capillary malformations resulting in venous stasis promoting neuronal loss and gliosis

(continued)

Table 10.5 (continued)

Characteristics	Tuberous sclerosis complex (TSC)	Neurofibromatosis type 1 (NF-1)	Sturge Weber syndrome (SWS)
Seizure	• Focal onset may generalize • Infantile spasms as a part of West syndrome • Onset of epilepsy before 3 years of age	• Focal onset may generalize • Incidence of epilepsy is 6–10%, which is remarkably less compared to TSC, which is 90%	• Focal onset may generalize • The onset of epilepsy is typically 2 years of age in 80% of cases
MRI findings	• Single/multiple cortical tubers typically at grey-white matter junction appearing hyperintense on T2-sequence [1]	• Focal lesions such as optic pathway glioma, or temporal lobe glioma with post-contrast enhancement on T1-sequence [14]	• Angiomatous malformation with leptomeningeal enhancement on post-contrast T1-sequence, typically seen in posterior temporo-occipital lobe ipsilateral to facial angioma [1]
Treatment	• Vigabatrin is the drug of choice for infantile spasms. • DRE occurs in up to 70% of patients with tuber resection resulting in seizure freedom in up to 59% of patients • mTOR inhibitors such as rapamycin/everolimus under investigation	• Good seizure control with anti-seizure medications (ASMs), although surgical resection is sometimes pursued	• Seizures sometimes respond to ASMs and ketogenic diet • DRE is common requiring focal resection or hemispherectomy

DRE drug-resistant epilepsy, *ASMs* anti-seizure medications, *TSC* tuberous sclerosis, *NF* neurofibromatosis, *GNAQ* gene for guanine nucleotide-binding protein (G protein), subunit alpha, q polypeptide, *Gαq* guanine nucleotide-binding protein (G protein), subunit alpha, q polypeptide, *GTP* guanosine-5′-triphosphate, *mTOR* mechanistic target of rapamycin

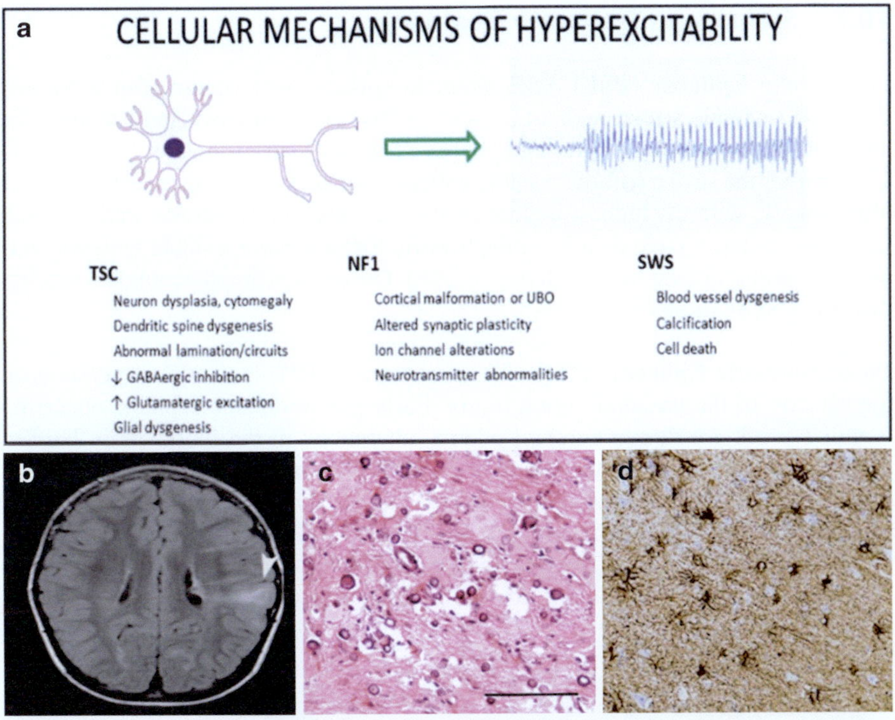

Fig. 10.4 (**a**) Possible contributors to cellular hyperexcitability and seizures in three common neurocutaneous disorders—tuberous sclerosis complex (TSC), neurofibromatosis type 1 (NF1), and Sturge–Weber syndrome (SWS); (**b**) cortical tuber (arrow) and effacement of the cortex gray-white matter junction; (**c**) balloon cells characteristic of a TSC tuber; and (**d**) silver (Bielschowsky)-stained section of a cortical tuber, showing cytoskeletal fibrillar aggregates. Images used with permission from Stafstrom et al. [23] (**a**), and Mühlebner et al. [24] (**b**–**d**) under CC-BY 4.0 license

10.6 Perinatal Insults

Non-traumatic destructive brain lesions in the perinatal period are usually attributed to hypoxia and vascular insults. These perinatal insults lead to severe hypoxia causing an alteration in calcium homeostasis and acutely increasing neurotropic factors such as BDNF, accelerating immature neuron formation [25]. The inability to eliminate the immature neurons in the remodeling process causes hyper-innervated circuits to become eileptogenic [26]. Epilepsy can also occur due to gliosis forming polymicrogyria, porencephaly, encephalomalacia, or ulegyria, visible as reduced gyral depth with a "mushroom" appearance on MRI [25].

10.7 Stroke and Traumatic Brain Injury

Post-Stroke Epilepsy (PSE) PSE refers to epilepsy that occurs after a hemorrhagic or ischemic stroke without a previous history of epilepsy and accounts for about 30–50% of newly diagnosed epilepsy [27]. Early onset seizures occur within 1 week after the stroke without a stable epileptic network. Late-onset seizures occur after 1 week, with the highest incidence at 6–12 months after stroke. Patients with late-onset epilepsy have definite epileptogenic foci and form a stable epileptic network, resulting in long-term epilepsy [27, 28]. Different pathophysiological mechanisms in PSE is shown in Fig. 10.5.

Post-Traumatic Epilepsy (PTE) The incidence of PTE is 2–50% depending on the severity of the traumatic brain injury. Early post-traumatic seizures occurring within 1 week are due to cerebral edema, alterations in the blood–brain barrier, excessive release of excitatory neurotransmitters, tissue damage by free radicals, and changes in the way cells produce energy. Late seizures occurring after one week are thought to indicate permanent changes resulting in neuronal loss and aberrant sprouting [29, 30]. Animal studies have demonstrated that cortical injuries cause "selective vulnerability" leading to loss of hilar and CA3 neurons (Table 10.6),

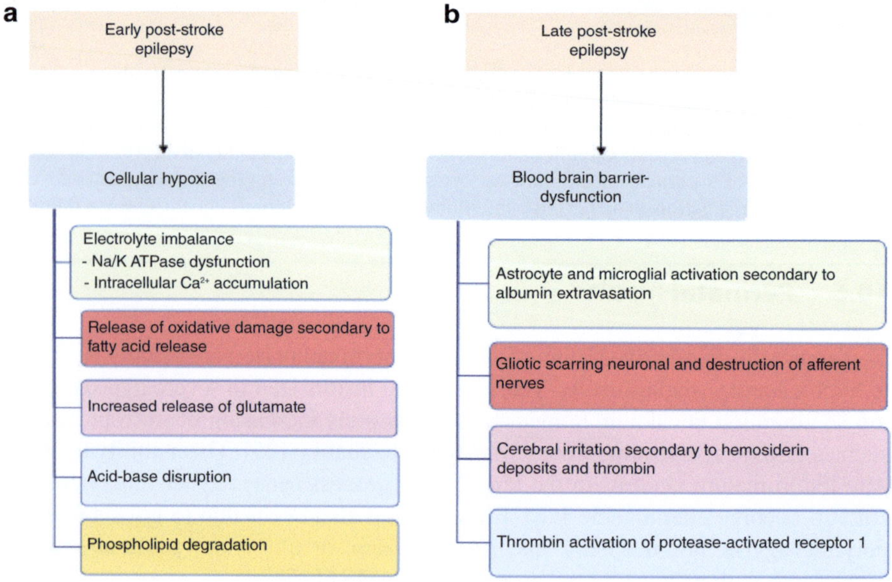

Fig. 10.5 (**a**) General pathophysiology of early poststroke epilepsy (PSE). Early PSE is often due to cellular hypoxia secondary to reduced blood supply caused by a hemorrhagic or ischemic stroke (**b**). General pathophysiology of late PSE. Late PSE is often a complication of blood-brain barrier dysfunction resulting in multiple glial cell dysfunction. Image from Phan et al. [28] under CC-BY 4.0 license

Table 10.6 Progression of gross cell loss in the ipsilateral hippocampal CA3 and hilum in temporal regions during the weeks after TBI (images from Golarai et al. [31] under CC-BY 4.0 license, Copyright [2001] Society for Neuroscience. https://www.jneurosci.org/content/21/21/8523.long)

3 weeks post-TBI Gross cell loss in the ipsilateral CA3 (between arrows) of the temporal hippocampus	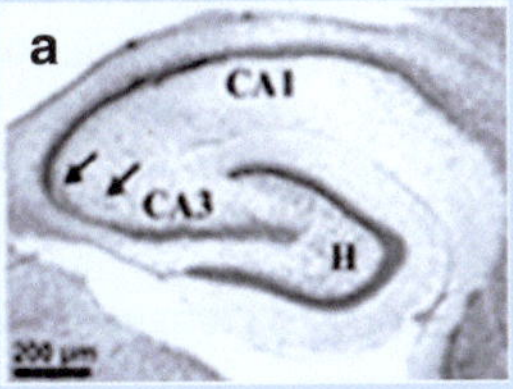	15 weeks post-TBI Widening of gross cell loss in CA3 (between arrows) compared with A	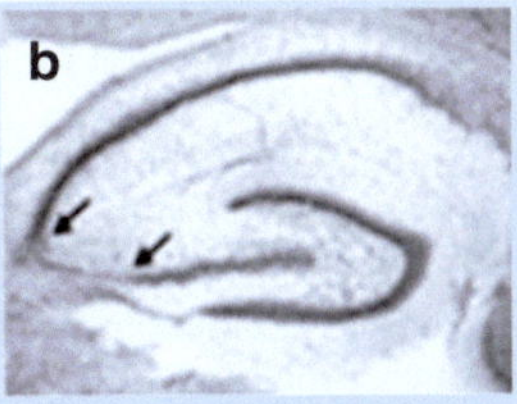
27 weeks post-TBI *(ipsilateral hippocampus)* Progression of cell loss across the entire CA3	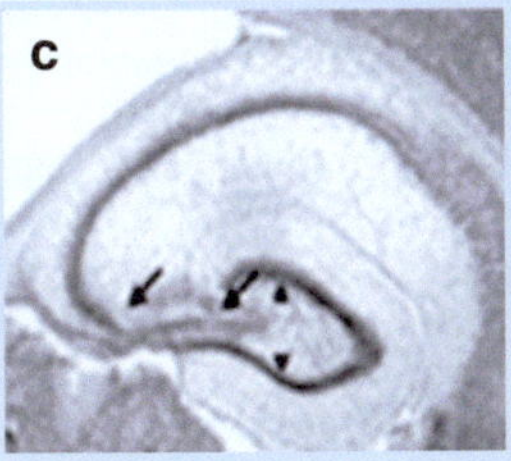	27 weeks post-TBI *(Contralateral hippocampus)* No evidence of gross cell loss	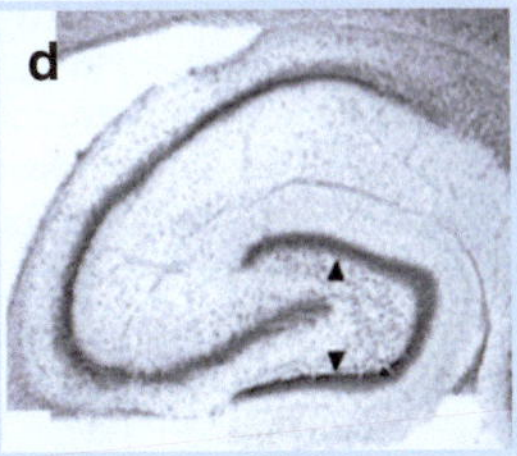

TBI traumatic brain injury

enhanced excitatory connectivity, and mossy fiber synaptic reorganization [32]. Initiation of prophylactic anti-seizure medications is a common practice with surgical treatment reserved for refractory PTE [32].

Acknowledgments We would like to thank Hunaid Hasan for his contributions to an earlier version of this chapter. We also thank Brian Hanrahan and Arun Antony for contributing to sections that were included in this chapter from a previous version of their chapter on Pathophysiology.

References

1. Passaro EA. Neuroimaging in adults and children with epilepsy. Contin Minneap Minn. 2023;29(1):104–55.
2. Lapalme-Remis S, Nguyen DK. Neuroimaging of epilepsy. Contin Minneap Minn. 2022;28(2):306–38.
3. Nayak CS, Bandyopadhyay S. Mesial temporal lobe epilepsy. In: StatPearls. Treasure Island: StatPearls Publishing; 2023. http://www.ncbi.nlm.nih.gov/books/NBK554432/.
4. Jardim AP, Neves RS, Caboclo LO, Lancellotti CL, Marinho MM, Centeno RS, et al. Temporal lobe epilepsy with mesial temporal sclerosis: hippocampal neuronal loss as a predictor of surgical outcome. Arq Neuropsiquiatr. 2012;70(5):319–24.
5. Wiebe S, Blume WT, Girvin JP, Eliasziw M, Effectiveness and Efficiency of Surgery for Temporal Lobe Epilepsy Study Group. A randomized, controlled trial of surgery for temporal-lobe epilepsy. N Engl J Med. 2001;345(5):311–8.
6. Al Sufiani F, Ang LC. Neuropathology of temporal lobe epilepsy. Epilepsy Res Treat. 2012;2012:624519.
7. Jaiswal V, Hanif M, Sarfraz Z, Nepal G, Naz S, Mukherjee D, et al. Hemimegalencephaly: a rare congenital malformation of cortical development. Clin Case Reports. 2021;9(12):e05238.

8. Oegema R, Barakat TS, Wilke M, Stouffs K, Amrom D, Aronica E, et al. International consensus recommendations on the diagnostic work-up for malformations of cortical development. Nat Rev Neurol. 2020;16(11):618–35.
9. Barkovich AJ, Guerrini R, Kuzniecky RI, Jackson GD, Dobyns WB. A developmental and genetic classification for malformations of cortical development: update 2012. Brain J Neurol. 2012;135(5):1348–69.
10. Berlangieri SU, Mito R, Semmelroch M, Pedersen M, Jackson G. Bottom-of-sulcus dysplasia: the role of 18F-FDG PET in identifying a focal surgically remedial epileptic lesion. Eur J Hybrid Imaging. 2020;4(1):23.
11. Matsuo T, Fujimoto S, Komori T, Nakata Y. Case report: the origin of transmantle-like features. Front Radiol. 2022;2:927764.
12. Guerrini R, Barba C. Focal cortical dysplasia: an update on diagnosis and treatment. Expert Rev Neurother. 2021;21(11):1213–24.
13. Beal J. Human brain frontal (coronal) section. 2005. https://commons.wikimedia.org/wiki/File:Human_brain_frontal_(coronal)_section.JPG.
14. Resta IT, Singh A, Gilbert BC, Rojiani MV, Alleyne C, Rojiani AM. Suprasellar ganglioglioma: expanding the differential diagnosis. Case Rep Pathol. 2018;2018:9486064.
15. Rahim S, Ud Din N, Abdul-Ghafar J, Chundriger Q, Khan P, Ahmad Z. Clinicopathological features of dysembryoplastic neuroepithelial tumor: a case series. J Med Case Rep. 2023;17(1):327.
16. Ferreira Furtado LM, Da Costa Val Filho JA, Rodrigues da Costa GA, Gouvea Braga PS. Pleomorphic xanthoastrocytoma of the frontal lobe in a child: a rare entity. Cureus. 2021;13(6):e15566.
17. Bodi I, Curran O, Selway R, Elwes R, Burrone J, Laxton R, et al. Two cases of multinodular and vacuolating neuronal tumour. Acta Neuropathol Commun. 2014;2:7.
18. Han W, Jiang C, Qi Z, Xiang W, Lin J, Zhou Y, et al. Adult-onset hypothalamic hamartoma: origin of epilepsy? Acta Epileptol. 2023;5(1):11.
19. Englot DJ, Chang EF, Vecht CJ. Epilepsy and brain tumors. Handb Clin Neurol. 2016;134:267–85.
20. Thom M, Blümcke I, Aronica E. Long-term epilepsy-associated tumors. Brain Pathol. 2012;22(3):350–79.
21. Soldozy S, Norat P, Yağmurlu K, Sokolowski JD, Sharifi KA, Tvrdik P, et al. Arteriovenous malformation presenting with epilepsy: a multimodal approach to diagnosis and treatment. Neurosurg Focus. 2020;48(4):E17.
22. Ogaki A, Ikegaya Y, Koyama R. Vascular abnormalities and the role of vascular endothelial growth factor in the epileptic brain. Front Pharmacol. 2020;11:20.
23. Stafstrom CE, Staedtke V, Comi AM. Epilepsy mechanisms in neurocutaneous disorders: tuberous sclerosis complex, neurofibromatosis type 1, and Sturge-Weber syndrome. Front Neurol. 2017;8:87.
24. Mühlebner A, van Scheppingen J, Hulshof HM, Scholl T, Iyer AM, Anink JJ, et al. Novel histopathological patterns in cortical tubers of epilepsy surgery patients with tuberous sclerosis complex. PLoS One. 2016;11(6):e0157396.
25. Kadam SD, Dudek FE. Neuropathogical features of a rat model for perinatal hypoxic-ischemic encephalopathy with associated epilepsy. J Comp Neurol. 2007;505(6):716–37.
26. Jensen FE, Wang C, Stafstrom CE, Liu Z, Geary C, Stevens MC. Acute and chronic increases in excitability in rat hippocampal slices after perinatal hypoxia in vivo. J Neurophysiol. 1998;79(1):73–81.
27. Chen J, Ye H, Zhang J, Li A, Ni Y. Pathogenesis of seizures and epilepsy after stroke. Acta Epileptol. 2022;4(1):2.
28. Phan J, Ramos M, Soares T, Parmar MS. Poststroke seizure and epilepsy: a review of incidence, risk factors, diagnosis, pathophysiology, and pharmacological therapies. Oxidative Med Cell Longev. 2022;2022:7692215.

29. Ding K, Gupta PK, Diaz-Arrastia R. Epilepsy after Traumatic Brain Injury. In: Laskowitz D, Grant G, editors. Translational research in traumatic brain injury. Boca Raton: CRC Press; 2016. http://www.ncbi.nlm.nih.gov/books/NBK326716/.
30. Lowenstein DH, Thomas MJ, Smith DH, McIntosh TK. Selective vulnerability of dentate hilar neurons following traumatic brain injury: a potential mechanistic link between head trauma and disorders of the hippocampus. J Neurosci. 1992;12(12):4846–53.
31. Golarai G, Greenwood AC, Feeney DM, Connor JA. Physiological and structural evidence for hippocampal involvement in persistent seizure susceptibility after traumatic brain injury. J Neurosci. 2001;21(21):8523–37.
32. Vespa P, Prins M, Ronne-Engstrom E, Caron M, Shalmon E, Hovda DA, et al. Increase in extracellular glutamate caused by reduced cerebral perfusion pressure and seizures after human traumatic brain injury: a microdialysis study. J Neurosurg. 1998;89(6):971–82.

Autoimmune Epilepsy

11

Shirin Jamal Omidi and Jay R. Gavvala

11.1 Introduction

Autoimmune epilepsies encompass a diverse group of conditions and may present acutely with seizures/status epilepticus or may have a more indolent and chronic course. This condition exhibits diverse clinical presentations, varying both in the rapidity of symptom onset and constellation of symptoms. Symptomatology of autoimmune epilepsy can include seizures, psychiatric symptoms, memory deficits, cognitive decline, abnormal movements, dysautonomia, and encephalopathy [1]. Focal seizures, especially in ages above 50 years, are a common semiology in autoimmune encephalitis [2–4], and it has been recognized as the most frequent cause of new-onset refractory status epilepticus (NORSE) [5]. Since the discovery of the ANNA1 (Anti-Hu) antibody (Ab), numerous other Abs have been recognized for their role in causing seizures. The newest members of this Ab family are KLH 11 and GFAP. Identification of Ab in the field of autoimmune encephalitis prompted the International League Against Epilepsy to include autoimmunity in the classification of epilepsies [6]. Although not as commonly appreciated, chronic epilepsies can also have an autoimmune-mediated etiology, most commonly GAD. In this chapter we will discuss clinical aspects, terminology, common Ab-positive autoimmune causes of seizures, diagnostic challenges, treatment, and prognosis of this group of disorders.

S. J. Omidi · J. R. Gavvala (✉)
Neurology, Department of Neurology, University of Texas Health Science Center at Houston, Houston, TX, USA
e-mail: jay.r.gavvala@uth.tmc.edu

Z. Haneef (ed.), *Epilepsy Fundamentals*,
https://doi.org/10.1007/978-3-031-77741-7_11

11.2 Clinical Aspects

Autoimmune epilepsy is typically suspected in patients experiencing a sudden and severe onset of seizures, often presenting as status epilepticus. Notably, it has been identified as the primary cause of NORSE. Initially characterized by limbic encephalopathy with symptoms of seizures, psychiatric manifestations, and memory deficits, the understanding of autoimmune epilepsy has evolved to encompass a broader spectrum, including cognitive decline, abnormal movements, dysautonomia, and encephalopathy. Recent advances in recognizing pathogenic antibodies and diverse symptomatology have led to the acknowledgment of subacute and chronic forms of autoimmune epilepsy, even without evident signs of encephalitis.

Despite advances in antibody detection and therapeutics, effective identification of patients who are presenting with autoimmune epilepsy remains a challenge. In acute settings for patients admitted with new-onset seizures, determining candidacy for autoimmune panel and immune therapy can be guided by criteria such as the APE2 (**A**ntibody **P**revalence in **E**pilepsy and **E**ncephalopathy) score [7]. APE2 incorporates history, clinical semiology, CSF, and MRI findings to assess the probability of an autoimmune cause [7]. Table 11.1 demonstrates important clinical data used in the APE2 scoring system. Figure 11.1 summarizes the approach to such patients using the APE2 score.

Navigating the challenge in the outpatient setting is crucial, especially when patients with an established epilepsy diagnosis are referred months after symptom onset. It is noteworthy that a positive autoimmune Ab has been reported in 10.5% of adults presenting to the clinic with epilepsy and no other signs of encephalitis, emphasizing the

Table 11.1 APE2 and ACES scoring

APE2 (maximum score 18)	
New, rapidly progressive mental status changes over 1–6 weeks OR new onset seizure	1
Neuropsychiatric changes: agitation, aggressiveness, emotional lability	1
Autonomic dysfunction	1
Viral prodrome	2
Facio-brachial dystonic seizures	3
Facial dyskinesias without facio-brachial dystonic seizures	2
Seizure refractory from at least two antiseizure medications	2
CSF findings of inflammation (protein >50 mg/dl and/or lymphocytes >5 cells/μL)	2
Brain MRI suggesting encephalitis	2
Systemic cancer diagnosed within 5 years of neurological symptom onset	2

ACES (maximum score 6)	
Cognitive symptoms	1
Behavioral changes	1
Autonomic symptoms	1
Autoimmune Diseases	1
Speech problems	1
Temporal MRI hyperintensities	1

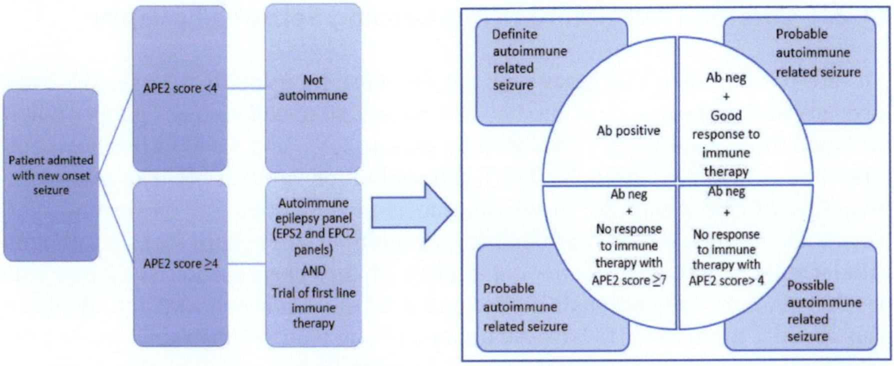

Fig. 11.1 Approach to a patient with new onset seizures admitted to hospital. *Abb-EPS2* epilepsy autoimmune/paraneoplastic evaluation serum, *EPC2* epilepsy autoimmune/paraneoplastic evaluation CSF

importance of broadening indications for further testing [8]. The 6-item ACES score (**A**ntibodies **C**ontributing to focal **E**pilepsy **S**igns and Symptoms score) recommends autoimmune tests when the ACES score is at least 2 [9]. This practical approach provides guidance for identifying individuals who may benefit from additional diagnostic evaluation. Table 11.1 summarizes clinical data used in ACES scoring.

11.2.1 Autoimmune-Associated Epilepsy (AAE) and Acute Symptomatic Seizures Secondary to Autoimmune Encephalitis (ASSAE)

Differentiating ASSAE from AAE is important as it distinguishes the long-term management of disease with antiseizure medications. Patients are ideally diagnosed with epilepsy when the immune process has subsided and without evidence of ongoing inflammatory activity. However, due to the absence of reliable noninvasive tools to confirm this, a suggested time point for using the term epilepsy is one year from the time of diagnosis [10]. While EEG can reliably demonstrate seizures, epileptiform or periodic discharges, and slowing, it has not shown any specific patterns in autoimmune encephalitis patients other than the well-described extreme delta brushes [11]. Nevertheless, it remains a valuable tool in predicting the progression to epilepsy in the presence of multifocal seizure onset and perisylvian involvement [12]. Other factors that portend increased risk for epilepsy in this group are GAD65 positivity, Rasmussen encephalitis, presentation with NORSE, [2] age over 54, seizures with autonomic piloerection, lowered self-reported mood, reduced attention, MRI limbic system changes, and the absence of conventional epilepsy risk factors [13]. Conversely, while mesial temporal hyperintensity in FLAIR MRI is recognized as a risk factor for epilepsy, volumetric studies on post-encephalitic MRIs have not proven successful in predicting refractory epilepsy in patients with focal volume loss [14].

11.2.2 Common Autoimmune Abs Causing Seizures/Epilepsy

Abs are positive in 1/3 of cases with autoimmune encephalitis. Classically, causative antibodies are classified into two categories: Ab to cell surface epitopes that is mediated through antibody production by plasma cells, and Ab to intracellular epitopes that is mediated through CD8 T cell infiltration in brain [4]. The most common Abs of this group are GAD 65, and paraneoplastic, aka onconeural Abs. Nevertheless, some patients can demonstrate multiple Abs against surface and intracellular epitopes. Adding to the complexity of this phenomenon are cases who develop autoimmune encephalitis after initiation of immune checkpoint inhibitors. This group is more likely to have Ab to intracellular epitopes than cell surface [15]. Tables 11.2 and 11.3 summarize clinical features of autoimmune-related seizures based on the type of Ab.

In **anti-LGI1 Ab encephalitis**, following seizures, symptoms of limbic encephalitis may develop in 75% of cases. Cognitive impairment parallels the onset of paraclinical abnormalities, such as SIADH, and MRI FLAIR hyperintensity extending from the basal ganglia to mesial temporal regions [20]. Progressive cognitive impairment poses a risk for hippocampal atrophy and subsequent permanent memory deficits, a risk mitigated by early immune therapy [21]. Importantly, LGI1 Ab titers are more sensitive in serum than CSF. Notably, these patients exhibit higher rates of Stevens-Johnson syndrome to carbamazepine and phenytoin [19]. This observation led to the recognition of an association between LGI1 Ab seropositivity and HLA class II, though such an association is not observed in tumor-associated LGI1 [19].

In **anti-NMDAR encephalitis**, approximately 80% of patients are women, with 40% being younger than 18 years and two-thirds of those aged 18–45 having a teratoma. In 55% of cases, no etiology is found [22]. NMDAR Ab has higher sensitivity in CSF than serum, and its increase in CSF is more likely to be detected during relapse. Titers tend to decrease in serum and CSF regardless of the outcome, though they may stay positive for a long time [1]. In 20% of cases of infectious encephalitis with HSV and less commonly with VZV and mycoplasma, occurrence or a worsening in symptoms is observed after a few weeks. This is due to Abs to NMDA and less frequently against GABAa [22]. Distinctive EEG finding of extreme delta brushes, generalized rhythmic delta slowing with superimposed 20–30 Hz beta activity [11], is seen in a small number of patients as shown in Fig. 11.2. Interpretation of this finding should consider clinical features, as this pattern can also be seen in mesial temporal lobe epilepsy with a good prognosis after surgery and, secondly, in poor prognosis ICU patients with hypoxic encephalopathy, brain tumor, stroke, and metabolic derangements [23].

Another autoimmune epilepsy syndrome in the younger age group is GAD 65 Ab, typically affecting patients in their 30s. A specific semiology that has been attributed to GAD 65 Ab is musicogenic seizures [18]. Diagnosis is confirmed by the detection of any titer of GAD 65 Ab in CSF or a titer of >20 nmol/L or >1000 U/mL in serum [4]. In the encephalitic phase of this disorder, which can last for about 6 years, CD8 T cells predominate in cerebral parenchyma, implying

Table 11.2 Clinical features of autoimmune-related seizures in disorders with Ab against neuronal cell surface antigens

Antibody to cell surface antigens	Seizure type	Prevalence of acute symptomatic seizures	Presence of mass/malignancy	MRI	Response to immune therapy	Additional features
NMDAR	Focal, generalized, EPC	High	Ovarian teratoma or carcinoma	Unremarkable or nonspecific changes	Good	Dyskinesias and behavioral changes Can be seen with use of ICI [15]
LGI1	FBDS, autonomic, focal with possible generalization	High, but 15% become epilepsy	Thymoma	T1 basal ganglia hyperintensity, mesial temporal FLAIR hyperintensity	Good	LE Association with HLA II, hyponatremia
CASPR2	Focal	Moderate	Thymoma	Unremarkable or mesial temporal FLAIR hyperintensity	Good	Morvan's syndrome, neuro-myotonia. Hyponatremia. Can be seen with the use of ICI [15]
GABAa	Status epilepticus, EPC	High	Thymoma	Multifocal GM and subcortical	Good	75% associated with underlying tumor
GABAb	Focal with possible generalization	High, but 20–30% become epilepsy	SCLC	Mesial temporal	Good if no concurrent onconeural Ab	LE 50% with malignancy, final outcome correlates with the stage and recurrence of SCLC
AMPA	Focal temporal, opsoclonus	Moderate	SCLC, Thymoma, Breast	Deep Gray nuclei FLAIR hyperintensity	Good if no concurrent onconeural Ab	Anterograde amnesia, LE 70% with malignancy

(continued)

Table 11.2 (continued)

Antibody to cell surface antigens	Seizure type	Prevalence of acute symptomatic seizures	Presence of mass/ malignancy	MRI	Response to immune therapy	Additional features
mGluR5	Myoclonic	Moderate	Hodgkin Lymphoma	Unremarkable or limbic/ cortical FLAIR changes	Good	LE
Glycine	Status epilepticus, EPC, myoclonus	Low	B lymphoma, Thymoma, Hodgkin lymphoma, Breast, Melanoma	Unremarkable or nonspecific	Good	Stiff person, PERM
DPPX		Rare	Lymphoma	Unremarkable or nonspecific		Months of diarrhea
IgLON5	Frontal lobe nocturnal	Rare		Unremarkable or nonspecific	Temporary, partial	Sleep or movement disorder
Neurexin 3a		Rare		Unremarkable		

Prevalence of acute symptomatic seizures: >60% high, 30–60% moderate, <30% low [4, 16]

NMDAR N-methyl-D-aspartate receptor, *LGI1* leucine-rich Glioma-inactivated 1, *CASPR2* contactin-associated protein-like 2, *GABA-A* gamma-aminobutyric acid A receptor, *GABA-B* gamma-aminobutyric acid B, *AMPA* α-amino-3-hydroxy-5-methyl-4-isoxazolepropionic acid, *mGluR3* metabotropic glutamate receptor subtype 3, *DPPX* dipeptidyl-peptidase–like protein 6, *ICI* immune checkpoint inhibitors, *FBD* faciobrachial dystonic seizures, *EPC* epilepsia partialis continua, *LE* limbic encephalitis, *SCLC* small cell lung cancer, *PERM* progressive encephalomyelitis with rigidity and myoclonus

Table 11.3 Clinical features of autoimmune-related seizures in disorders with Ab against intracellular antigens

Antibody to intra-cellular antigens	Seizure type	Prevalence of acute symptomatic seizures	Presence of mass/ malignancy	MRI	Response to immune therapy	Additional features
ANNA-1 (Hu)	Temporal/extra temporal lobe, EPC, multifocal and migrating myoclonic jerks, opsoclonus myoclonus [17]	Moderate, but 60% become epilepsy	SCLC	Normal or cortical/limbic FLAIR hyperintensities	Poor	Bulbar symptoms, sensory neuropathy, autonomic neuropathy, LE Can be seen with ICI [15]
ANNA-2 (Ri)		Rare	Breast cancer, SCLC	Brainstem FLAIR changes		Stridor, laryngospasm, jaw dystonia, opsoclonus myoclonus, ataxia Can be seen with ICI [15]
GAD 65	Temporal lobe, musicogenic [18]	Low to moderate, but 80% become epilepsy	Rare adenocarcinoma, thymoma, neuroendocrine tumors	Normal or mesial temporal FLAIR hyperintensity	Poor	Stiff person, ataxia, PERM Can be seen with ICI [15]
Ma-2	Temporal lobe	Moderate, but 60% become epilepsy	Germ cell tumor, breast cancer, SCLC	Brainstem or medial temporal FLAIR hyperintensity	Poor	Ataxia, brainstem encephalitis, LE Can be seen with ICI [15]
Amphiphysin	Focal	Low	Breast cancer	Normal or nonspecific		Stiff person, PERM
CRMP5	Focal		SCLS, Thymoma	Normal or mesial temporal FLAIR hyperintensity		Optic neuritis, retinitis, myelitis

(continued)

Table 11.3 (continued)

Antibody to intra-cellular antigens	Seizure type	Prevalence of acute symptomatic seizures	Presence of mass/malignancy	MRI	Response to immune therapy	Additional features
KLH-11 [19]	EPC	Rare	Testicular germ cell tumor or teratoma		Moderate	Rhombencephalitis
GFAP [19]	Focal	Rare	Ovarian teratoma	Radial perivascular periventricular enhancement, or leptomeningeal enhancement, or longitudinally extensive myelitis with central enhancement	Good	Meningoencephalitis or myelitis, 30% anti-NMDAR encephalitis or AQP4 positive, 30% malignancy

Prevalence of acute symptomatic seizures: >60% high, 30–60% moderate, <30% low [4, 16]

ANNA1/Anti Hu anti-neural nuclear Ab type 1, *ANNA2/Anti Ri* anti-neural nuclear Ab type 2, *GAD 65* anti-glutamic acid decarboxylase 65, *CRMP5* collapsin response-mediator protein-5, *KLH11* Kelch-like 11 protein, *GFAP* glial fibrillary acidic protein, *EPC* epilepsia partialis continua, *ICI* immune checkpoint inhibitors, *LE* limbic encephalitis, *SCLC* small cell lung cancer, *PERM* progressive encephalomyelitis with rigidity and myoclonus

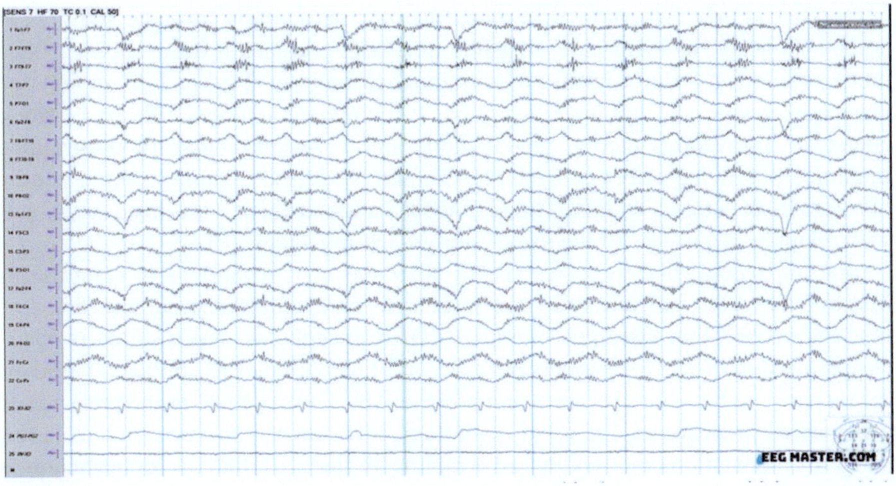

Fig. 11.2 Extreme delta brushes in patients with anti-NMDAR encephalitis, bipolar montage. Sensitivity, 7 μV; high pass filter, 70 Hz. EEG from EEGmaster.com under CC-BY license

responsibility for refractory epilepsy. This inflammatory response cools down later, only for the seizures to continue due to secondary temporal lobe structural damage [24]. In this group of patients, even epilepsy surgery might not result in long-term seizure freedom [25]; however, brain stimulation may provide meaningful seizure reduction [26].

Onconeural antibodies like Anti ANNA1 (Anti-Hu) and Anti Ma-2 are also likely associated with CD8 T cells, implying more refractory course than disorders related to plasma cell activation. In anti-Hu encephalitis, focal findings, seizures, and multifocal epileptiform discharges are common EEG findings. Interestingly, 21% of cases exhibit focal EEG findings from areas other than MRI changes [27].

11.3 Management

Patients with positive antibodies require cancer screening, including skin examination, chest, abdomen and pelvic CT scans with contrast, gender-appropriate imaging including mammogram, transvaginal sonography and/or pelvic MRI in females, and testicular ultrasound in males. In onconeural cases, in the absence of malignancy in the aforementioned workup, FDG-PET is recommended. In cases with negative results, cancer surveillance needs to be repeated every 6 months for 4 years [28].

The treatment approach involves a triad of immune therapy, removal of the underlying tumor if applicable, and antiseizure medications. Sodium channel blocking agents like phenytoin, carbamazepine, oxcarbazepine and lacosamide have been shown to be more effective in autoimmune-related seizures [29]. In contrast to traditional forms of epilepsy, the importance of immunomodulatory therapy cannot be understated. Early initiation of immune therapy, particularly in acute settings, is crucial. In acute inpatient care, first-line treatment consists of intravenous methyl prednisolone for 5 days, intravenous immunoglobulin at 2 g/kg, 5–7 sessions of plasma exchange and immunoadsorption. In outpatient settings, a 6- or 12-week course of intravenous methyl prednisolone and a 6-week course of intravenous immunoglobulin have been utilized [4]. Caution is advised in rapid tapering of corticosteroids in patients with LGI1 and CASPR2 antibodies, as it has been associated with relapses, regardless of the use of steroid-sparing agents [8].

Starting second-line therapies shortly after the first, regardless of response to first-line therapies, has ensured good results. As an example, the combined use of first and second-line therapies in anti-NMDAR encephalitis has resulted in long-term remission in 95% of patients [19]. Second-line therapies are initiated based on the response to first-line therapies and positive Abs. For example, in the presence of a suspected or proven CD8 T cell pathology, such as Abs to intracellular neural antigens, medications modulating all lymphocyte lineages, like cyclophosphamide and mycophenolate, are preferred. In cases with Abs to cell surface antigens, B cell-targeted therapies like rituximab and azathioprine are more commonly used [4]. Notably, medications affecting the interleukin system, such as tocilizumab (anti-IL6) and anakinra (anti-IL1), have shown efficacy in NORSE and chronic febrile infection-related epilepsy syndrome (FIRES) [30]. Surveillance includes monitoring for infections and periodic evaluation of blood counts, liver function, and renal function [28].

Clinical trials are ongoing for additional treatment options in autoimmune encephalitis, including inebilizumab (anti-CD19) for anti-NMDAR encephalitis, satralizumab (anti-IL6) for anti-NMDAR and anti-LGI1 encephalitides, bortezomib (a proteasome inhibitor) and monoclonal Abs against fragment crystallizable (FcRN) for anti-LGI1. Epilepsy surgery is another treatment option in cases of AAE refractory to treatment, including surgical resection, radiofrequency ablation, and brain-responsive neurostimulation. Reports on surgical approaches in patients with AEE are scarce, which stems from challenges in diagnosing these patients. Still, in general, surgical outcomes are less favorable than non-immune-mediated epilepsies. Available studies have reported favorable Engel 1 and 2 outcomes in 38% of 13 patients with onconeural AEE [25] and more than 50% seizure reduction in three of four patients with GAD 65 [26].

11.4 Prognosis

Although autoimmune encephalitis is mainly described as a monophasic disease, recurrence has been reported in up to 35% of patients with positive Ab, more commonly with Ab to surface epitopes [1]. While cases with Ab to intracellular antigens generally have a more resistant and progressive course [31]. Timely administration of immune therapy is linked to high rates of seizure reduction and prevention of cognitive impairment [21, 32]. Conversely, predictors of worse outcomes include presentation with NORSE [5], CD8 T cell-mediated mechanisms including onconeural Abs or GAD 65 Ab [16, 28], and lack of response to initial immunotherapy [19].

Although anti-NMDAR encephalitis generally has favorable neurological outcomes, in children under 12 years of age, it is more associated with worse long-term adaptive behavior [33]. In patients over 45 years old, milder symptoms with lower seizure incidence are noted, while lower rates of recovery and higher rates of association with carcinoma (less teratoma) are observed [34]. On EEG, extreme delta brushes indicate a prolonged course of the disease in anti-NMDAR encephalitis [11]. Moreover, relapses and poor response to treatment are associated with prolonged or secondary elevation of CXCL13 in CSF [35]. Imaging markers of decreased hippocampal connectivity in resting-state functional connectivity MRI may portend adverse outcomes [36].

Outcomes in onconeural Ab syndromes are dependent on tumor treatment. As an example, in patients with anti-Ma-2, better outcomes are seen in younger patients with testicular tumors with a complete response to treatment and negative Ma-1 Ab [4]. However, in general, onconeural Abs portend poorer responses to immune therapy [28, 31]. In cases where Abs are positive in the treatment setting with immune checkpoint inhibitors, the main determinant of outcome is clinical findings prior to treatment with immune checkpoint inhibitors [37].

11.5 Rasmussen Encephalitis

Rasmussen encephalitis typically involves children, but adults can also be affected with a milder and slower pace. Pathology slides often reveal the presence of CD8 T cells and CD3 T cells. In these patients, immune pathways are notably altered, particularly at the crosstalk between dendritic and natural killer cells [38].

The condition progresses through three stages:

1. Stage 1 (prodromal): Patients experience focal motor seizures that are not frequent, without other focal neurologic abnormalities. MRI evidence of insular and peri-insular atrophy may begin within the first 4 months.
2. Stage 2 (acute): Seizures become intractable, including epilepsia partialis continua involving different areas on the same side of the body. EEG shows loss of background activity on one side and migrating repetitive focal rhythmic discharges on the same side. Progressive hemiparesis, aphasia, hemianopia, and cognitive impairment occur, with clear brain atrophy visible on MRI. Motor exam fluctuates with seizure frequency.
3. Stage 3 (residual): Seizure frequency decreases with stable, permanent neurologic deficits. EEG may reveal epileptiform discharges on the contralateral side as well [39].

Early initiation of immunotherapy in the early stages helps prevent refractory epilepsy and improve cognitive and motor outcomes. Therefore, intravenous methylprednisolone should be started at the time of diagnosis, followed by Tacrolimus [40].

References

1. Dalmau J, Graus F. Antibody-mediated encephalitis. N Engl J Med. 2018;378(9):840–51.
2. Spatola M, Dalmau J. Seizures and risk of epilepsy in autoimmune and other inflammatory encephalitis. Curr Opin Neurol. 2017;30(3):345–53.
3. Kaaden T, Madlener M, Angstwurm K, Bien CG, Bogarin Y, Doppler K, et al. Seizure semiology in antibody-associated autoimmune encephalitis. Neurol Neuroimmunol Neuroinflamm. 2022;9(6):e200034.
4. Husari KS, Dubey D. Autoimmune epilepsy. Neurotherapeutics. 2019;16(3):685–702.
5. Gaspard N, Foreman BP, Alvarez V, Cabrera Kang C, Probasco JC, Jongeling AC, et al. New-onset refractory status epilepticus: etiology, clinical features, and outcome. Neurology. 2015;85(18):1604–13.
6. Scheffer IE, Berkovic S, Capovilla G, Connolly MB, French J, Guilhoto L, et al. ILAE classification of the epilepsies: position paper of the ILAE Commission for Classification and Terminology. Epilepsia. 2017;58(4):512–21.
7. Dubey D, Kothapalli N, McKeon A, Flanagan EP, Lennon VA, Klein CJ, et al. Predictors of neural-specific autoantibodies and immunotherapy response in patients with cognitive dysfunction. J Neuroimmunol. 2018;323:62–72.
8. Brenner T, Sills GJ, Hart Y, Howell S, Waters P, Brodie MJ, et al. Prevalence of neurologic autoantibodies in cohorts of patients with new and established epilepsy. Epilepsia. 2013;54(6):1028–35.
9. de Bruijn MAAM, Bastiaansen AEM, Mojzisova H, van Sonderen A, Thijs RD, Majoie MJM, et al. Antibodies contributing to focal epilepsy signs and symptoms score. Ann Neurol. 2021;89(4):698–710.
10. Geis C, Planagumà J, Carreño M, Graus F, Dalmau J. Autoimmune seizures and epilepsy. J Clin Invest. 2019;129(3):926–40.
11. Schmitt SE, Pargeon K, Frechette ES, Hirsch LJ, Dalmau J, Friedman D. Extreme delta brush: a unique EEG pattern in adults with anti-NMDA receptor encephalitis. Neurology. 2012;79(11):1094–100.
12. Gillinder L, Papacostas J, McCombe P, Chauvel P. Effect of immunotherapy on intracranial EEG in patients with seronegative autoimmune-associated epilepsy. Epileptic Disord. 2022;24(6):1081–6.
13. McGinty RN, Handel A, Moloney T, Ramesh A, Fower A, Torzillo E, et al. Clinical features which predict neuronal surface autoantibodies in new-onset focal epilepsy: implications for immunotherapies. J Neurol Neurosurg Psychiatry. 2021;92(3):291–4.
14. Steriade C, Patel PS, Haynes J, Desai N, Daoud N, Yuan H, et al. Predictors of seizure outcomes of autoimmune encephalitis: a clinical and morphometric quantitative analysis study. Clin Neurol Neurosurg. 2023;231:107854.
15. Nersesjan V, McWilliam O, Krarup L-H, Kondziella D. Autoimmune encephalitis related to cancer treatment with immune checkpoint inhibitors: a systematic review. Neurology. 2021;97(2):e191–202.
16. Chen B, Lopez Chiriboga AS, Sirven JI, Feyissa AM. Autoimmune encephalitis-related seizures and epilepsy: diagnostic and therapeutic approaches. Mayo Clin Proc. 2021;96(8):2029–39.
17. Frazzini V, Nguyen-Michel V-H, Habert M-O, Pichit P, Apartis E, Navarro V. Focal status epilepticus in anti-Hu encephalitis. Autoimmun Rev. 2019;18(11):102388.
18. Smith KM, Zalewski NL, Budhram A, Britton JW, So E, Cascino GD, et al. Musicogenic epilepsy: expanding the spectrum of glutamic acid decarboxylase 65 neurological autoimmunity. Epilepsia. 2021;62(5):e76–81.
19. Varley JA, Strippel C, Handel A, Irani SR. Autoimmune encephalitis: recent clinical and biological advances. J Neurol. 2023;270(8):4118–31.
20. Flanagan EP, Kotsenas AL, Britton JW, McKeon A, Watson RE, Klein CJ, et al. Basal ganglia T1 hyperintensity in LGI1-autoantibody faciobrachial dystonic seizures. Neurol Neuroimmunol Neuroinflamm. 2015;2(6):e161.

21. Thompson J, Bi M, Murchison AG, Makuch M, Bien CG, Chu K, et al. The importance of early immunotherapy in patients with faciobrachial dystonic seizures. Brain. 2018;141(2):348–56.
22. Dalmau J, Geis C, Graus F. Autoantibodies to synaptic receptors and neuronal cell surface proteins in autoimmune diseases of the central nervous system. Physiol Rev. 2017;97(2):839–87.
23. Baykan B, Gungor Tuncer O, Vanli-Yavuz EN, Baysal Kirac L, Gundogdu G, Bebek N, et al. Delta brush pattern is not unique to NMDAR encephalitis: evaluation of two independent long-term EEG cohorts. Clin EEG Neurosci. 2018;49(4):278–84.
24. Tröscher AR, Mair KM, Verdú de Juan L, Köck U, Steinmaurer A, Baier H, et al. Temporal lobe epilepsy with GAD antibodies: neurons killed by T cells not by complement membrane attack complex. Brain. 2023;146(4):1436–52.
25. Carreño M, Bien CG, Asadi-Pooya AA, Sperling M, Marusic P, Elisak M, et al. Epilepsy surgery in drug resistant temporal lobe epilepsy associated with neuronal antibodies. Epilepsy Res. 2017;129:101–5.
26. Feyissa AM, Mirro EA, Wabulya A, Tatum WO, Wilmer-Fierro KE, Won SH. Brain-responsive neurostimulation treatment in patients with GAD65 antibody-associated autoimmune mesial temporal lobe epilepsy. Epilepsia Open. 2020;5(2):307–13.
27. Rudzinski LA, Pittock SJ, McKeon A, Lennon VA, Britton JW. Extratemporal EEG and MRI findings in ANNA-1 (anti-Hu) encephalitis. Epilepsy Res. 2011;95(3):255–62.
28. Devine MF, Kothapalli N, Elkhooly M, Dubey D. Paraneoplastic neurological syndromes: clinical presentations and management. Ther Adv Neurol Disord. 2021;14:1756286420985323.
29. Feyissa AM, López Chiriboga AS, Britton JW. Antiepileptic drug therapy in patients with autoimmune epilepsy. Neurol Neuroimmunol Neuroinflamm. 2017;4(4):e353.
30. Aledo-Serrano A, Hariramani R, Gonzalez-Martinez A, Álvarez-Troncoso J, Toledano R, Bayat A, et al. Anakinra and tocilizumab in the chronic phase of febrile infection-related epilepsy syndrome (FIRES): effectiveness and safety from a case-series. Seizure. 2022;100:51–5.
31. Binks S, Uy C, Honnorat J, Irani SR. Paraneoplastic neurological syndromes: a practical approach to diagnosis and management. Pract Neurol. 2022;22(1):19–31.
32. Dubey D, Samudra N, Gupta P, Agostini M, Ding K, Van Ness PC, et al. Retrospective case series of the clinical features, management and outcomes of patients with autoimmune epilepsy. Seizure. 2015;29:143–7.
33. Yeshokumar A, Gordon-Lipkin E, Arenivas A, Rosenfeld M, Patterson K, Blum R, et al. Younger age at onset is associated with worse long-term behavioral outcomes in anti-NMDA receptor encephalitis. Neurol Neuroimmunol Neuroinflamm. 2022;9(5):e200013.
34. Bastiaansen AEM, de Bruijn MAAM, Schuller SL, Martinez-Hernandez E, Brenner J, Paunovic M, et al. Anti-NMDAR encephalitis in the Netherlands, focusing on late-onset patients and antibody test accuracy. Neurol Neuroimmunol Neuroinflamm. 2022;9(2):e1127.
35. Leypoldt F, Höftberger R, Titulaer MJ, Armangue T, Gresa-Arribas N, Jahn H, et al. Investigations on CXCL13 in anti-N-methyl-D-aspartate receptor encephalitis: a potential biomarker of treatment response. JAMA Neurol. 2015;72(2):180–6.
36. Kuchling J, Jurek B, Kents M, Kreye J, Geis C, Wickel J, et al. Impaired functional connectivity of the hippocampus in translational murine models of NMDA-receptor antibody associated neuropsychiatric pathology. Mol Psychiatry. 2023;29(1):85–96.
37. Sechi E, Markovic SN, McKeon A, Dubey D, Liewluck T, Lennon VA, et al. Neurologic autoimmunity and immune checkpoint inhibitors: autoantibody profiles and outcomes. Neurology. 2020;95(17):e2442–52.
38. Leitner DF, Lin Z, Sawaged Z, Kanshin E, Friedman D, Devore S, et al. Brain molecular mechanisms in Rasmussen encephalitis. Epilepsia. 2023;64(1):218–30.
39. Bien CG, Granata T, Antozzi C, Cross JH, Dulac O, Kurthen M, et al. Pathogenesis, diagnosis and treatment of Rasmussen encephalitis: a European consensus statement. Brain. 2005;128(3):454–71.
40. Takahashi Y, Yamazaki E, Mine J, Kubota Y, Imai K, Mogami Y, et al. Immunomodulatory therapy versus surgery for Rasmussen syndrome in early childhood. Brain and Development. 2013;35(8):778–85.

Neuropsychiatric Comorbidities 12

Yosefa Modiano and Erin Sullivan-Baca

Throughout much of history, epilepsy was understood as a mental health disorder, with early medical texts positing that epilepsy reflected possession or a divine curse. As the field of neurology began to distinguish itself from psychiatry during the mid-eighteenth century, there was increasing recognition that epilepsy resulted from aberrant neurologic functioning. Still, the boundaries between neural and mental remain murky. Evidence of dysregulated limbic activation underlying epilepsy and observations of distinct psychiatric ictal semiology highlight a complex interplay among epilepsy and psychiatric conditions and establish epilepsy as an important natural model for understanding brain-behavior relations.

Interictal or chronic mental health conditions are highly prevalent among people with epilepsy (PWE) and are associated with a range of adverse outcomes that impact patient care and quality of life. The following chapter focuses on common neuropsychiatric comorbidities known to affect PWE. This chapter will adhere to the most updated psychiatric diagnostic tool, the DSM-5, to promote consistency across disciplines. As part of the update from DSM-IV-TR, the DSM-5 organizes diagnoses into classes that share etiological mechanisms and/or symptoms to promote conceptual clarity.

Y. Modiano (✉)
Clinical Neuropsychologist, Vivian L. Smith Department of Neurosurgery, McGovern Medical School at UTHealth, Houston, TX, USA

E. Sullivan-Baca
Clinical Neuropsychologist, Neurocognitive Specialty Group, Dallas, TX, USA

Baylor College of Medicine, Houston, TX, USA

Michael E. DeBakey VA Medical Center, Houston, TX, USA

Z. Haneef (ed.), *Epilepsy Fundamentals*,
https://doi.org/10.1007/978-3-031-77741-7_12

12.1 Neurodevelopmental Disorders

Neurodevelopmental disorders (NDDs) are conditions that begin early in development and result in a range of deficits impacting personal, social, academic, or occupational functioning. There is a preponderance of epilepsy syndromes across early development with varied etiologies, including perinatal complications, malformations of cortical development, early neuronal insult, and genetic conditions. Many epilepsy syndromes commonly present with comorbid intellectual impairment and developmental delay.

Intellectual disability (ID) is characterized by the following diagnostic criteria:

1. Intellectual functioning falling two or more standard deviations below normative means (IQ $\leq$ 70) in combination with deficits in adaptive functioning across domains of communication, self-care, home living, social skills, use of community resources, self-direction, functional academic skills, work, leisure, health issues, or safety.
2. Onset of deficits before age 18.
3. A lifelong condition with no expectation of marked improvement in intellectual functioning over time.

Some conditions with comorbid epilepsy, such as Fragile X syndrome, are in fact marked by deterioration of IQ over time.

Autism spectrum disorder (ASD) shows high overlap with epilepsy perhaps related to shared mechanisms of under- or over-connectivity. ASD is characterized by persistent deficits in social communication and social interaction in combination with restricted or repetitive patterns of behavior, interests, or activities. Intellectual functioning can vary widely in ASD, but impairment in executive functions, such as fluid reasoning and cognitive flexibility, is common.

Attention deficit/hyperactivity Disorder (ADHD) is characterized by symptoms of inattention and hyperactivity/impulsivity that are inconsistent with developmental level and negatively impact social, academic, and occupational functioning. Overlap between ADHD and epilepsy is high and may reflect the ictal semiology of inattention characteristic of some childhood epilepsies (e.g., absence seizures), adverse effects of antiseizure medications (ASMs), or shared underlying brain dysfunction.

Learning disabilities (LDs) are common among children with epilepsy and can impact reading, mathematics, and writing skills. Academic challenges may be more pronounced among PWE due to other factors related to epilepsy, such as missed school for medical appointments or seizure recovery, inattention related to ictal and/or interictal discharges, or adverse reactions to ASMs.

Epidemiology The epidemiology of neurodevelopmental disorders in children and adults with epilepsy is outlined in Table 12.1. Prevalence of epilepsy in children with ID ranges from 5.5% to 35% [1], with roughly one in four adults with epilepsy demonstrating ID [2]. Prevalence of epilepsy in children with ASD ranges from

Table 12.1 Epidemiology of neurodevelopmental disorders in epilepsy

	Adults	Children
ID	1 in 4 adults demonstrate ID [2].	5.5–35% [1]
ASD	6–19% [3, 4].	2.4–37% [3] with higher rates of ASD presenting in specific epilepsy syndromes (e.g., Dravet, infantile spasms).
ADHD	Odds of ADHD are 3 times higher among PWE [5]	12–70% [6–8]
LD		31–41% [9], 50% of children with epilepsy demonstrate academic difficulties [10]

2.4% to 37% [3] with higher rates of ASD presenting in specific epilepsy syndromes (e.g., Dravet, Infantile spasms). Among adults with epilepsy, rates of ASD range from 6% to 19% [3, 4]. The odds of an ADHD diagnosis are 3 times higher among PWE [5], and the prevalence of ADHD in children with epilepsy ranges from 12% to 70% [6–8]. Prevalence of LDs in children with epilepsy ranges from 31–41% [9], and upwards of 50% of children with epilepsy demonstrate academic difficulties [10].

Treatment Among children with epilepsy and ID, adaptive functioning can be markedly improved via academic accommodations, curricula focusing on the development of life skills, and supported vocational training. Nevertheless, given the potential chronicity and medical complexity of epilepsy, caregivers may wish to establish guardianship and/or medical power of attorney, and advanced planning about long-term care is advised. Management of ASD includes early intervention for speech or motor deficits, behavioral skills training, such as applied behavioral analysis (ABA), social skills training, and academic accommodations focused on functional communication and supports. Regulatory agencies have cautioned against psychostimulant use for individuals with ADHD and epilepsy due to purported concerns about reduced seizure threshold. Still, large population studies have failed to identify a link between stimulant use and worsening seizures among PWE [11], arguing that a diagnosis of epilepsy should not automatically preclude stimulant treatment. LDs are managed through a combination of additional academic supports and accommodations that may already be afforded to children with epilepsy due to adverse impacts of seizures and ASMs on learning; these may include extended time on testing, additional breaks, note-taking supports, and proximity to medical assistance.

12.2 Mood Disorders

Depressive disorders are defined by the presence of emotional dysregulation, including sadness, a feeling of emptiness, or irritability, that influence an individual's functional ability. Within this category, major depressive disorder (MDD) is the most prevalent condition, defined by cardinal symptoms of depressed mood or loss of interest or pleasure (anhedonia) along with weight changes, changes in sleep patterns, psychomotor agitation or retardation, fatigue, feelings of worthlessness, difficulty concentrating, or recurrent thoughts of death. In PWE, depressive symptoms may present as part of a prodromal, ictal, or peri-ictal phenomena [12]; however, interictal depression is significantly more prevalent, representing the most common type of mood disturbance in PWE [13].

The presentation of depressive symptoms in PWE may be unique when compared to depression in people in the general population. Specifically, PWE frequently exhibit more prominent paranoia, irritability, psychotic symptoms (peri-ictally), and a chronic dysthymic state [12]. There is some evidence of common physiological risk factors between depression and epilepsy. This is supported by literature describing approximately 3.7 times higher rates of pre-existing diagnosis of depression in PWE when compared to controls [14]. Furthermore, psychosocial factors such as social isolation, reduced self-efficacy, stress, stigma, financial strain, and work and transportation safety restrictions may compound physiologic factors to exacerbate the risk of depression in this population [15–17]. In PWE, depression can have a negative influence on clinical outcomes, as demonstrated in Table 12.2.

Epidemiology The lifetime prevalence rate of depression in PWE has been reported in the range of 25–60% [18–20] compared to a lifetime prevalence rate of 17% in the general U.S. population (with a median of approximately 9% in an international sample) [21]. When compared to people with other chronic illnesses (e.g., cancer, asthma, or diabetes), the rate of depression remains higher in PWE, suggesting a unique psychosocial and physiologic influence from this particular disease [22].

Treatment The treatment of depression in PWE will depend on several factors, including symptomatology, duration of symptoms, and relationship with seizure semiology. Psychotropic medication, psychotherapeutic intervention, and lifestyle modifications may all play a role in reducing depressive symptoms. For more information on treatment considerations, see Table 12.6.

Bipolar and related disorders represent the modern terminology for previously described "manic depression." Bipolar I Disorder and Bipolar II Disorder are characterized by the presence of manic or hypomanic episodes (respectively), defined as distinct periods of abnormally and persistently elevated, expansive, or irritable mood and persistently increased activity or energy lasting at least 4 days (hypomania) or 1 week (mania). Major depressive episodes present between manic periods

Table 12.2 Epilepsy outcomes associated with depression

Greater severity of epilepsy (as measured by 1-year seizure freedom)
Low health-related quality of life in epilepsy
Increased risk of suicide (10% of deaths in PWE vs. 1% in the general population)

in the disease course. Epilepsy and bipolar and related disorders frequently co-occur, and there are several shared features between the two conditions. For example, there is an overlap in symptoms such that mania can present in patients with temporal lobe epilepsy (TLE) in the context of seizure aura, ictal excitement, or post-ictal symptoms [23]. Furthermore, both epilepsy and bipolar disorders present with an episodic nature, often persist chronically, and may respond to the same antiseizure medications (ASMs), suggestive of a potential shared pathogenesis [24].

Epidemiology Rates of bipolar disorders are elevated in PWE, although they remain less prevalent than depressive disorders. It is estimated that between 4.5% and 12% of people with epilepsy have a bipolar or related disorder, compared to 1–2% of the general population [25]. It is possible, however, that this number does not provide a full picture of the prevalence of sub-clinical hypomanic symptoms presenting outside the context of a diagnosed disorder.

Treatment ASMs provide an interface between epilepsy management and management of bipolar and related disorders. While medication management is the front-line treatment for each of these conditions, supplementary psychotherapy may be beneficial in further managing mood symptoms.

12.3 Schizophrenia Spectrum Disorders/Psychotic Symptoms in Epilepsy

Schizophrenia spectrum disorders are characterized by abnormalities in perception, including delusions, which are fixed beliefs providing a false sense of reality (i.e., persecutory delusions, somatic delusions, or grandiose delusions); hallucinations, defined as perception-like experiences without an external stimulus (i.e., visual, auditory, or olfactory hallucinations); disorganized thinking; disorganized motor behavior; and negative symptoms, such as reduced emotional expression or avolition. In epilepsy, psychotic symptoms can occur either within the context of a comorbid schizophrenia spectrum disorder or in the context of the epilepsy itself, which may in turn meet the criteria for **psychotic disorder due to another medical condition (epilepsy)**. DSM-5 criteria for such a diagnosis are described in Table 12.3.

While this is one classification system available, epileptologists are more likely to describe psychotic symptoms in relation to seizure location and/or timing of symptoms. For seizures involving the limbic or lateral (neocortical) temporal areas, psychotic symptoms such as visual or auditory hallucinations, depersonalization, derealization, and autoscopy can occur. Post-ictal psychosis (PIP) usually does not ensue immediately after the end of the last seizure but emerges after a 2.5–48 hour (rarely up to 1 week) interval of cognitive lucidity. Chronic interictal psychosis (CIP) does not show any direct correlation with seizure activity. The onset is often insidious, with paranoid delusions and hallucinations. Table 12.4 compares epilepsy-related psychoses to primary psychoses.

Table 12.3 DSM-5 criteria for psychotic disorder due to another medical condition (epilepsy)

Prominent hallucinations or delusions which are a direct pathophysiological consequence of epilepsy
Symptoms are not better explained by another mental disorder
Symptoms do not occur exclusively in the course of a delirium
Symptoms cause clinically significant distress or impairment

Table 12.4 Features differentiating epilepsy-related psychoses and primary psychoses

Epilepsy-related psychoses	Differences from primary psychoses
Ictal psychotic symptoms	More often superimposed by mood disturbances Often present with altered states of consciousness Brief (<3 min)
Post-ictal psychosis (PIP)	Frequently associated with affective disturbances
Chronic interictal psychosis (CIP)	Less deterioration of premorbid personality Fewer negative symptoms Less severe psychotic episodes More variable course of illness.

Epidemiology Estimates of epilepsy with comorbid schizophrenia are typically low, in the 1–2% range [26]; however, it may present more frequently in certain epilepsy populations, such as patients with drug-resistant epilepsy [27]. Regarding psychotic symptoms outside a formal diagnosis of schizophrenia, the pooled estimate of the prevalence of psychosis in epilepsy is approximately 5%, although there is heterogeneity dependent upon seizure location and timing of psychotic symptoms [28]. In TLE, the prevalence of any type of psychosis is estimated to be around 7%. PIP can emerge in approximately 2–7.8% of PWE. PWE with a long history of intractability have a 5% prevalence of CIP, particularly in those who already experience PIP.

Treatment Comorbid schizophrenia and epilepsy can result in a high rate of emergency care and hospitalizations [27], highlighting the importance of monitoring and routine intervention. For psychotic symptoms presenting in the context of epilepsy, such as PIP and CIP, clinical vigilance and prevention are critical, warranting aggressive optimization of seizure burden, as well as neuroleptic management of PIP. The symptoms of CIP among some PWE have been described to paradoxically worsen after the apparent successful intervention of epilepsy (i.e., with ASMs or epilepsy surgery)—a phenomenon described as "forced normalization." The existence of this rare but noteworthy phenomenon once again underscores the importance of clinical vigilance in recognizing that psychosis, in some cases, could be the adverse effects of the intervention.

12.4 Anxiety Disorders

Anxiety is characterized by excessive fear of a real or perceived threat or anticipation of a future threat, with behavioral disturbances. In PWE, anxiety may be present either as a comorbid condition, such as panic disorder or generalized anxiety disorder or as a symptom indicative of an anxiety disorder due to another medical condition (epilepsy). Panic disorder is marked by recurrent unexpected panic attacks and persistent concern about having more panic attacks. While ictal auras of anxiety/fearfulness can resemble a panic attack, there are several key differentiating features (see Table 12.5). Regarding anxiety more broadly, ictal anxiety, with accompanying perception of fear or sense of doom, can represent a manifestation of focal seizures, particularly those involving the limbic system. Anxiety may also result from avoidance behaviors in response to the unpredictable nature of epilepsy (i.e., fear of when or where the next seizure will take place) or from a comorbid generalized anxiety disorder, which is prevalent in PWE.

Epidemiology The estimated prevalence of anxiety disorders is around 22% in PWE, approximately twice that of the general population [29]. In DRE, comorbid anxiety disorders may be as high as 38% [27].

Treatment A combination of seizure control, psychopharmacology, and cognitive behavioral psychotherapeutic interventions may be helpful for individuals who experience either epilepsy-related anxiety symptoms or comorbid anxiety disorders. Specific interventions are described in detail in Table 12.6.

Table 12.5 Comparison of panic disorder vs. ictal anxiety

Panic disorder	Ictal anxiety
Slower onset (minutes)	Faster (peaks in 2 min)
Remains conscious	Can lose consciousness
Quick recovery	Post-ictal confusion

Table 12.6 Synopsis of treatment approaches to comorbid psychiatric syndromes

Mood disorders	• *Ictal or peri-ictal mood disturbances:* Usually self-limited and resolve without psychotropic treatment. Optimization of seizure control through ASM adjustments. • *Interictal mood disturbances:* Investigate the temporal relationship between earlier ASM adjustments and the onset or exacerbation of mood disorder. • Antidepressant therapy (SSRI) to be continued for 6 months after recovery from the first episode. However, upon subsequent episodes, antidepressants should continue for at least 2 years. • Amoxapine, clomipramine, and bupropion have high risks of seizure exacerbation. • Psychotherapy to help understand/accept illness, correct negative core beliefs, cope more effectively with stressful life events and develop an improved sense of agency/self-esteem.

(continued)

Table 12.6 (continued)

Psychosis	• *Ictal or peri-ictal psychosis:* Usually self-limited and resolves without psychotropic treatment. Optimization of seizure control through ASM adjustments. • *Post-ictal psychosis:* (1) for very brief episodes (lasting a few days), neuroleptics can be tapered *starting* about 5 days after symptom remission. For more sustained episodes (lasting more than a few days), neuroleptics should be maintained longer. A period of 1–2 months of careful observation after symptom remission is advised before neuroleptics taper. (2) can resolve following successful epilepsy surgery, although these patients carry a higher risk of mood problems following surgery. • *Chronic interictal psychosis:* (1) similar to treatment for primary psychotic disorders. Neuroleptics should be maintained long-term following remission. (2) chlorpromazine and clozapine have a high risk for seizure exacerbation. Proconvulsive risks elevated by rapid escalation of dose, high dosages, and combination therapy with multiple neuroleptic drugs.
Anxiety disorders	• *Electro-clinical seizures, demonstrating ictal panic symptoms:* Timely recognition of this condition is critical. Optimization of seizure control through ASM adjustments. • *Co-existing diagnoses of epilepsy as well as panic disorder* SSRIs as first-line agents. High-potency benzodiazepines, such as clonazepam and alprazolam can also be useful, but patients should be carefully selected to minimize the risk of addiction. • *Behavioral interventions* behavioral interventions, such as relaxation techniques as a countermeasure against negative states to help disrupt the potential self-reinforcing cycle.
PNES	• Critical that patients and caregivers learn to identify, distinguish, and quantify the different seizure types, if possible, so as to direct treatment targets. Review videos of captured seizures (from video-EEG) with the patient/family to help consolidate diagnostic insights. • ASMs should be adjusted to the minimum required dose to achieve satisfactory control of the patient's epileptic seizures. • Cognitive behavioral therapy.
Neurocog.	• Particular attention should be paid to identifying cognitive challenges that impact the management of daily activities with important consequences, such as medication compliance. • Among those with identified cognitive impairments, cognitive rehabilitation can be considered to promote neural recovery and develop ongoing compensatory strategies.

12.5 Psychogenic Nonepileptic Seizures

Within the DSM-5, **psychogenic nonepileptic seizures (PNES)** are defined within broader categories. Categorization and diagnostic criteria are described in Table 12.7. Psychological comorbidity in PNES is high and exposure to traumatic events, especially childhood sexual trauma, is a prominent risk factor for the development of PNES [30, 31]. The diagnoses of epilepsy and PNES are not mutually exclusive, and individuals with epilepsy can also be diagnosed with PNES. Among many patients with such mixed disorders, PNES and epileptic seizures usually occur independently across separate times. However, they may also occur closely together

Table 12.7 PNES classification in the DSM-5

Categorization	Criteria
Category: Somatic symptoms and related disorders	Symptoms of altered voluntary motor or sensory functions
Diagnosis: Conversion disorder (functional neurological symptom disorder)	Clinical findings that provide evidence of incompatibility between the symptom and recognized neurological or medical conditions.
Specifier: "With attacks or seizures"	

as in reported cases of epileptic seizures evolving into PNES during the same episode. In such cases, it has been postulated that the psychological changes induced by electrical seizures may activate the emergence of conversion symptoms. When comparing the ictal manifestations of PNES versus epileptic seizures among patients with mixed disorders, only some patients (36–60%) demonstrate PNES that are readily distinguishable from their epileptic seizures [32]. In other words, many patients with mixed disorders may manifest PNES that are challengingly similar to their epileptic seizures. This observation highlights the diagnostic challenges that confront clinicians evaluating patients with suspected mixed disorders.

Epidemiology It has been reported that approximately 10% of patients with PNES also have co-existing epilepsy [33]. Notably, among patients with learning disability, the percentage of mixed PNES with epilepsy cases can be up to 30%. In most cases, the onset of PNES is after the onset of epilepsy, with hypotheses that PNES is driven by emotional benefits and symptom modeling gained from the prior illness experiences associated with epilepsy.

Treatment Identification and differentiation between seizures with epileptic versus nonepileptic etiology is crucial to direct treatment. Cognitive behavioral therapy is a gold standard treatment for addressing PNES symptoms.

12.6 Neurocognitive Disorders

Cognitive impairment is common among individuals with epilepsy. Deficits can be seen across domains of processing speed, attention, executive functioning, language, visuospatial skills, and learning/memory, with unique patterns reflecting seizure focus, underlying pathophysiology, and medication effects. See Fig. 12.1 for an overview of the assessment of cognition in epilepsy. Overlapping psychiatric comorbidities and psychosocial factors among PWE can also influence cognition. As the third most prevalent neurologic disorder among the elderly [34], the intersection between aging and epilepsy renders this population particularly susceptible to cognitive sequelae.

The DSM-5 includes a diagnosis of neurocognitive disorders (NCD), which replaces the DSM-IV's term of "dementia or other debilitating conditions." NCD can be classified into "mild" or "major" categories, both of which involve a decline

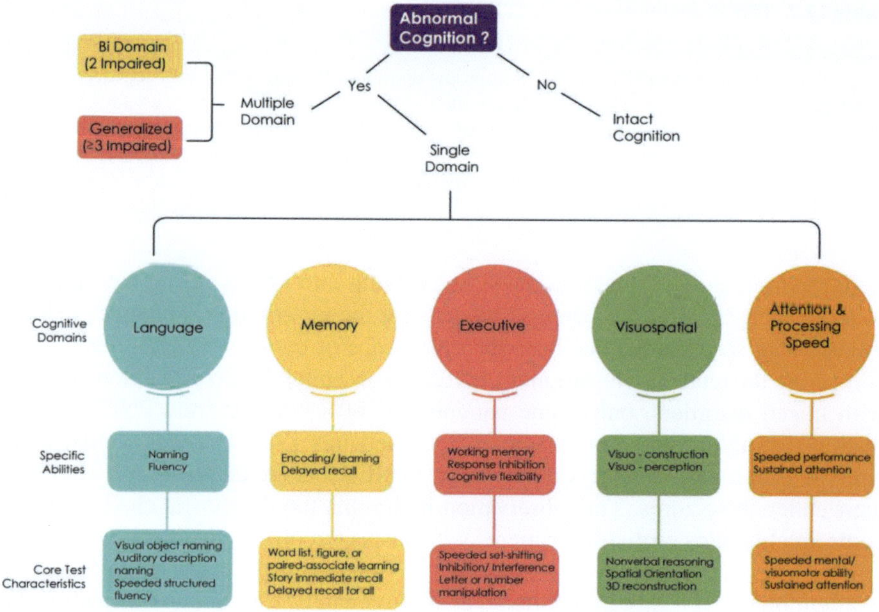

Fig. 12.1 Assessment of cognition in epilepsy. (Adapted from McDonald et al. [35])

in cognition across one or more domains; major NCD is diagnosed when cognitive decline impacts the capacity to safely and independently manage instrumental activities of daily living and is synonymous with the term “dementia.” DSM-5 specifies a range of etiologies (e.g., Alzheimer’s disease), each of which has a unique course, risk factors, and prognoses. Epilepsy is not included as a known etiology of NCD, and diagnosing NCD due to epilepsy would fall under the broad, catch-all category of “NCD due to another medical condition.” The International Classification of Cognitive Disorders in Epilepsy (IC-CoDE) initiative has aimed to better characterize cognitive phenotypes across epilepsy syndromes [35, 36]. This diagnostic approach is largely research-based but can better codify cognitive sequelae from epilepsy. Characterizing cognitive profiles among PWE is an important first step in identifying areas of deficit that warrant intervention. Neuropsychological evaluation is advised as part of a multidisciplinary approach to epilepsy management for assessment of cognition, mood, and behavior [37, 38].

12.7 Considerations for Diverse Populations

While the above statistics and descriptions are provided based on the general population, special considerations exist in the identification and treatment of diverse populations. Regarding gender differences, women with epilepsy represent a unique group for several reasons, including differential impact from psychological risk factors such as interpersonal trauma, variability in mood in relation to reproductive stages such as the postpartum period and perimenopause, and psychosocial influences including gender roles and differential stigma. Such influences lead to a higher rate of nearly all psychological conditions in women compared to men with epilepsy, including anxiety, depression, PTSD, and PNES [27, 31]. Such vast differences in prevalence rates highlight the need for symptom monitoring in women with epilepsy. In racial and ethnic minorities, psychological comorbidity in epilepsy has been underexplored and further investigation is paramount to creating more individualized screening and treatment procedures. From the nascent literature, it appears that sociocultural characteristics of racial and ethnic minority groups, including health literacy, attitudes toward epilepsy, and level of family support, may influence both psychological and seizure-related outcomes [39, 40].

Pearls: neuro-psychiatric comorbidity

- Lifetime prevalence rate of depression in PWE is 25–60%.
- Suicide contributes to ~5% of deaths in PWE. The odds of suicide in epilepsy is 3 times that of the general population.
- Prevalence of chronic interictal psychosis in intractable epilepsy is 5%.
- Comorbid anxiety occurs in PWE at approximately twice the rate of the general population.
- Approximately 10% of patients with PNES have co-existing epilepsy.
- Individuals with drug-resistant epilepsy, women, and racial/ethnic minorities may be especially impacted by comorbid psychiatric conditions.

Acknowledgments David Chen, M.D. for authoring a previous version of this chapter.

References

1. Robertson J, Hatton C, Emerson E, Baines S. Prevalence of epilepsy among people with intellectual disabilities: a systematic review. Seizure. 2015;29:46–62.
2. McGrother CW, Bhaumik S, Thorp CF, Hauck A, Branford D, Watson JM. Epilepsy in adults with intellectual disabilities: prevalence, associations and service implications. Seizure. 2006;15:376–86.
3. Strasser L, Downes M, Kung J, Cross JH, De Haan M. Prevalence and risk factors for autism spectrum disorder in epilepsy: a systematic review and meta-analysis. Dev Med Child Neurol. 2018;60:19–29.
4. Liu X, Sun X, Sun C, Zou M, Chen Y, Huang J, Wu L, Chen WX. Prevalence of epilepsy in autism spectrum disorders: a systematic review and meta-analysis. Autism. 2022;26:33–50.
5. Brikell I, Ghirardi L, D'Onofrio BM, Dunn DW, Almqvist C, Dalsgaard S, Kuja-Halkola R, Larsson H. Familial liability to epilepsy and attention-deficit/hyperactivity disorder: a Nationwide cohort study. Biol Psychiatry. 2018;83:173–80.
6. Neto FK, Noschang R, Nunes ML. The relationship between epilepsy, sleep disorders, and attention deficit hyperactivity disorder (ADHD) in children: a review of the literature. Sleep Sci. 2016;9:158–63.
7. Ott D, Caplan R, Guthrie D, Siddarth P, Komo S, Shields WD, Sankar R, Kornblum H, Chayasirisobhon S. Measures of psychopathology in children with complex partial seizures and primary generalized epilepsy with absence. J Am Acad Child Adolesc Psychiatry. 2001;40:907–14.

8. Reilly CJ. Attention deficit hyperactivity disorder (ADHD) in childhood epilepsy. Res Dev Disabil. 2011;32:883–93.
9. Pavlou E, Gkampeta A. Learning disorders in children with epilepsy. Childs Nerv Syst. 2011;27:373–9.
10. Beghi M, Cornaggia CM, Frigeni B, Beghi E. Learning disorders in epilepsy. Epilepsia. 2006;47:14–8.
11. Brikell I, Chen Q, Kuja-Halkola R, D'Onofrio BM, Wiggs KK, Lichtenstein P, Almqvist C, Quinn PD, Chang Z, Larsson H. Medication treatment for attention-deficit/hyperactivity disorder and the risk of acute seizures in individuals with epilepsy. Epilepsia. 2019;60:284–93.
12. Lambert MV, Robertson MM. Depression in epilepsy: etiology, phenomenology, and treatment. Epilepsia. 1999;40:s21–47.
13. Jackson MJ. Depression and anxiety in epilepsy. J Neurol Neurosurg Psychiatry. 2005;76:i45–7.
14. Hesdorffer DC, Ishihara L, Mynepalli L, Webb DJ, Weil J, Hauser WA. Epilepsy, suicidality, and psychiatric disorders: a bidirectional association. Ann Neurol. 2012;72:184–91.
15. Gandy M, Sharpe L, Perry KN. Psychosocial predictors of depression and anxiety in patients with epilepsy: a systematic review. J Affect Disord. 2012;140:222–32.
16. Karakis I, Boualam N, Moura LM, Howard DH. Quality of life and functional limitations in persons with epilepsy. Epilepsy Res. 2023;190:107084.
17. Reisinger EL, DiIorio C. Individual, seizure-related, and psychosocial predictors of depressive symptoms among people with epilepsy over six months. Epilepsy Behav. 2009;15:196–201.
18. LaFrance WC, Kanner AM, Hermann B. Psychiatric comorbidities in epilepsy. Int Rev Neurobiol. 2008;83:347–83.
19. Rai D, Kerr MP, McManus S, Jordanova V, Lewis G, Brugha TS. Epilepsy and psychiatric comorbidity: a nationally representative population-based study. Epilepsia. 2012;53:1095–103.
20. Selassie AW, Wilson DA, Martz GU, Smith GG, Wagner JL, Wannamaker BB. Epilepsy beyond seizure: a population-based study of comorbidities. Epilepsy Res. 2014;108:305–15.
21. Kessler RC, Bromet EJ. The epidemiology of depression across cultures. Annu Rev Public Health. 2013;34:119–38.
22. Fuller-Thomson E, Brennenstuhl S. The association between depression and epilepsy in a nationally representative sample. Epilepsia. 2009;50:1051–8.
23. Mazza M, Di Nicola M, Della Marca G, Janiri L, Bria P, Mazza S. Bipolar disorder and epilepsy: a bidirectional relation? Neurobiological underpinnings, current hypotheses, and future research directions. Neurosci Rev J Bringing Neurobiol Neurol Psychiatry. 2007;13:392–404.
24. Knott S, Forty L, Craddock N, Thomas RH. Epilepsy and bipolar disorder. Epilepsy Behav. 2015;52:267–74.
25. Li J, Ledoux-Hutchinson L, Toffa DH. Prevalence of bipolar symptoms or disorder in epilepsy: a systematic review and meta-analysis. Neurology. 2022;98:e1913–22.
26. Lu E, Pyatka N, Burant CJ, Sajatovic M. Systematic literature review of psychiatric comorbidities in adults with epilepsy. J Clin Neurol Seoul Korea. 2021;17:176–86.
27. Sullivan-Baca E, Rehman R, Towne AR, Haneef Z. Psychiatric comorbidity of drug-resistant epilepsy in veterans. Epilepsy Behav. 2023;139:109059.
28. Clancy MJ, Clarke MC, Connor DJ, Cannon M, Cotter DR. The prevalence of psychosis in epilepsy; a systematic review and meta-analysis. BMC Psychiatry. 2014;14:75.
29. Kanner AM. Anxiety disorders in epilepsy: the forgotten psychiatric comorbidity: anxiety disorders in epilepsy: the forgotten psychiatric comorbidity. Epilepsy Currents. 2011;11(3):90–1. https://doi.org/10.5698/1535-7511-11.3.90.
30. Myers L, Trobliger R, Bortnik K, Zeng R, Saal E, Lancman M. Psychological trauma, somatization, dissociation, and psychiatric comorbidities in patients with psychogenic nonepileptic seizures compared with those in patients with intractable partial epilepsy. Epilepsy Behav. 2019;92:108–13.
31. Sullivan-Baca E, Weitzner DS, Choudhury TK, Fadipe M, Miller BI, Haneef Z. Characterizing differences in psychiatric profiles between male and female veterans with epilepsy and psychogenic non-epileptic seizures. Epilepsy Res. 2022;186:106995.

32. Mari F, Di Bonaventura C, Vanacore N, Fattouch J, Vaudano AE, Egeo G, Berardelli A, Manfredi M, Prencipe M, Giallonardo AT. Video-EEG study of psychogenic nonepileptic seizures: differential characteristics in patients with and without epilepsy. Epilepsia. 2006;47(Suppl 5):64–7.
33. Benbadis SR, Agrawal V, Tatum WO. How many patients with psychogenic nonepileptic seizures also have epilepsy? Neurology. 2001;57:915–7.
34. World Health Organization. Epilepsy: a public health imperative. Geneva: World Health Organization; 2019.
35. McDonald CR, Busch RM, Reyes A, Arrotta K, Barr W, Block C, Hessen E, Loring DW, Drane DL, Hamberger MJ, Wilson SJ, Baxendale S, Hermann BP. Development and application of the International Classification of Cognitive Disorders in Epilepsy (IC-CoDE): initial results from a multi-center study of adults with temporal lobe epilepsy. Neuropsychology. 2023;37:301–14.
36. Norman M, Wilson SJ, Baxendale S, Barr W, Block C, Busch RM, Fernandez A, Hessen E, Loring DW, CR MD, Hermann BP. Addressing neuropsychological diagnostics in adults with epilepsy: introducing the international classification of cognitive disorders in epilepsy: the IC CODE initiative. Epilepsia Open. 2021;6:266–75.
37. Morrison CE, MacAllister WS, Barr WB. Neuropsychology within a tertiary care epilepsy center. Arch Clin Neuropsychol. 2018;33:354–64.
38. Wilson SJ, Baxendale S, Barr W, Hamed S, Langfitt J, Samson S, Watanabe M, Baker GA, Helmstaedter C, Hermann BP, Smith ML. Indications and expectations for neuropsychological assessment in routine epilepsy care: report of the ILAE neuropsychology task force, diagnostic methods commission, 2013–2017. Epilepsia. 2015;56:674–81.
39. Trujillo S, Wetmore JB, Camarillo IA, Misiewicz S, May H, Choi H, Siegel K, Chung WK, Phelan JC, Yang LH, Leu CS, Bergner AL, Ottman R. Knowledge and beliefs about epilepsy genetics among Hispanic and non-Hispanic patients. Epilepsia. 2023;64:2443–53.
40. Chong J, Drake K, Atkinson PB, Ouellette E, Labiner DM. Social and family characteristics of Hispanics with epilepsy. Seizure. 2012;21:12–6.

Social Considerations in Epilepsy

13

Jennifer Haynes, Kamakshi Patel, and Christine M. Baca

Beyond seizures, persons with epilepsy (PWE) face significant social challenges. Management of epilepsy extends beyond the medical and surgical treatment of seizures and requires recognition and consideration of such challenges in the development of a treatment plan that incorporates counseling and education of PWE, family members and caretakers, in addition to the broad community that includes employers and educators. This chapter reviews the social implications of epilepsy, including patient and family counseling, lifestyle modifications related to seizure triggers and injury avoidance, support services and regulatory guidelines or legislation related to sports participation, education, employment, disability and driving.

13.1 Patient and Family Education

Patients and family members need to have a strong understanding of epilepsy, seizure triggers, and emergency response procedures to seizures. They also need to have strategies to increase adherence to anti-seizure medications (ASM) to minimize injury secondary to seizures. Numerous online educational resources are available for patients and families (Table 13.1).

J. Haynes · C. M. Baca (✉)
Neurology, Virginia Commonwealth University, Richmond, VA, USA
e-mail: Christine.Baca@vcuhealth.org

K. Patel
Department of Neurology, Michael DeBakey VA Medical Center, Baylor College of Medicine, Houston, TX, USA

Z. Haneef (ed.), *Epilepsy Fundamentals*,
https://doi.org/10.1007/978-3-031-77741-7_13

Table 13.1 Resources for PWE, families and caretakers

Epilepsy Foundation www.epilepsy.com	• Offers educational material, seizure first aid training, and seizure action plan templates. Support groups, both virtual and regional in-person. 24/7 hotline available.
Centers for Disease Control—Epilepsy Page www.cdc.gov/epilepsy	• Provides educational materials, data and statistics on epilepsy, current studies. • Links to other resources (Epilepsy Foundation and Managing Epilepsy Well Network).
Managing Epilepsy Well (MEW) Network https://managingepilepsywell.org	• Promotes epilepsy self-management tools and research. • Links to evidence-based self-management programs for patients.
National Associations of Epilepsy Center www.naec-epilepsy.org	• Information for patients regarding epilepsy centers and locator to find epilepsy centers.
Danny Did Foundation https://www.dannydid.org/	• Nonprofit organization focusing on SUDEP education. • Grants available for seizure detection devices.

13.1.1 Treatment Adherence

Adherence to anti-seizure medication is particularly important for PWE given the risk of seizure exacerbation associated with non-adherence. The most common cause of ASM non-adherence is that patients forget to take their medication [1]. Studies have found that adherence can be affected by numerous different potential factors negatively (ASM polytherapy, frequency of dosing, side effects and feelings of stigma) and positively (effective medical follow-up, employment, educational attainment, and medication reminder strategies, e.g., pillbox, displayed medication schedule). Since the development of neuromodulation for the treatment of epilepsy, treatment plans can involve device interrogation and downloading data, such as with NeuroPace® Responsive Neurostimulator (RNS). Interrogation and data uploading are integral to RNS treatment to refine pattern detection and monitor response.

Self-management is an integral component of long-term care for patients with chronic diseases such as epilepsy. Self-management strategies for PWE include strategies to (1) identify and manage seizure triggers, (2) increase adherence to ASMs, (3) minimize risks of injury due to seizures, and (4) improve patient/family response to seizures. Such strategies should be reviewed with patients and family members [2]. The Center for Disease Control (CDC) Managing Epilepsy Well network provides several different online self-management tools for PWE (Tables 13.1 and 13.2) [3].

13.1.2 Sudden Unexpected Death in Epilepsy (SUDEP)

SUDEP is defined as sudden death in a patient with epilepsy with no other identifiable cause and normal autopsy. The incidence of SUDEP is estimated to be 1.2 cases per 1000 persons with epilepsy in the epilepsy population; however, it is

Table 13.2 Managing epilepsy well (MEW) network self-management programs

Project UPLIFT (Using Practice and Learning to Increase Favorable Thoughts) for Epilepsy	• Self-management program for PWE and depression. • Provides home-based telephonic group sessions of mindfulness and cognitive therapies. • Anonymity is optional.
HOBSCOTCH (Home-Based Self-management and Cognitive Training Changes Lives)	• Behavioral program focusing on cognitive issues in adults with epilepsy. • Virtual and phone-based sessions teach participants self-awareness and memory strategies.
PACES (Program of Active Consumer Engagement in Self-management of Epilepsy)	• Self-management program for adults with epilepsy with education on epilepsy and then coping strategies for depression and cognitive impacts of epilepsy. • Offering in-person and telephone-based sessions.
SMART (Self-Management for People with Epilepsy and a History of Negative Health Events)	• Self-management program for adults with epilepsy, sub-optimally controlled and from disadvantaged subgroups. • Utilizes virtual group sessions and then telephone maintenance sessions facilitated by a nurse educator and per educator.

higher in epilepsy centers (1.1–5.9 per 1000) and in refractory patients and those referred for receiving surgery (6.3–9.3 per 1000) [4, 5]. SUDEP often occurs during sleep, and patients are often found in prone positions. Risk factors for SUDEP include poor control of primary or secondary tonic-clonic seizures, use of polytherapy, duration of epilepsy and young age at onset [6–8].

Self-management tools may help reduce the risk of SUDEP by improving adherence and addressing lifestyle issues that can provoke seizures (Table 13.3). As most SUDEP cases occur during sleep, seizure detection devices may help reduce the risk, particularly in those that sleep or live alone, although this has not been studied extensively. Nocturnal supervision has been shown to be protective for SUDEP [8]. Further, there are some studies that suggest that seizure detection devices may reduce anxiety, improve quality of life and increase independence [9, 10]. Seizure detection devices can also aid in providing more accurate seizure counts. Seizure detection devices vary in their mechanism, and the choice of seizure detection device varies based on seizure type and patient and provider goals [11]. As there is often an associated cost, seizure detection devices may not be affordable for all patients, and many lack peer-reviewed performance data [11]. Some foundations, such as the Danny Did Foundation offer grants to patients and families. False alarms may be a consideration with seizure detection devices when discussing with patients and families [11].

Table 13.3 Seizure tracking & medication reminder tools

Applications[a]	• Examples: Epsy, Seizure tracker. • Medication reminders, video recordings. • Ability to transmit electronically to the care team. • Requires computer, tablet or smartphone.
Paper-based	• Templates available on http://epilepsy.com/ • Option for those without access to technology.

[a] Not all applications are available in the U.S [11]

13.2 Safety Issues

13.2.1 Seizure Triggers and Lifestyle Modifications

Tailoring lifestyle modifications to specific seizure types, severity, frequency, and the underlying cause of epilepsy is instrumental in preventing seizures. Successfully minimizing and steering clear of seizure triggers constitutes a valuable approach to seizure prevention [12] as outlined in Table 13.4.

Sports There are no data to suggest that sports exacerbate seizures. Recommendations should be individualized to each specific clinical scenario taking into consideration the risk of injury should a seizure occur during the activity. Any sports that could cause injury to self or others should be discouraged in patients with active seizures (e.g., motorcycling, auto racing, driving an all-terrain vehicle, snowmobiling, jet skiing, skydiving, hang gliding, free climbing, and scuba diving). Patients whose seizures are triggered by hyperventilation should monitor pace and respirations [13].

Driving In the United States, the privilege to drive a motor vehicle is governed by individual states or territorial governments. Six states require Mandatory Reporting to DMV: California, Delaware, Nevada, New Jersey, Oregon, and Pennsylvania [14] Table 13.5 outlines state based driving restrictions for patients with seizures.

A 1994 consensus conference on epilepsy and driver licensing, with the participation of the American Academy of Neurology, American Epilepsy Society and the Epilepsy Foundation of America, agreed on recommendations for resumption or maintenance of driving privileges for PWE that included: (1) seizure freedom of 3 months, (2) allowances for patients with seizures that occur exclusively nocturnally, and (3) allowances for patients with established prolonged and consistent auras [15]. Requirements for commercial licenses vary per state as well, and updated information should be obtained whenever necessary [14].

Piloting The Federal Aviation Administration (FAA) regulates commercial and noncommercial piloting medical certifications and uses certified examiners to assess and make final determinations regarding the impact of medical conditions on the

Table 13.4 General lifestyle modifications for PWE

Sleep	• Sleep deprivation can exacerbate seizures in some patients. • Modification of sleep habits to achieve a consistent and sufficient sleep schedule. • Patients with seizures that generalize should avoid sleeping in a prone position.
Water activities	• Epilepsy patients are encouraged not to swim or take baths alone. • PWE should be visually supervised by someone with both swimming and seizure safety skills when swimming; this person should be knowledgeable regarding safe seizure response actions.
Alcohol and Illicit drugs	• The use of recreational drugs should be avoided due to the potential for lowering the seizure threshold and possible ASM interactions. • Alcohol consumption can lower the threshold for seizures for certain types of epilepsy (e.g., JME) and interact with some ASMs; it should be avoided in such circumstances. • If alcohol will be consumed, moderation and small quantities are encouraged.
Flashing lights	• Flashing lights can trigger seizures in some seizure patients; such patients should avoid activities involving strobe lights or flashing bulbs. • When encountered with a flickering pattern, PWE should try to focus on distant objects and/or wear polarized sunglasses.
Menses	• Some women with epilepsy might have cyclical seizure exacerbations related to the menstrual cycle (catamenial epilepsy). • Women with catamenial epilepsy are encouraged to keep seizure and menstrual cycle calendars. • Special care should be taken when oral contraceptives are used, given significant interactions that could alter the effectiveness of certain ASMs or contraceptive medications.
Stress reduction	• Relaxation techniques such as progressive muscle relaxation (PMR), yoga, mindfulness, biofeedback and deep breathing exercises can reduce seizure frequency.
Exercise	• Appropriate protective equipment and appropriative selection of exercise has been shown to reduce seizure frequency.

ability of a person to pilot an aircraft. Federal regulations, in effect, bar anyone with a history of epilepsy, who has been diagnosed with epilepsy or who has experienced a "*disturbance of consciousness without a satisfactory medical explanation of the cause*," from obtaining any type of pilot's license [16]. Additional information from the FAA can also be found on their website regarding seizure and epilepsy guides for medical examiners [17, 18].

Unpredictability of Seizures The unpredictability of seizures can limit the ability of a patient with epilepsy to participate in certain activities, especially if consciousness might be impaired during the event. When engaging in a situation, patients and caretakers need to understand all aspects of the specific activity and must try to anticipate any injury that might develop if a seizure were to occur during that respective activity. Factors including seizure behavior, medication adherence, ASM side-effects, age and concomitant medical problems must be taken into

Table 13.5 State-based seizure driving restriction regulations

3 months	6 months		12 months	18 months	Undetermined
Arizona	Alabama	New Jersey	Arkansas	Rhode Island	Colorado
California	Alaska	New Mexico	District of Columbia		Connecticut
Kentucky	Florida	North Dakota	New Hampshire		Delaware
Maine	Georgia	North Carolina	New York		Idaho
Maryland	Hawaii	Oklahoma			Illinois
Minnesota	Iowa	Pennsylvania			Indiana
Nevada	Kansas	South Carolina			Louisiana
Oregon	Massachusetts	South Dakota			Montana
Texas	Michigan	Tennessee			Nebraska
Utah	Mississippi	Virginia			Ohio
Wisconsin	Missouri	Washington			Vermont
					Wyoming

Source: Epilepsy Foundation of America: State driving laws database. https://www.epilepsy.com/lifestyle/driving-and-transportation/laws

Table 13.6 Seizure rescue medications

Medication	Dose for age 12+ years	½ life	Peak concentration
Diazepam rectal gel (>2 years)	0.2 mg/kg, max 20 mg	45–46 h	1.5 h
Diazepam Nasal spray (>6 years)	0.2 mg/kg, max 20 mg	~49 h	1.5 h
Midazolam (>12 years)	5 mg	2.1–6.2 h	10–12 min

consideration before determining how safe it is for a given PWE to participate in a desired activity.

If a seizure does occur, seizure action plans developed with a neurologist can be helpful in seizure termination early on [19]. Currently, three rescue medications as outlined in Table 13.6 are available and should be optimally selected depending on patient's preference and levels of functioning [20].

13.3 Employment Challenges for PWE

PWE have higher unemployment rates compared to the general population and persons with other chronic medical conditions [21, 22]. When employed, PWE can be faced with underemployment and reduced income. High seizure frequency, ASM side effects, earlier age of seizure onset, education, cognitive impairment, job types and discrimination are several factors that have been associated with unemployment in PWE. Some jobs have restrictions that limit the ability of PWE to qualify for given regulatory agencies (e.g., commercial truck driver, air traffic controller). Several jobs have inherent dangers for PWE that could increase the risk of injury in the setting of a seizure (e.g., scuba diver, welder).

Given the heterogeneity of epilepsy with respect to severity, seizure types, and treatment, the decision to disclose epilepsy to employers should be made on an individual basis with review and consideration of job requirements, risk of seizure and injury and any need for accommodations.

Table 13.7 provides information and resources regarding the Americans with Disabilities Act, the U.S. Equal Employment Opportunity Commission (EEOC), the Job Accommodation Network (JAN), Social Security Administration and Disability, in addition to general resources for employment in PWE.

Table 13.7 PWE and employment accommodation and disability resources

General resources	• Epilepsy Foundation of America—Employment Help with Epilepsy: https://www.epilepsy.com/lifestyle/employment
Americans with Disabilities Act (ADA)	• Federal law prohibits discrimination against qualified individuals with disabilities. • Individuals with disabilities include those who have impairments that substantially limit a major life activity, have a record (or history) of a substantially limiting impairment, or are regarded as having a disability. • Title I of the ADA covers employment by private employers with 15 or more employees as well as state and local government employers. • Section 501 of the Rehabilitation Act provides similar protections related to federal employment. See U.S.C. §12,102(2); 29 C.F.R. §1630.2(g).
U.S. Equal Employment Opportunity & Commission (EEOC)	• The federal agency in charge of enforcing title I of the ADA "envisioned an interactive process that requires participation by both parties." • Accommodation(s) need only be "effective," and this could be different from the preference of the employee. • Physicians can play a role in the accommodation process.https://www.eeoc.gov/laws/guidance/epilepsy-workplace-and-ada
Job Accommodation Network (JAN)	• The leading source of free, expert, and confidential guidance on workplace accommodations and disability employment issues. • One of several services provided by the U.S. Department of Labor's Office of Disability Employment Policy (ODEP). • Accommodation & Compliance: Epilepsy/Seizure Disorder: https://askjan.org/disabilities/Epilepsy-Seizure-Disorder.cfm?
Disability: Social Security Administration (SSA)	• SSA requires PWE to meet requisite parameters to be eligible for disability benefits. • PWE can receive SSA benefits for both convulsive and non-convulsive epilepsy and non-epileptic events (if debilitating). • SSA regulations require patients to provide detailed certified medical documentation regarding seizure type, frequency, and response to treatment to qualify.https://www.ssa.gov/disability/professionals/bluebook/11.00-Neurological-Adult.htm#11_02

13.4 Education Challenges for PWE

PWE can face educational challenges. Recurrent seizures and ASM side effects can negatively impact learning in PWE through their effects on attention, memory, or alertness. Children with epilepsy may experience social isolation and preliminary evidence suggests that they are more prone to being victims of bullying compared to healthy and chronic disease controls [23]. Healthcare professionals must be proficient in the laws and regulations that safeguard educational and social integrative processes for PWE. Children and adolescents are protected by a series of laws that safeguard the educational process in PWE and other disabilities. Table 13.8 provides an overview of resources provided by Section 504 of the Rehabilitation Act of 1973 and Individuals with Disabilities Education Act to develop and implement Individualized Education Plans (IEP).

With nearly half a million children with epilepsy in the United States, teachers and other school staff need education and training about epilepsy and seizure management. The 2014 School Health Profile survey demonstrated a gap in teaching our

Table 13.8 Educational resources for PWE

General Education/ School Resources	• Centers for Disease Control (CDC): Epilepsy in Schools:https://www.cdc.gov/epilepsy/groups/schools.htm • Epilepsy Foundation of America—School and Childcare.https://www.epilepsy.com/parents-and-caregivers/kids/school-child-care
Section 504 of the Rehabilitation Act of 1973	• Prohibits disability discrimination in federally funded programs and activities, including public childcare, school systems and postsecondary education programs, even in adults. • Extends to state and local government activities in the ADA.
504 Plan	• Office of Civil Rights. • Eligible if they have a disability that significantly impacts a major life function. • Includes accommodations, modifications, and related services. • No age limits. • Used in school, work, and college. • Creation of each plan at each respective institution.
Individuals with Disabilities Education Act (IDEA)	• Helps guarantee early intervention services to infants and toddlers, as well as special education services to children until adulthood with the goal of creating and following individualized educational needs.
Individualized Educational Plans (IEP)	• Department of Education. • Eligible if has a disability that meets criteria under IDEA, significant impacts education performance and requires specialized services. • Includes specialized education services, accommodations, and related services. • Offered through 12th grade or until age 21 years. • Used through 12th grade and does not transfer to college. • Qualified personnel from school district agencies evaluate children academically, functionally, and developmentally to determine disability requirements. • IEP includes seizure action plans.

teachers; a median of 49% of lead health education teachers in secondary schools wanted professional development on epilepsy and seizure disorders, while a mean of 18.2% received such education [24]. Additional recommendations from the survey included education of nurses, staff, teachers, and students about treatment, first aid and stigma, in addition to following individualized seizure response plans, including administration of rescue medications (see prior safety section).

13.5 Quality of Life

When caring for PWE, it is important to investigate the impact seizures have on all aspects of the individual. Unfortunately, misperceptions about epilepsy by the public can propagate stigma and discrimination. Such feelings of stigma and discrimination can lead to stress, lower self-esteem, and decreased social functioning in PWE, including those with well-controlled seizures [25]. Counseling, self-management tools and good social support can help ameliorate the negative effects of stigma [26]. Developing networks of social support and relationships can sometimes be challenging for PWE [27]; lower self-esteem and social isolation may limit opportunities for dating and intimate relationships [28]. Until 1957, 17 states in the U.S. prohibited PWE from marrying. While progress has been made with discriminatory legislation, long-term social outcomes for young PWE demonstrate concerning outcomes related to decreased employment, lower marriage rates, poor social relationships, and decreased likelihood of living independently [29]. A concomitant diagnosis of anxiety or depression is common in PWE. Depression and anxiety have substantive negative impacts on quality of life and have been associated with the risk of suicide in PWE [30]. It is, therefore, imperative for health professionals to screen for mood disorders in PWE to ensure appropriate identification, treatment and referral if needed [31].

13.6 Conclusion

Beyond seizures, epilepsy has broad social and psychological impacts on PWE and families. Such impacts need to be identified and taken into consideration by health professionals when developing comprehensive treatment plans and identifying educational and employment-based resources to ensure safety, reduce SUDEP risk and optimize quality of life.

Acknowledgments Rafael Lopez-Baquero for co-authoring a previous version of this chapter.

References

1. Paschal AM, Rush SE, Sadler T. Factors associated with medication adherence in patients with epilepsy and recommendations for improvement. Epilepsy Behav. 2014;31:346–50.

2. Kralj-Hans I, Goldstein LH, Noble AJ, Landau S, Magill N, McCrone P, et al. Self-management education for adults with poorly controlled epILEpsy (SMILE (UK)): a randomised controlled trial protocol. BMC Neurol. 2014;14:69.
3. Control CfD. Managing Epilepsy Well (MEW) network. U.S. Department of Health & Human Services; 2022. [Available from: https://www.cdc.gov/epilepsy/research/MEW-network.htm.
4. Thurman DJ, Hesdorffer DC, French JA. Sudden unexpected death in epilepsy: assessing the public health burden. Epilepsia. 2014;55(10):1479–85.
5. Tomson T, Nashef L, Ryvlin P. Sudden unexpected death in epilepsy: current knowledge and future directions. Lancet Neurol. 2008;7(11):1021–31.
6. Devinsky O, Hesdorffer DC, Thurman DJ, Lhatoo S, Richerson G. Sudden unexpected death in epilepsy: epidemiology, mechanisms, and prevention. Lancet Neurol. 2016;15(10):1075–88.
7. Hesdorffer DC, Tomson T, Benn E, Sander JW, Nilsson L, Langan Y, et al. Combined analysis of risk factors for SUDEP. Epilepsia. 2011;52(6):1150–9.
8. Langan Y, Nashef L, Sander JW. Case-control study of SUDEP. Neurology. 2005;64(7):1131–3.
9. Chiang S, Moss R, Patel AD, Rao VR. Seizure detection devices and health-related quality of life: a patient- and caregiver-centered evaluation. Epilepsy Behav. 2020;105:106963.
10. Thompson ME, Langer J, Kinfe M. Seizure detection watch improves quality of life for adolescents and their families. Epilepsy Behav. 2019;98(Pt A):188–94.
11. Shum J, Friedman D. Commercially available seizure detection devices: a systematic review. J Neurol Sci. 2021;428:117611.
12. Osborne Shafter P, Schachter S. Managing triggers. Bowie, MD: Epilepsy Foundation of America; 2014. Available from: https://www.epilepsy.com/manage/managing-triggers.
13. Knowles BD, Pleacher MD. Athletes with seizure disorders. Curr Sports Med Rep. 2012;11(1):16–20.
14. Wheless J, Ji S, Osborne Shafter P. Driving and transportation. Bowie, MD: Epilepsy Foundation of America; 2013. [Available from: https://www.epilepsy.com/lifestyle/driving-and-transportation.
15. Consensus statements, sample statutory provisions, and model regulations regarding driver licensing and epilepsy. American Academy of Neurology, American Epilepsy Society, and Epilepsy Foundation of America. Epilepsia. 1994;35(3):696–705.
16. Osborne Shafter P. Pilots and epilepsy. Bowie, MD: Epilepsy Foundation of America; 2014. Available from: https://www.epilepsy.com/lifestyle/employment/pilot.
17. Association FA. Epilepsy (Seizure disorder); 2023. Available from: https://www.faa.gov/ame_guide/media/Epilepsy.pdf.
18. Association FA. Guide for aviation medical examiners—version 1/25/2023—seizure (all classes); 2023. Available from: https://www.faa.gov/ame_guide/media/Seizure.pdf.
19. Penovich P, Glauser T, Becker D, Patel AD, Sirven J, Long L, et al. Recommendations for development of acute seizure action plans (ASAPs) from an expert panel. Epilepsy Behav. 2021;123:108264.
20. Gidal B, Detyniecki K. Rescue therapies for seizure clusters: pharmacology and target of treatments. Epilepsia. 2022;63 Suppl 1(Suppl 1):S34–44.
21. Smeets VM, van Lierop BA, Vanhoutvin JP, Aldenkamp AP, Nijhuis FJ. Epilepsy and employment: literature review. Epilepsy Behav. 2007;10(3):354–62.
22. Sung C, Muller V, Jones JE, Chan F. Vocational rehabilitation service patterns and employment outcomes of people with epilepsy. Epilepsy Res. 2014;108(8):1469–79.
23. Hamiwka LD, Yu CG, Hamiwka LA, Sherman EM, Anderson B, Wirrell E. Are children with epilepsy at greater risk for bullying than their peers? Epilepsy Behav. 2009;15(4):500–5.
24. Prevention. CfDCa. School health policies and practices brief: epilepsy and seizure disorder. In: Services USDoHaH, editor; 2017.
25. Jacoby A. Epilepsy and stigma: an update and critical review. Curr Neurol Neurosci Rep. 2008;8(4):339–44.
26. Szaflarski JP, Szaflarski M. Seizure disorders, depression, and health-related quality of life. Epilepsy Behav. 2004;5(1):50–7.

27. Batchelor R, Taylor MD. Young adults with epilepsy: relationships between psychosocial variables and anxiety, depression, and suicidality. Epilepsy Behav. 2021;118:107911.
28. Suurmeijer TP, Reuvekamp MF, Aldenkamp BP. Social functioning, psychological functioning, and quality of life in epilepsy. Epilepsia. 2001;42(9):1160–8.
29. Berg AT, Baca CB, Rychlik K, Vickrey BG, Caplan R, Testa FM, et al. Determinants of social outcomes in adults with childhood-onset epilepsy. Pediatrics. 2016;137(4):e20153944.
30. Fiest KM, Dykeman J, Patten SB, Wiebe S, Kaplan GG, Maxwell CJ, et al. Depression in epilepsy: a systematic review and meta-analysis. Neurology. 2013;80(6):590–9.
31. Fiest KM, Patten SB, Jette N. Screening for Depression and Anxiety in Epilepsy. Neurol Clin. 2016;34(2):351–61.

14 Women with Epilepsy

Kamakshi Patel

Care of Women with Epilepsy (WWE) is unique due to the influence of hormones on seizures and anti-seizure medications (ASMs) and the effects of seizures and ASMs on fertility and pregnancy, including teratogenicity.

Estradiol, the bioactive form of estrogen, has a proconvulsant effect through the modulation of glutamate receptors and promotion of kindling. Allopregnanolone, the bioactive form of progesterone, has an anticonvulsant effect through the modulation of gamma amino Butyric acid (GABA) receptors. These effects can alter seizure control during the menstrual cycle, pregnancy, and other periods of hormonal changes [1].

14.1 Catamenial Epilepsy

This affects about one-third of women with epilepsy and is defined as a doubling of seizures or seizures occurring exclusively during certain phases of the menstrual cycle [2]. Phases of the menstrual cycle associated with the highest estrogen-progesterone ratio are associated with an increase in seizures. The main patterns seen with catamenial epilepsy are shown in Table 14.1. Currently, there is no FDA-approved treatment for catamenial epilepsy. Figure 14.1 shows the treatment algorithm based on the available data from the progesterone treatment trial and other studies [3–6].

K. Patel (✉)
Department of Neurology, Michael DeBakey VA Medical Center,
Baylor College of Medicine, Houston, TX, USA
e-mail: Kamakshi.Patel@bcm.edu

Z. Haneef (ed.), *Epilepsy Fundamentals*,
https://doi.org/10.1007/978-3-031-77741-7_14

Table 14.1 Types of catamenial epilepsy (PR = progesterone, ES = Estrogen)

Patterns	Characteristics (days of menstrual cycle)
C1-Perimenstrual (decrease in PR)	Most seizures on days 3 to –3
C2-Periovulatory (increase in ES)	Most seizures on days 10 to 13
C3-anovulatory (inadequate luteal phase; increased ES/PR ratio)	Most seizures on days 10 to 3

Adapted from Ref. [2]

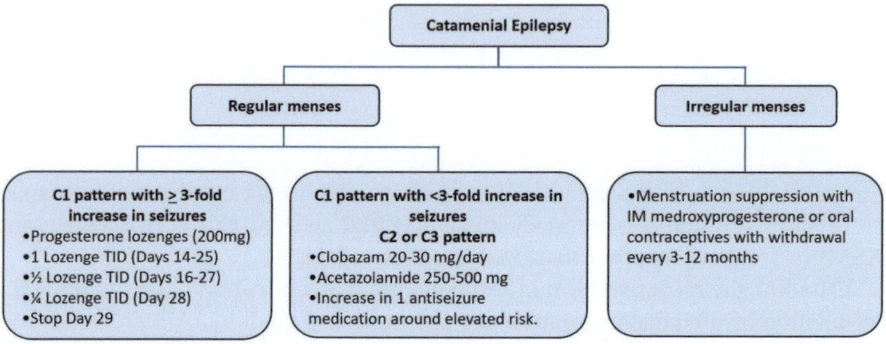

Fig. 14.1 Management of catamenial epilepsy (Inc: Increase, TID: three times a day)

14.2 Sexual Dysfunction (SD) and Fertility

Sexual dysfunction is commonly observed in WWE., with a multifactorial pathogenesis that makes it difficult to identify a single contributing factor [7, 8]. Research has not found a significant association between sexual dysfunction and factors such as seizure control, age of epilepsy onset, type of epilepsy, or use of enzyme-inducing antiepileptic drugs. [8, 9] However, some comorbid mood disorders and ASMs-related reductions in libido and sexual dysfunction may play a role. Medications frequently associated with SD include topiramate, valproate, pregabalin, and gabapentin, with less frequent associations reported for lamotrigine, levetiracetam, and oxcarbazepine [10].

WWE, who have no history of infertility or related disorders, experience similar rates of achieving pregnancy, time to conception, and live birth outcomes as women without epilepsy [11] A web-based study of the pregnancy registry reported a self-reported infertility rate of 9.2% (95% confidence interval [CI], 6.7% to 12.4%) compared to the observed 6% infertility rate among married women in the general population [12] Risk factors for infertility included the use of multiple ASMs, older age, and lower educational levels [13]. Encouragingly, the success rates for infertility treatments including live birth rates per embryo transfer were similar in women with and without epilepsy [14].

14.3 Contraception

Contraceptive counseling should be provided to all WWE of childbearing potential due to the teratogenic effects of ASMs early on in pregnancy. Moreover, the choice of contraceptive method can have an influence on seizure frequency. There is a 6.75 increased risk of seizures in women using hormonal birth control compared to women using the barrier method [15]. Conversely, some ASMs may interact with contraceptives and reduce the effectiveness of contraception. Table 14.2 outlines the interactions between ASMs and contraceptives. Neurologists play an important role in counseling about the ideal contraceptive in WWE. The following approach may be useful in contraception counseling in WWE [16]. (Fig. 14.2).

Table 14.2 Interactions between ASMs and contraceptives

ASMs that increase contraceptives' metabolism	ASMs that *do not* increase contraceptives' metabolism	ASMs whose metabolism is altered by contraceptives
Strong effect: PHT, CBZ, PB, OXC *Weak effect:* TPM, CLB, LTG	LEV, VPA, ESX, GBP, PGB, ZNS	LTG, VPA

ASMs Anti-Seizure Medications, *PHT* Phenytoin, *CBZ* Carbamazepine, *PB* Phenobarbital, *OXC* Oxcarbazepine, *TPM* Topiramate, *CLB* Clobazam, *LTG* Lamotrigine, *LEV* Levetiracetam, *GBP* Gabapentin, *VPA* Valproic acid, *PGB* Pregabalin, *ZNS* Zonisamide, *ESX* Ethosuximide

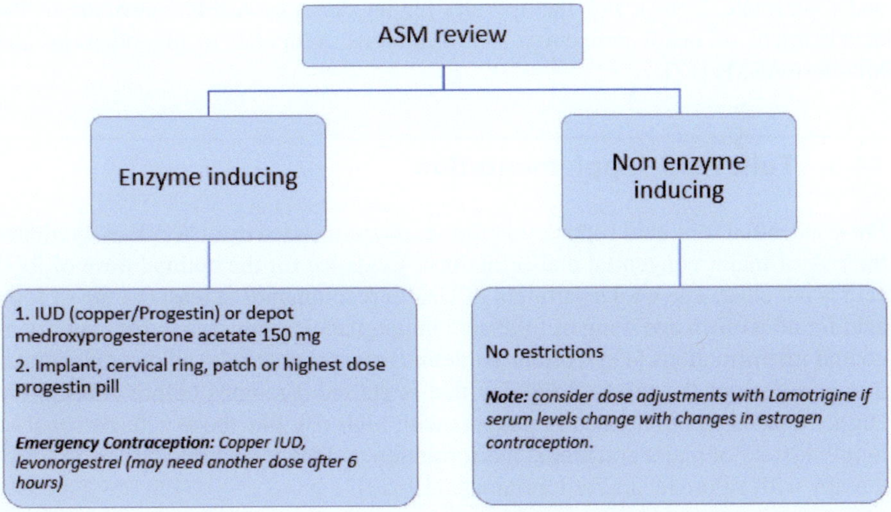

Fig. 14.2 Approach to contraception in women with epilepsy. (Adapted from Ref. [16])

Table 14.3 Teratogenicity profiles with different ASMs

ASMs (exposed numbers-n)	Cardiac defects	Neural tube defects	Cleft lip/cleft palate	Hypospadias
CBZ (n=7308)	0.71%	0.28%	0.32%	0.35%
LTG (n=7100)	0.71%	0.09%	0.19%	0.28%
Barbiturates (n=852)	3.20%	0.23%	1.29%	0.23%
PHT (n=1218)	0.73%	0%	0.24%	0.41%
VPA (n=4270)	1.40%	1.38%	0.72%	1.31%
LEV (n=754)	0.13%	0.13%	0%	0%
TPM (n=470)	0.42%	–	1.27%	0.63%

The total number of exposed fetuses/infants in parenthesis (n) and a specific percentages of malformations as indicated in the table

Data from Tomson, T., Battino, D. & Perucca, E. Lancet Neurol. 15, 210–218

CBZ Carbamazepine, *LTG* Lamotrigine, *PHT* Phenytoin, *VPA* Valproic acid, *LEV* Levetiracetam, *TPM* Topiramate

The US pregnancy registry has given comparative data for major malformations for over 5000 pregnancies. Any pregnant woman in the U.S. and Canada, who is taking any anticonvulsant drug for any reason, is eligible to enroll toll-free by calling 1-888-233-2334

14.4 Teratogenicity of ASMs

In utero exposure to ASMs is associated with teratogenicity (Table 14.3). The risk is highest (10.3%) with valproate monotherapy and polytherapy. Other high-risk ASMs include phenobarbital (6.5%), phenytoin (6.4%), carbamazepine (4.5–6.6%) and topiramate (3.9%). Polytherapy and higher doses of ASMs contribute to the development of major congenital malformations compared to monotherapy and low-dose ASMs [17].

14.5 Folic Acid Supplementation

Pre-conceptual folic acid supplementation is recommended in all WWE as it reduces the risk of major congenital malformations. Evidence for the optimal dose of folic acid is not clear. The US Department of Health recommends at least 0.4 mg of folic acid for all women and 4 mg of folic acid in pregnancies at high risk for major congenital malformations [18] Current guidelines recommend folic acid supplementation at a dose of 0.4 mg to 4 mg/day that is started 3 months before conception. Higher dose is recommended in patients with high risk like those with personal or family history of major congenital malformation and those on monotherapy or polytherapy with valproate and/or topiramate [19, 20].

14.6 Pregnancy

All WWE who are considering pregnancy should receive preconception counseling due to the teratogenic effects of ASMs, which can occur early on in pregnancy, often before a woman becomes aware of their pregnancy. Additional topics to address include prenatal screening methods and their implications, the genetic aspects of epilepsy, supplementation with folic acid and vitamin K, considerations for labor, breastfeeding, and childcare planning.

During pregnancy, seizure patterns can vary: roughly one-third of patients have more seizures, one-third have less, and the rest remain stable. Prepregnancy seizure control is the best predictor of seizure stability during pregnancy [21]. Physiological changes in pregnancy such as increased volume of distribution and hormonal fluctuations, can alter serum levels of ASMs as outlined in Table 14.4 Noncompliance is also a major issue for worsening seizures in pregnancy. WWE should be screened for anxiety and depression during and post pregnancy. Therapeutic drug monitoring is an important tool for ASM dose optimization in pregnancy, and most experts recommend checking these levels monthly to ensure effective management.

The lowest effective dose of the most appropriate ASM should be used, with aim of monotherapy where possible during pregnancy. At least, one documented preconception drug level that achieved seizure freedom/control should be obtained to establish a target level during pregnancy. Increased obstetric-related complications have also been observed in WWE, and multidisciplinary care with obstetricians, pediatricians and neurologists is important. Management of WWE in different stages of pregnancy is outlined in Table 14.5.

Table 14.4 Changes in serum concentration of ASMs during pregnancy

ASM	Percent decrease in serum concentration	Recommendations for monitoring
Phenobarbital	Up to 55%	Yes
Phenytoin	60–70%	Yes (free levels)
Carbamazepine	0–12%	Optional
Valproate	Up to 23%	Optional (free level)
Oxcarbazepine	32–62%	Yes
Lamotrigine	69% in 0.77 of population 17% in 0.23 of population	Yes
Gabapentin	Insufficient data	Yes
Topiramate	Up to 30%	Yes
Levetiracetam	40–60% (maximum decrease in first trimester)	Yes
Zonisamide	Up to 35%	Yes

Data from Ref. [19]

Table 14.5 Management of WWE during various stages of pregnancy

Preconception	1st trimester
Establish the epilepsy syndrome (EEG, MRI, clinical review) Review risks and benefits of current ASM Consider valproate only in extremely limited circumstances. Monotherapy with the lowest effective dose If possible, control seizures prior to pregnancy and counsel patients regarding the importance of maintaining compliance with ASMs during pregnancy. Discuss risks of increased seizures and maternal and fetal complications if ASMs stopped abruptly. Recommend supplementary folic acid 0.4–4 mg daily. Determine and get baseline levels of prepregancy ASM at doses that have been optimal in controlling seizures. Discuss the contraception choice, its initiation and discontinuation.	Get monthly ASM levels. Adjust dosing to maintain prepregnancy levels. Redose ASMs if emesis occurs. Screen for anxiety and depression Counsel compliance with ASMs
2nd trimester	**3rd trimester**
Get monthly ASM levels. Adjust dosing to maintain prepregnancy levels. History and exam for medication side effects Review results from prenatal screening Screen for anxiety and depression Counsel compliance with ASMs	Monthly ASM levels Discuss the birth plan. Discuss the possible risk of seizure worsening peripartum. Discuss breastfeeding and the data to support it. Discuss strategies to ensure sleep. Discuss postpartum ASM taper. Screening for anxiety and depression

Adapted from Ref. [19]

14.7 Post Pregnancy and Lactation

Postpartum depression is higher in WWE on polytherapy and those who are multiparous [22]. During the postpartum period, with return to pre-pregnancy physiology and metabolism, anticipatory ASM readjustments should be considered. The tapering schedule for ASMs as outlined in Table 14.6 is largely determined by the metabolic pathway of ASMs and may need to be done immediately after delivery to avoid toxicity [23].

Table 14.6 Guide to determining ASM tapering post-partum

ASM Metabolism pathway	Time of return to pre-pregnancy levels
Glucuronidation (e.g., LTG)	10–21 days (start taper as soon as postpartum day 3)
Renal clearance (e.g., LEV)	2–3 weeks
Cytochrome P450 (e.g., CBZ)	4–8 weeks

Data from Ref. [23] (LTG: Lamotrigine, LEV: Levetiracetam, CBZ: Carbamazepine)

Table 14.7 ASM accumulation in breast milk

ASMs that accumulate in breast milk	ASMs that DO NOT accumulate in breast milk
Primidone, levetiracetam, gabapentin, lamotrigine, and topiramate	Phenobarbital, phenytoin, carbamazepine and valproate do not accumulate in breast milk

Data adapted from Ref. [20]

14.8 Breastfeeding

All women with epilepsy should be encouraged to breastfeed their babies as it is beneficial for both mother and child. Concerns about ASM accumulation and transfer to infants via breast milk can be addressed using information outlined in Table 14.7 [20].

The total amount of drug transferred to infants via breast milk is usually much smaller than the amount transferred via the placenta during pregnancy. Multiple studies have shown that breastfeeding is safe in women taking ASMs and does not result in any adverse consequences [24–26]. WWE taking benzodiazepines and barbiturates should be counseled to monitor for sedation in infants.

14.9 Menopause

WWE may experience early menopause. Patients with a higher seizure burden have been reported to experience menopause onset at a young age [27]. For most WWE, perimenopause does not significantly affect seizure frequency, but those with catamenial epilepsy may experience worsening seizure frequency. Hormone replacement therapy is associated with an increase in seizure frequency during menopause, and especially in women with a history of catamenial epilepsy. Treatment for menopausal symptoms that can be considered in WWE include non–estrogen-based therapies such as clonidine, selective serotonin reuptake inhibitors (SSRIs), serotonin norepinephrine reuptake inhibitors (SNRIs), and vaginal lubricants for symptomatic treatment [28].

14.10 Bone Health

Women with epilepsy are at increased risk of fractures, osteoporosis, and osteomalacia due to accelerated bone loss with the use of ASMs. Phenytoin, phenobarbital, primidone, valproate, carbamazepine and oxcarbazepine, gabapentin, and topiramate are more frequently associated with bone loss [29, 30]. Baseline and annual DEXA scans, Vitamin D, Calcium and phosphorus levels should be obtained. Moreover, osteoporosis prevention strategies like exercise, nutrition, calcium and high-dose Vitamin D supplementation should be recommended.

Pearls

- Management of Catamenial Epilepsy will depend on whether WWE is experiencing regular or irregular menses. For WWE with regular menses, progesterone supplementation or an increase in ASM is indicated. For WWE with irregular menses, menstrual suppression may be needed.
- WWE of childbearing potential should be offered preconception contraception counseling due to the teratogenic effects of some ASMs.
- There is no difference in achieving pregnancy in WWE and women without epilepsy.
- Care of WWE during pregnancy requires periodic checking of ASM levels, ensuring that the levels are similar to prepregnancy levels and safely tapering the dose of ASM to prepregnancy levels after childbirth.
- WWE are at an increased risk of bone loss, and appropriate screening with DEXA should be done 1–2 yearly. Management of abnormalities seen on DEXA scan is recommended. In addition, baseline Vitamin D, Calcium, and phosphorus with appropriate low-level supplementation is recommended.

References

1. Harden CL, Pennell PB. Neuroendocrine considerations in the treatment of men and women with epilepsy. Lancet Neurol. 2013;12:72–83.
2. Herzog AG, Klein P, Ransil BJ. Three patterns of catamenial epilepsy. Epilepsia. 1997;38:1082–8.
3. Lim LL, Foldvary N, Mascha E, Lee J. Acetazolamide in women with catamenial epilepsy. Epilepsia. 2001;42:746–9.
4. Herzog AG, Fowler KM, Smithson SD, Kalayjian LA, Heck CN, Sperling MR, Liporace JD, Harden CL, Dworetzky BA, Pennell PB, Massaro JM, Progesterone Trial Study Group. Progesterone vs placebo therapy for women with epilepsy: a randomized clinical trial. Neurology. 2012;78:1959–66.
5. Mattson RH, Cramer JA, Caldwell BV, Siconolfi BC. Treatment of seizures with medroxyprogesterone acetate: preliminary report. Neurology. 1984;34:1255–8.
6. Feely M, Calvert R, Gibson J. Clobazam in catamenial epilepsy. A model for evaluating anticonvulsants. Lancet Lond Engl. 1982;2:71–3.
7. Rathore C, Henning OJ, Luef G, Radhakrishnan K. Sexual dysfunction in people with epilepsy. Epilepsy Behav EB. 2019;100:106495.
8. Henning OJ, Nakken KO, Træen B, Mowinckel P, Lossius M. Sexual problems in people with refractory epilepsy. Epilepsy Behav. EB. 2016;61:174–9.
9. Henning O, Johannessen Landmark C, Traeen B, Svendsen T, Farmen A, Nakken KO, Lossius M. Sexual function in people with epilepsy: similarities and differences with the general population. Epilepsia. 2019;60:1984–92.

10. Yang Y, Wang X. Sexual dysfunction related to antiepileptic drugs in patients with epilepsy. Expert Opin Drug Saf. 2016;15:31–42.
11. Pennell PB, French JA, Harden CL, Davis A, Bagiella E, Andreopoulos E, Lau C, Llewellyn N, Barnard S, Allien S. Fertility and birth outcomes in women with epilepsy seeking pregnancy. JAMA Neurol. 2018;75:962–9.
12. MacEachern DB, Mandle HB, Herzog AG. Infertility, impaired fecundity, and live birth/pregnancy ratio in women with epilepsy in the USA: findings of the epilepsy birth control registry. Epilepsia. 2019;60:1993–8.
13. Sukumaran SC, Sarma PS, Thomas SV. Polytherapy increases the risk of infertility in women with epilepsy. Neurology. 2010;75:1351–5.
14. Larsen MD, Jølving LR, Fedder J, Nørgård BM. The efficacy of assisted reproductive treatment in women with epilepsy. Reprod Biomed Online. 2020;41:1015–22.
15. Herzog AG, Mandle HB, Cahill KE, Fowler KM, Hauser WA. Differential impact of contraceptive methods on seizures varies by antiepileptic drug category: findings of the epilepsy birth control registry. Epilepsy Behav. EB. 2016;60:112–7.
16. Bui E. Women's issues in epilepsy. Contin Minneap Minn. 2022;28:399–427.
17. Tomson T, Battino D, Perucca E. Teratogenicity of antiepileptic drugs. Curr Opin Neurol. 2019;32:246–52.
18. US Preventive Services Task Force, Bibbins-Domingo K, Grossman DC, Curry SJ, Davidson KW, Epling JW Jr, García FA, Kemper AR, Krist AH, Kurth AE, Landefeld CS, Mangione CM, Phillips WR, Phipps MG, Pignone MP, Silverstein M, Tseng CW. Folic acid supplementation for the prevention of neural tube defects: US preventive services task force recommendation statement. JAMA. 2017;317:183–9.
19. Tomson T, Battino D, Bromley R, Kochen S, Meador K, Pennell P, Thomas SV. Management of epilepsy in pregnancy: a report from the international league against epilepsy task force on women and pregnancy. Epileptic Disord Int Epilepsy J Videotape. 2019;21:497–517.
20. Harden CL, Pennell PB, Koppel BS, Hovinga CA, Gidal B, Meador KJ, Hopp J, Ting TY, Hauser WA, Thurman D, Kaplan PW, Robinson JN, French JA, Wiebe S, Wilner AN, Vazquez B, Holmes L, Krumholz A, Finnell R, Shafer PO, Le Guen C, American Academy of Neurology; American Epilepsy Society. Practice parameter update: management issues for women with epilepsy—focus on pregnancy (an evidence-based review): vitamin K, folic acid, blood levels, and breastfeeding: report of the quality standards subcommittee and therapeutics and technology assessment Subcommittee of the American Academy of neurology and American Epilepsy Society. Neurology. 2009;73:142–9.
21. Thomas SV, Syam U, Devi JS. Predictors of seizures during pregnancy in women with epilepsy. Epilepsia. 2012;53:e85–8.
22. Galanti M, Newport DJ, Pennell PB, Titchner D, Newman M, Knight BT, Stowe ZN. Postpartum depression in women with epilepsy: influence of antiepileptic drugs in a prospective study. Epilepsy Behav EB. 2009;16:426–30.
23. Pennell PB, Hovinga CA. Antiepileptic drug therapy in pregnancy I: gestation-induced effects on AED pharmacokinetics. Int Rev Neurobiol. 2008;83:227–40.
24. Veiby G, Engelsen BA, Gilhus NE. Early child development and exposure to antiepileptic drugs prenatally and through breastfeeding: a prospective cohort study on children of women with epilepsy. JAMA Neurol. 2013;70:1367–74.
25. Birnbaum AK, Meador KJ, Karanam A, Brown C, May RC, Gerard EE, Gedzelman ER, Penovich PE, Kalayjian LA, Cavitt J, Pack AM, Miller JW, Stowe ZN, Pennell PB, MONEAD Investigator Group. Antiepileptic drug exposure in infants of breastfeeding mothers with epilepsy. JAMA Neurol. 2020;77:441–50.
26. Meador KJ, Baker GA, Browning N, Cohen MJ, Bromley RL, Clayton-Smith J, Kalayjian LA, Kanner A, Liporace JD, Pennell PB, Privitera M, Loring DW, Neurodevelopmental Effects of Antiepileptic Drugs (NEAD) Study Group. Breastfeeding in children of women taking antiepileptic drugs: cognitive outcomes at age 6 years. JAMA Pediatr. 2014;168:729–36.

27. Harden CL, Koppel BS, Herzog AG, Nikolov BG, Hauser WA. Seizure frequency is associated with age at menopause in women with epilepsy. Neurology. 2003;61:451–5.
28. Erel T, Guralp O. Epilepsy and menopause. Arch Gynecol Obstet. 2011;284:749–55.
29. Heo K, Rhee Y, Lee HW, Lee SA, Shin DJ, Kim WJ, Song HK, Song K, Lee BI. The effect of topiramate monotherapy on bone mineral density and markers of bone and mineral metabolism in premenopausal women with epilepsy. Epilepsia. 2011;52:1884–9.
30. Pack AM, Walczak TS. Bone health in women with epilepsy: clinical features and potential mechanisms. Int Rev Neurobiol. 2008;83:305–28.

Semiology

15

Zulfi Haneef, Mohamed Hegazy, and Jay R. Gavvala

Semiology is the study and interpretation of seizure manifestations, including the subjective seizure experience and objective seizure presentation. Seizure manifestations arise from activation of an area of cortex by the ictal epileptiform discharge (the so-called symptomatogenic zone). It is important to realize that the ictal epileptiform discharge may originate in one cortical area (seizure onset zone), which may be "silent" before spreading to the symptomatogenic zone to produce the seizure phenomena. In other words, initial seizure manifestations may present the spread of the ictal discharge and not necessarily where the seizure originates. A detailed and accurate understanding of the patient's semiology is critical and complements the EEG and imaging when establishing the epilepsy classification and localization [1–3]. Seizures are divided into prodrome, aura, ictal, and postictal phases, and each may contribute to an accurate diagnosis and localization. Diurnal predilection can inform diagnosis. Myoclonus of JME is typically early in the day or soon after awakening from sleep. Nocturnal seizures are more common in benign rolandic epilepsy (BRE/BECTS/SeLECTS), frontal lobe seizures, and the tonic seizures in Lennox-Gastaut syndrome. A seizure semiology glossary has been developed by the international league against epilepsy (ILAE), which is a good resource for studying semiology and will be referenced in the following text [4].

Z. Haneef · M. Hegazy
Neurology, Kellaway Section of Neurophysiology, Baylor College of Medicine, Epilepsy Center of Excellence, DeBakey VA Medical Center, Houston, TX, USA

J. R. Gavvala (✉)
Neurology, Texas Comprehensive Epilepsy Program, University of Texas Health Science Center at Houston, Houston, TX, USA
e-mail: jay.r.gavvala@uth.tmc.edu

Z. Haneef (ed.), *Epilepsy Fundamentals*,
https://doi.org/10.1007/978-3-031-77741-7_15

15.1 Aura

Although the dramatic generalized tonic–clonic phase captures the attention of the patient, family, and much of the medical community, the epileptologist recognizes the aura as a highly informative aspect of the seizure for localization. As the patient sometimes may not give much importance to this stage, directed questioning may be necessary to elicit a description of auras ("Have you sometimes almost had a seizure, but not quite?" or "Do you have any warning before your seizures?").

Although auras are sometimes truly "indescribable," careful questioning can often elicit a wealth of diagnostic information (e.g., "Is the sensation in the body, head, or mind?").

According to the 2017 ILAE seizure classification [4], an aura is a focal aware seizure without external manifestations. Depending on the symptomatogenic zone activated, an aura can be sensory, emotional (e.g. fear), cognitive (e.g., Déjà vu), or autonomic (e.g., palpitations). Sensory auras can be somatosensory (e.g., tingling sensation), olfactory (e.g., foul smell), gustatory (e.g., metallic taste), auditory (e.g., ringing noise), visual (e.g., flashing lights) or psychic (limbic cortex).

15.2 Prodrome

Prodromal symptoms include headache, irritability, and behavioral/personality change and can precede the ictus by several minutes, hours, a day, and sometimes longer. The prodrome precedes the ictal onset and does not have an epileptiform EEG correlate.

15.3 Seizure Semiology (Table 15.1)

15.3.1 Motor (Fig. 15.1)

When assessing motor symptomatology, it is important to consider the spatial and temporal distribution of symptoms. Motor semiology can be broadly categorized into simple and complex semiologies [4].

Simple motor semiology includes myoclonic and clonic jerks, versive, tonic, and tonic–clonic seizures. Classically, focal activation results in somatotopically organized unilateral motor symptoms, whereas bilateral symmetric symptoms (particularly myoclonic, tonic, and tonic–clonic) tend to be more consistent with generalized seizures. *Dystonia* refers to sustained agonist–antagonist muscle co-contractions, causing twisting movements and unnatural postures. In temporal lobe epilepsy, unilateral upper limb dystonia strongly indicates a contralateral seizure onset. Some of these hand movements, including grasping, clenching, or opening-closing movements, often evolve into full dystonic posturing and have sometimes been referred to by the acronym RINCH (rhythmic ictal non-clonic hand motions). *Unilateral simple motor semiology* tends to arise from activation of contralateral primary

Table 15.1 Summary of common semiology in clinical practice, including lateralizing and localizing value of each

Semiology	Clinical relevance	IL/CL	Lobe
Motor			
Akinetic	Reflects involvement of "negative motor" areas	CL	F
Astatic	Helps identify sz type but lacks specific localizing value	–	–
Atonic	Indicates potential involvement of motor and premotor cortex; significant in specific syndromes	CL	F
Myoclonic/clonic	Suggests cortical involvement and is typically contralateral to the hemispheric of onset	CL	F, P
Dystonic	Often has lateralizing value; can indicate sz spread and the hemisphere of origin	CL	F, T, P
Eye blinking	May indicate frontal or anterior temporal lobe involvement; unilateral blinking suggests IL focus	IL	F, T, O
Gyratory	Indicates the hemisphere of sz onset; often CL to the direction of body movement, especially when preceded by head version in same direction	CL	F
Epileptic nystagmus	Typically indicates the CL hemisphere involvement relative to the fast phase	CL	P, O
Paresis	Ictal paresis suggests CL motor cortex involvement	CL	F
Spasm	Generalized or focal onset; localization can be challenging without additional context	–	Variable
Chapeau de gendarme	Ictal pouting. Indicates frontal lobe involvement, especially anterior prefrontal and anterior cingulate	–	F
Asymmetric clonic ending	Last clonic jerk is typically ipsilateral to the site of seizure onset	IL	Any
Tonic	Simple tonic/dystonic: Unilateral tonic activity suggests CL cortical involvement	CL	F, P
	Fencing: Lateralizes to the hemisphere opposite the extended arm	CL	F
	Figure of 4: Lateralizing sign, pointing to the CL hemisphere of the extended limb	CL	F
Versive	Forced head turning, particular in context of neck extension indicates sz activity in the hemisphere opposite the direction of version	CL	F, T, P, O
Automatisms	Can indicate the side of the sz onset, especially when unilateral and distal in temporal lobe epilepsies; more complex in bilateral cases	IL/CL	T, F
Sensory			
Auditory	Auditory phenomena can localize to temporal lobes, especially the superior temporal gyrus and temporal operculum	CL	T
Visual	Visual disturbances may point to occipital or temporal lobe involvement	CL	O, T
Somatosensory	Helps localize to the parietal lobe or specific sensory cortex areas	CL	P
Olfactory	May indicate temporal lobe involvement, particularly the mesial structures	IL	T, F

(continued)

Table 15.1 (continued)

Semiology	Clinical relevance	IL/ CL	Lobe
Gustatory	Suggests insular or opercular involvement, pointing to frontal or temporal lobes	IL/ CL	F, T
Depersonalization	Often associated with temporal lobe sz, particularly the non-dominant hemisphere	IL	T
Body-perception illusion	Indicates parietal lobe involvement, especially CL to the distortion	CL	P
Forced thinking	Often associated with dominant hemisphere temporal lobe sz		T
Vestibular	Non-lateralizing sign but suggestive of posterior temporal peri-Sylvian or TPO epilepsies	–	T, P, O
Affective			
Fear	Emotional response can suggest temporal or frontal lobe involvement	–	F, T
Anger	Might indicate frontal lobe engagement	–	F, T
Anxiety	Can be associated with temporal or frontal lobe sz	–	F, T
Guilt	Can be associated with temporal or frontal lobe sz, particularly the right hemisphere	–	F, T
Ecstasy	Often associated with right temporal lobe involvement	–	T
Sadness	May suggest involvement of the non-dominant temporal lobe or frontal lobe	–	F, T
Mystic	Suggests temporal lobe involvement, often with a religious or transcendental quality	–	T
Sexual	Rare; can be associated with temporal or frontal lobe sz, particularly the dominant hemisphere	–	F, T
Cognitive			
Dysmnesia	Impaired memory function suggests involvement of the temporal lobe		T
Aphasia	Indicates involvement of the dominant hemisphere, particularly the frontal and temporal lobes	IL	F, T
Autonomic			
Cardiovascular	Changes like tachycardia or bradycardia can indicate sz spread but are less specific for localization	–	T
Cutaneous	Flushing	–	–
	Pallor	L	T
	Piloerection	–	T, I, A
Respiratory	Hyperventilation or apnea		Limbic, Bs, T
	Laryngeal constriction, choking		I
Gastrointestinal	Epigastric aura		T > F, I
	Hypersalivation		T, FO, I
	Ictal spitting	ND	T
	Ictal vomiting is often occipital in children (e.g., Panayiotopoulos syndrome) and temporal in adults		O, T
Genitourinary	Urinary urgency, sexual automatisms	ND	T
Pupillary	Unilateral pupillary dilatation	IL	T, OT
	Postictal		
Confusion	Suggests widespread brain involvement or significant sz impact	–	–

(continued)

Table 15.1 (continued)

Semiology	Clinical relevance	IL/ CL	Lobe
Todd's paralysis	Indicates the region of sz onset, often with lateralizing value	CL	F
Postictal nose wiping	Could have lateralizing value, suggesting the side of sz focus	IL	T
Postictal headache	General symptom, indicating recent sz activity without specific localizing value	–	–
Postictal psychosis	Reflects significant brain impact, potentially indicating bilateral or temporal lobe involvement	–	T, F

A amygdala, *Bs* brainstem, *CL* contralateral, *F* frontal, *FO* frontal operculum, *I* insula, *IL* ipsilateral, *L* left, *ND* non-dominant, *O* occipital, *P* parietal, *Sz* seizure, *T* temporal

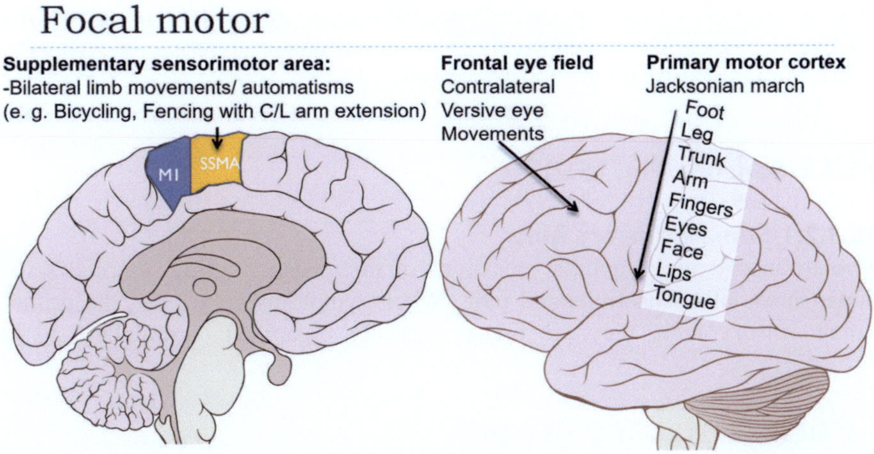

Fig. 15.1 Focal motor seizures. (Modified from images by Patrick J. Lynch, medical illustrator CC-BY-2.5 via Wikimedia Commons)

motor or sensori-motor cortices [4, 5]. Head version can be an effective lateralizing sign when seen immediately prior to progression into focal to bilateral tonic–clonic seizure (seizure onset contralateral to side of version). Head version as an initial semiology of an epileptic seizure is suggestive of a premotor onset. *Bilateral asymmetric tonic posturing* is commonly seen with activation of the supplementary sensorimotor area (SSMA), with the quintessential feature being the fencing posturing with contralateral arm extension and ipsilateral arm flexion. It is important to note that posturing could be unilateral or restricted to a single limb, but proximal limb and axial muscle involvement is always prominent. Furthermore, seeing this presentation implies involvement of the SSMA but may not imply an onset there as there is rich connectivity from other regions to the SSMA [4].

Complex motor semiology can include hyperkinetic seizures and seizures with automatisms. Automatisms can be further divided by their clinical presentation into gestural, distal, genital, ictal grasping, proximal, mimic, oroalimentary, verbal, and vocal [4]. In temporal lobe epilepsy, unilateral *distal manual automatisms* lateralize to the ipsilateral hemisphere when there is contralateral hand dystonia; however, automatisms in the left hand may also be lateralizing to the ipsilateral hemisphere even without contralateral dystonia [6]. Common *mimic automatisms* are laughing (gelastic) and crying (dacrystic) behaviors which are commonly associated with hypothalamic lesions such as hamartoma but also may localize to frontal (anterior cingulate) and temporal lobes. *Oroalimentary automatisms*, including chewing, lip smacking, lip pursing, licking, and swallowing, are typically seen in mesial temporal lobe epilepsy with the presence of oral automatisms related to ictal oscillatory activity in the operculo-insular region based on SEEG data [7].

Negative motor phenomena can range from generalized seizures, resulting in atonic/astatic seizures, also known as drop attacks, to focal seizures, resulting in focal paralysis. These phenomena do not have clear localization. There are also akinetic seizures defined as paroxysmal inability to move with preservation of awareness of focal onset, commonly from mesial premotor cortex, as well as the inferior frontal gyrus and pre-SSMA regions.

15.3.2 Sensory (Fig. 15.2)

15.3.2.1 Somatosensory Aura

Somatosensory auras include sensations of numbness, tingling, thermal sensations, shock like sensations and pain. When symptoms arise from the primary somatosensory area, they are somatotopically organized contralateral to the

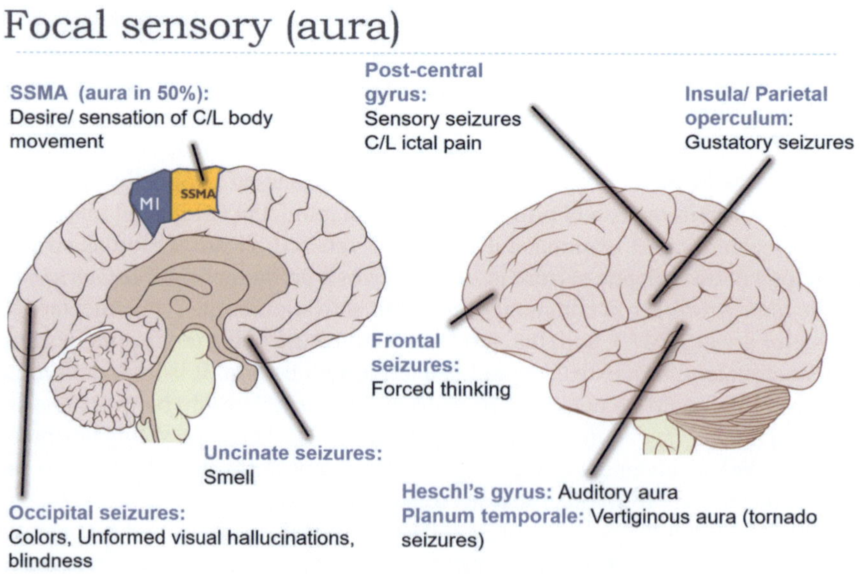

Fig. 15.2 Focal sensory seizures (auras). (Modified from images by Patrick J. Lynch, medical illustrator CC-BY-2.5 via Wikimedia Commons)

hemisphere of onset. Sensory illusions include distortions of sensation and movement and are typically associated with activation of somatosensory association cortices, including inferior parietal lobe and temporo-parieto-occipital junction [4]. While somatosensory auras are commonly thought of as being unilateral, there are several regions that may produce bilateral somatosensory symptoms, including the secondary somatosensory area (SII) in the parietal operculum, the insula, and the SSMA [3, 5]. The characteristic of the somatosensory aura can also be valuable as painful somatosensory auras are typically associated with operculo-insular epilepsies (Fig. 15.3) [8].

15.3.2.2 Olfactory Auras

Olfactory auras can be difficult to distinguish from gustatory symptoms and in certain focal epilepsies can occur in tandem. Classically, olfactory auras are unpleasant, including an odor of burning, garbage, or sulfur, but neutral or even pleasant smells have been described. Most commonly, spontaneous olfactory symptoms have been reported in seizures arising from the medial temporal structures. Cortical stimulation studies have evoked olfactory responses from not only the amygdala, piriform cortex, and the uncus but also the medial orbitofrontal cortex, mid-dorsal insula, and insular central sulcus (Fig. 15.3) [4]. There has been no demonstrable lateralizing value to olfactory auras.

15.3.2.3 Gustatory Auras

Gustatory symptoms reported during focal seizures are rare. As above, they can be difficult to distinguish from olfactory auras. Typically reported gustatory auras are unpleasant and commonly are described as metallic, salty, bitter, or acidic tastes. Gustatory auras have been reported in seizures from peri-rolandic, insular, opercular, and less frequently from the medial temporal structures [4, 5, 8]. Electrical stimulation studies implicate the posterior dorsal insula (Fig. 15.3) and parietal or

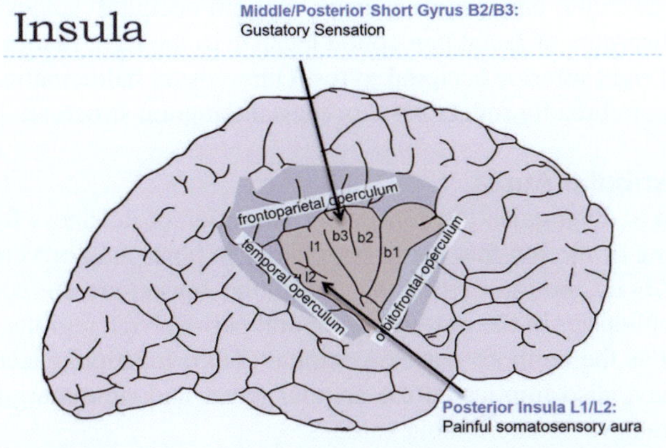

Fig. 15.3 Insular seizures: The overlying opercula have been retracted exposing the five principal insular gyri. *b1* anterior short gyrus, *b2* middle short gyrus, *b3* posterior short gyrus, *l1* anterior long gyrus, *l2* posterior long gyrus

rolandic operculum. As with olfactory auras, gustatory auras do not have any clear lateralizing value.

15.3.2.4 Auditory Auras

Auditory auras range from simple auditory hallucinations such as buzzing and ringing that typically arise in proximity to the Heschl's gyrus. Complex auditory auras such as music or voices and auditory illusions such as alterations in the perception of sound (intensity, tone, distance) arise from auditory association cortex, including anterior superior temporal gyrus and the planum temporale [4].

15.3.2.5 Visual Auras

There can be a spectrum of visual symptoms seen during a seizure that can be classified as negative or positive phenomena as well as simple and complex visual symptoms [4]. Positive visual symptoms can be static, flashes of lights, or spots that can be black, white, or colorful, a localized part of, or throughout, the visual field. Negative visual symptoms are loss of some or all of a visual field, while complex visual symptoms can include formed visual hallucinations ranging from geometric forms to fully formed scenes, illusions of movement, size, and shape, among others. While classically visual auras are associated with occipital seizures, they are not specific for the occipital lobe and can also be seen in parietal, medial temporal, and occipitotemporal seizures [4].

Elementary visual hallucinations typically localize to the calcarine sulcus but can also be seen in the lingula, cuneus, fusiform gyrus, and less frequently from the basal or medial temporal lobe [4]. Geometric visual hallucinations typically arise from the anterior lingula and ventral temporal cortex. Visual blurring and changes in color perception are more commonly associated with ventral and medial temporal structures where illusions of movement reflect lateral occipital and lateral temporo-occipital sources [4]. Changes in size, shape, and number reflect activation of the precuneus and parieto-occipital sulcus. Hallucinations of the face reflect involvement of right cuneus, precuneus, parieto-occipital sulcus, or fusiform gyrus while disorders of facial perception localize to the right fusiform, right basal temporal, and right inferior occipital gyrus. Other visual hallucinations not involving the face more broadly reflect basal or mesial temporal structures [4].

15.3.2.6 Vestibular Auras

It is important to distinguish the nonspecific sensation of dizziness from the sensation of spinning or motion that could be rotatory in horizontal or vertical direction or a sense of body motion or floating. Vestibular sensations are commonly seen arising from epilepsies in the posterior temporal peri-Sylvian regions, including the insula as well as the temporo-parieto-occipital (TPO) junction. Electrical stimulation studies have also implicated the angular gyrus and supramarginal gyrus [4]. Vestibular aura have no lateralizing value.

15.3.3 Affective

Emotional symptoms during a seizure are more commonly unpleasant sensations, typically anxiety and fear. About 20% of temporal lobe seizures include an affective aura. The presence of emotional symptoms suggests activation of the limbic system, mostly commonly the amygdala for fear and anxiety. However, the anterior cingulate and orbitofrontal cortex can be involved in the production of fear; anterior dorsal insula can result in feelings of ecstasy and bliss; mesial temporal lobe, orbitofrontal and prefrontal cortex for anger; and mesial temporal region, orbitofrontal cortex, and anterior insula for sadness [4, 9].

15.3.4 Cognitive

Cognitive changes in seizures include various forms of dysmnesia and aphasia.

15.3.4.1 Dysmnesia

This encompasses memory disturbances such as ictal amnesia, déjà vu (incorrect sense of familiarity with an unfamiliar situation), jamais vu (incorrect sense of unfamiliarity with a familiar situation), and dream-like states with vivid hallucinations, often pointing to mesial temporal structures- in particular the amygdala, hippocampus, and rhinal cortices [10]. Derealization, the perception of surroundings as unreal, although common in psychiatric conditions, can also occur in temporal lobe epilepsy. Ictal amnesia involves the inability to recall events during a seizure, aiding in the classification of focal seizures based on patient awareness.

15.3.4.2 Ictal Aphasia

This is the inability to understand or produce language during seizures, and may include Wernicke's, Broca's, and global aphasia, highlighting involvement of the dominant hemisphere's language areas. Despite different patterns correlating with specific brain regions, ictal aphasia shows limited localizing value within the dominant hemisphere, often in the parieto-occipital regions, and could be spread from the non-dominant hemisphere [11].

15.3.5 Autonomic

Autonomic seizure semiology encompasses a variety of autonomic symptoms that can manifest during both focal and generalized seizures. These symptoms are further described below. Autonomic signs are particularly notable when they occur early in a focal seizure, classifying the event as a focal autonomic seizure, which is a subset of focal aware seizures with non-motor features. Some autonomic phenomena, like ictal vomiting or urinary urge, may have localizing or lateralizing value. The differentiation between subjective autonomic auras, which are experienced by the patient, and objective autonomic seizures, which are observable by clinicians, is essential for accurate diagnosis and effective management.

15.3.5.1 Cardiovascular Manifestations

These are among the most prominent ictal autonomic signs and can include alterations in heart rate and rhythm, blood pressure changes, and sensations of chest pain. Sinus tachycardia is the most common ictal autonomic sign, present in approximately 82% of patients with epilepsy. It can precede EEG changes and is particularly frequent in temporal lobe seizures. In contrast, bradycardia and ictal asystole are rare, occurring in 0–2% of cases, predominantly associated with temporal lobe seizures, although they rarely can be from other areas such as the left cingulate [12]. Cardiovascular changes are important as they can impact the overall management and treatment strategies for patients with epilepsy [13].

15.3.5.2 Cutaneous Phenomena

Flushing is associated with both temporal and extratemporal lobe epilepsies without a clear lateralizing value [14]. Pallor is mostly observed in the context of syncope, although it can occur with seizures and has been said to be associated with left mesial temporal onset [15]. Ictal sweating, which may occur alongside symptoms such as shivering and abdominal pain, lacks systematic studies to determine its lateralizing value but remains a relevant clinical observation during seizures. Piloerection is thought to be associated with temporal lobe, insular, or amygdalar seizures, with inconsistently reported lateralization, and some reports suggesting association with autoimmune encephalitis and glioma-associated seizures [16].

15.3.5.3 Respiratory Manifestations

Hyperventilation often accompanies other autonomic signs. Ictal hypoventilation is thought to be centrally mediated and is often associated with oxygen desaturation [17]. This can proceed to ictal apnea. There is no clear lateralizing/localizing value to hypoventilation, although localization may be related to mesial temporal [18] or brainstem involvement. Laryngeal constriction and ictal choking can occur with insular seizures [19].

15.3.5.4 Gastrointestinal Symptoms

Epigastric aura is the most common autonomic symptom in adults with epilepsy, occurring most commonly in temporal lobe epilepsy [20], although they have been reported in frontal or insular seizures. Ictal hypersalivation is thought to be mesial temporal, although it is non-lateralizing. Ictal spitting, meanwhile, is more consistently non-dominant mesial temporal [21]. Ictal vomiting is common in childhood occipital epilepsies, but can also be seen in adults in mesial temporal seizures, likely from insular propagation [22], without a clear lateralizing value.

15.3.5.5 Genitourinary Manifestations

Urinary urgency has been reported with non-dominant temporal seizures [23]. Sexual arousal or genital automatisms including fondling or scratching the genitals are typically thought to be non-dominant temporal in localization [24].

15.3.5.6 Pupillary Findings

Unilateral pupillary dilatation (mydriasis) is thought to be associated with ipsilateral temporal or occipitotemporal seizures [25].

15.4 Semiology in Lobar Epilepsy

15.4.1 Temporal Lobe

Temporal lobe epilepsy is the most common focal epilepsy in adults and can present with varied semiology, including ipsilateral and contralateral findings of localizing significance, as noted in Fig. 15.4. Features suggestive of non-dominant temporal seizures including gastro-intestinal/genito-urinary symptoms are also mentioned.

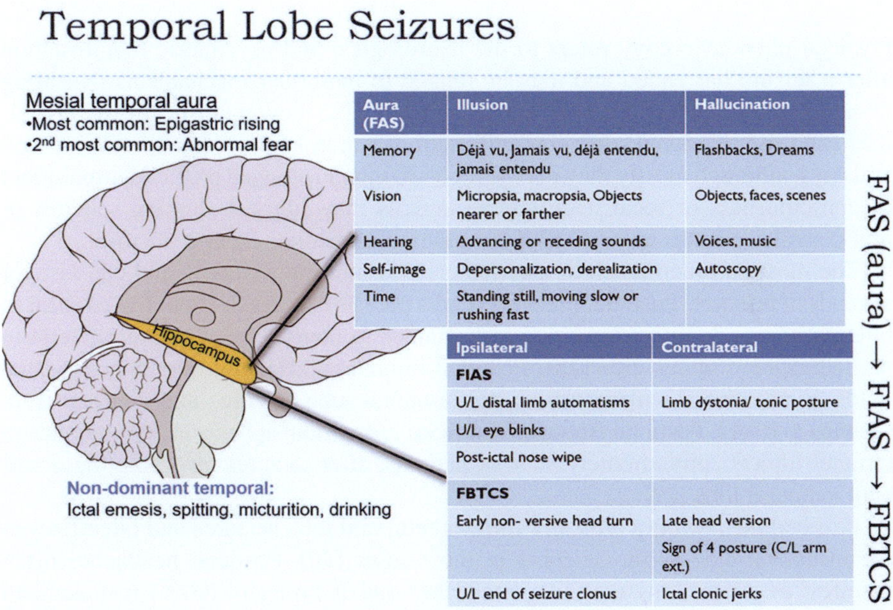

Fig. 15.4 Semiology in of temporal lobe seizures and spread from focal to bilateral involvement. *FAS* focal aware seizure, *FIAS* focal impaired awareness seizure, *FBTCS* focal to bilateral tonic–clonic seizure. (Modified from images by Patrick J. Lynch, medical illustrator CC-BY-2.5 via Wikimedia Commons)

Dystonia is typically contralateral to the focus and is thought to result from involvement of the ipsilateral basal ganglia. Automatisms are not lateralizing in the absence of dystonia. Contralateral dystonia restricts automatisms in that side and makes ipsilateral (in the non-dystonic limb) automatisms more prominent.

15.4.2 Frontal Lobe

Frontal lobe epilepsies can be conceptualized as precentral, premotor, and prefrontal [26, 27]. *Precentral cortex* epilepsies demonstrating elementary motor behaviors such as clonic jerks or tonic posturing. *Premotor* epilepsies include symptoms such as proximal symmetric or asymmetric tonic posturing, head/eye version, and axial/proximal complex motor behaviors. *Prefrontal* epilepsies include more complex patterns of gestural motor behaviors, which can have naturalistic appearances, such as bicycling movements and can also have emotional features. There is commonly alteration in awareness with vocalization. More anterior prefrontal structures will also typically include distal more than proximal stereotypies.

15.5 Postictal Phase

Postictal phenomenology refers to the neurological and psychiatric manifestation after a seizure has ended and may be related to inhibitory mechanisms that bring about ictal termination. These often accompany EEG slowing or postictal generalized EEG suppression. Postictal manifestations can last from a few minutes to several days, although mostly they persist for 5–30 min. Prolonged postictal impairment of consciousness or focal neurological deficits may suggest ongoing seizures or non-convulsive status epilepticus, necessitating immediate EEG evaluation.

The most common postictal phenomenon is unresponsiveness and can last for seconds to minutes. Postictal paresis (Todd's paresis) occurs in about 6% of focal to bilateral tonic–clonic seizures and last minutes to hours, typically contralateral to the hemisphere of seizures [28]. Postictal blindness, affecting one or both visual fields, is associated with occipital and temporal lobe seizures and can last from seconds to hours. Postictal Broca's, Wernicke's, or global aphasia lasting seconds to minutes affects approximately 40% of patients, after seizures involving the dominant temporal lobe [29].

Postictal nose wiping typically follows temporal lobe seizures and lateralizes to the ipsilateral side as the seizures in most cases [30]. Postictal headaches often manifest as severe, migraine-like headaches and is typically (90%) ipsilateral to seizure involving the temporal lobe but not following extratemporal seizures [31].

Postictal psychiatric symptoms, including delirium, psychosis, catatonia, impaired cognition, and amnesia, can last from hours to days. Postictal psychosis, occurring in 2–7% of patients, often follows a lucid interval of 6 h to a week and can last from 12 h to 3 months without treatment. About half the patients with postictal psychosis may have an underlying psychiatric disorder [32]. Postictal psychosis is treatable and can lead to morbidity and mortality if left untreated [33].

Postictal autonomic signs include tachycardia, bradycardia, hypersalivation, apnea, and changes in blood pressure, with respiratory depression and cardiac dysfunction being critical factors in SUDEP [34].

Pearls: Semiology

- Motor semiology can be simple or complex. Simple motor seizures can be tonic, clonic, tonic–clonic, atonic, and myoclonic.
- Myoclonic jerks differ from clonic motor activity in being arrhythmic. Clonic seizures are rhythmic.
- Cortical stimulation studies have shown auras to arise from areas beyond the conventional cortical locations for the sensations [35].
- About 14% neurons in the trigger zone increase their firing rate with an aura, compared to 7% with a subclinical seizure and 36% with a focal impaired awareness seizure (FIAS) [36].
- Tonic seizures are thought to arise from simultaneous bilateral activation of Brodmann area 6 (rostral to precentral gyrus) [37].
- Ictal/postictal headache occurs with benign epilepsy of childhood with occipital paroxysms and with occipital seizures of Lafora disease [37].
- "Dialeptic seizures" is a term sometimes used to describe seizures with both loss of awareness and hypomotor activity. It serves as an umbrella category for both generalized absence and focal dyscognitive seizures. However, this term is not part of the ILAE seizure classification scheme (Chap. 1).

Acknowledgements We would like to thank Atul Maheshwari for authoring a previous version of this chapter.

References

1. Jan MMS, Girvin JP. Seizure semiology: value in identifying seizure origin. Can J Neurol Sci. 2008;35:22–30.
2. Engel J, Pedley TA, Aicardi J. Epilepsy: a comprehensive textbook. Lippincott Williams & Wilkins; 2008.
3. Foldvary-Schaefer N, Unnwongse K. Localizing and lateralizing features of auras and seizures. Epilepsy Behav. 2011;20:160–6.
4. Beniczky S, et al. Seizure semiology: ILAE glossary of terms and their significance. Epileptic Disord. 2022;24:447–95.
5. Tufenkjian K, Lüders HO. Seizure semiology: its value and limitations in localizing the epileptogenic zone. J Clin Neurol. 2012;8:243.
6. Janszky J, et al. Unilateral hand automatisms in temporal lobe epilepsy. Seizure. 2006;15:393–6.
7. Aupy J, et al. Insulo-opercular cortex generates oroalimentary automatisms in temporal seizures. Epilepsia. 2018;59:583–94.

8. Isnard J, Guenot M, Sindou M, Mauguiere F. Clinical manifestations of insular lobe seizures: a stereo-electroencephalographic study. Epilepsia. 2004;45:1079–90.
9. Singh R, et al. Characteristics and neural correlates of emotional behavior during prefrontal seizures. Ann Neurol. 2022;92:1052–65.
10. Gloor P, Olivier A, Quesney LF, Andermann F, Horowitz S. The role of the limbic system in experiential phenomena of temporal lobe epilepsy. Ann Neurol. 1982;12:129–44.
11. Loesch AM, et al. Seizure-associated aphasia has good lateralizing but poor localizing significance. Epilepsia. 2017;58:1551–5.
12. Leung H, Schindler K, Kwan P, Elger C. Asystole induced by electrical stimulation of the left cingulate gyrus. Epileptic Disord. 2007;9:77–81.
13. Eggleston KS, Olin BD, Fisher RS. Ictal tachycardia: the head-heart connection. Seizure. 2014;23:496–505.
14. Fogarasi A, Janszky J, Tuxhorn I. Autonomic symptoms during childhood partial epileptic seizures. Epilepsia. 2006;47:584–8.
15. Fogarasi A, Janszky J, Tuxhorn I. Ictal pallor is associated with left temporal seizure onset zone in children. Epilepsy Res. 2005;67:117–21.
16. Franco AC, Noachtar S, Rémi J. Ictal ipsilateral sweating in focal epilepsy. Seizure. 2017;50:4–5.
17. Pavlova M, et al. Advantages of respiratory monitoring during video-EEG evaluation to differentiate epileptic seizures from other events. Epilepsy Behav. 2014;32:142–4.
18. Lacuey N, et al. Ictal central apnea is predictive of mesial temporal seizure onset: an intracranial investigation. Ann Neurol. 2024;95:998–1008.
19. Geevasinga N, Archer JS, Ng K. Choking, asphyxiation and the insular seizure. J Clin Neurosci. 2014;21:688–9.
20. Henkel A, Noachtar S, Pfänder M, Lüders HO. The localizing value of the abdominal aura and its evolution: a study in focal epilepsies. Neurology. 2002;58:271–6.
21. Kellinghaus C, Loddenkemper T, Kotagal P. Ictal spitting: clinical and electroencephalographic features. Epilepsia. 2003;44:1064–9.
22. Catenoix H, et al. The role of the anterior insular cortex in ictal vomiting: a stereotactic electroencephalography study. Epilepsy Behav. 2008;13:560–3.
23. Gurgenashvili K, et al. Intracranial localisation of ictal urinary urge epileptogenic zone to the non-dominant temporal lobe. Epileptic Disord. 2011;13:430–4.
24. Aull-Watschinger S, Pataraia E, Baumgartner C. Sexual auras: predominance of epileptic activity within the mesial temporal lobe. Epilepsy Behav. 2008;12:124–7.
25. Masjuan J, García-Segovia J, Barón M, Alvarez-Cermeño JC. Ipsilateral mydriasis in focal occipitotemporal seizures. J Neurol Neurosurg Psychiatry. 1997;63:810–1.
26. Bonini F, et al. Frontal lobe seizures: from clinical semiology to localization. Epilepsia. 2014;55:264–77.
27. McGonigal A. Frontal lobe seizures: overview and update. J Neurol. 2022;269:3363–71.
28. Rolak LA, Rutecki P, Ashizawa T, Harati Y. Clinical features of Todd's post-epileptic paralysis. J Neurol Neurosurg Psychiatry. 1992;55:63–4.
29. Privitera M, Kim KK. Postictal language function. Epilepsy Behav. 2010;19:140–5.
30. Leutmezer F, et al. Postictal nose wiping: a lateralizing sign in temporal lobe complex partial seizures. Neurology. 1998;51:1175–7.
31. Yankovsky AE, Andermann F, Bernasconi A. Characteristics of headache associated with intractable partial epilepsy. Epilepsia. 2005;46:1241–5.
32. Kanner AM, Soto A, Gross-Kanner H. Prevalence and clinical characteristics of postictal psychiatric symptoms in partial epilepsy. Neurology. 2004;62:708–13.
33. Devinsky O. Postictal psychosis: common, dangerous, and treatable. Epilepsy Curr. 2008;8:31–4.
34. Ryvlin P, et al. Incidence and mechanisms of cardiorespiratory arrests in epilepsy monitoring units (MORTEMUS): a retrospective study. Lancet Neurol. 2013;12:966–77.
35. Schulz R, et al. Localization of epileptic auras induced on stimulation by subdural electrodes. Epilepsia. 1997;38:1321–9.

36. Babb TL, Wilson CL, Isokawa-Akesson M. Firing patterns of human limbic neurons during stereoencephalography (SEEG) and clinical temporal lobe seizures. Electroencephalogr Clin Neurophysiol. 1987;66:467–82.
37. Wyllie E, Cascino GD, Gidal BE, Goodkin HP. Wyllie's treatment of epilepsy: principles and practice. Lippincott Williams & Wilkins; 2012.

Neuroradiology

16

Zulfi Haneef and David Chen

Both structural and functional imaging techniques are important in the diagnosis and treatment of epilepsy. Structural imaging techniques include CT and MRI, but these sometimes are not as informative in identifying the epileptogenic zone as functional imaging techniques. Functional imaging techniques include positron emission tomography (PET), ictal single photon emission computed tomography (ictal SPECT), functional MRI (fMRI), and magnetoencephalography/magnetic source imaging (MEG/MSI). A combination of DTI (a structural MRI technique) and fMRI has been found useful in many centers for identifying abnormalities and assessing the risk of deficits from surgical treatment. Other promising techniques include functional connectivity magnetic resonance imaging (fcMRI), magnetic resonance spectroscopy (MRS), arterial spin labeling (ASL), near infra-red spectroscopy (NIRS) and magentonanoparticle (MNP) imaging [1].

16.1 Magnetic Resonance Imaging (MRI)

Along with video-EEG, structural MRI is a cornerstone in the evaluation of epilepsy surgery. Identifying an MRI abnormality benefits the determination of surgical candidacy and also affects the likelihood of surgical success. When MRI is concordant with EEG in TLE, the surgical success rate is 82%, while it is only 56% when MRI is non-concordant [2].

A high-resolution MRI with thin sections through the area of interest is optimal for identifying the sometimes subtle abnormalities that can cause epilepsy. Hence, negative results from low-field strength scanners (e.g., some open MRI machines) and suboptimal imaging sequences should generally be repeated. Contrast-enhanced

Z. Haneef (✉) · D. Chen
Department of Neurology, Baylor College of Medicine & Neurology Care Line, Houston VA Medical Center, Houston, TX, USA
e-mail: Zulfi.Haneef@bcm.edu

Z. Haneef (ed.), *Epilepsy Fundamentals*,
https://doi.org/10.1007/978-3-031-77741-7_16

MRI should be performed for patients with new-onset epilepsy for acute/subacute lesions that cause seizures.

16.1.1 Temporal Lobe Epilepsy

Coronal MRI slices in mTLE are acquired perpendicular to the long axis of the hippocampus (Fig. 16.1). The cardinal MRI features of MTS are hippocampal atrophy (best in T1 volumetric sequences) and hippocampal hyperintensity in T2/FLAIR with loss of the internal architecture. Secondary features of hippocampal atrophy include loss of hippocampal head digitations, enlarged temporal horn, and atrophy of Papez circuit regions, including fornix, amygdala, mammillary body and thalamus. Quantitative measurement using hippocampal volumetry and T2 relaxometry may have increased sensitivity over visual analysis but have not found their way into routine clinical practice due to the complexity of such analyses [3].

16.1.2 Dual Pathology

Dual pathology (Fig. 16.2) refers to the dual findings of (1) hippocampal sclerosis and (2) another extra-hippocampal pathology, the most common of which are cortical malformations (e.g., cortical dysplasia) [4, 5]. Dual pathology occurs in 5–20% of patients with refractory TLE and can be diagnosed using MRI [6]. In the setting of dual pathology, the best resective surgical outcome is obtained by removing both the hippocampus and the extra-hippocampal lesion [6]. Neuromodulation may also be appropriate in this situation.

16.1.3 Neocortical Epilepsy

In frontal lobe epilepsy, the surgical success rate of 72% when MRI is concordant falls to 41% when MRI is non-concordant [7]. Lesional epilepsies are often due to

Fig. 16.1 Hippocampal coronal slice direction (solid lines) perpendicular to its long axis (dashed line)

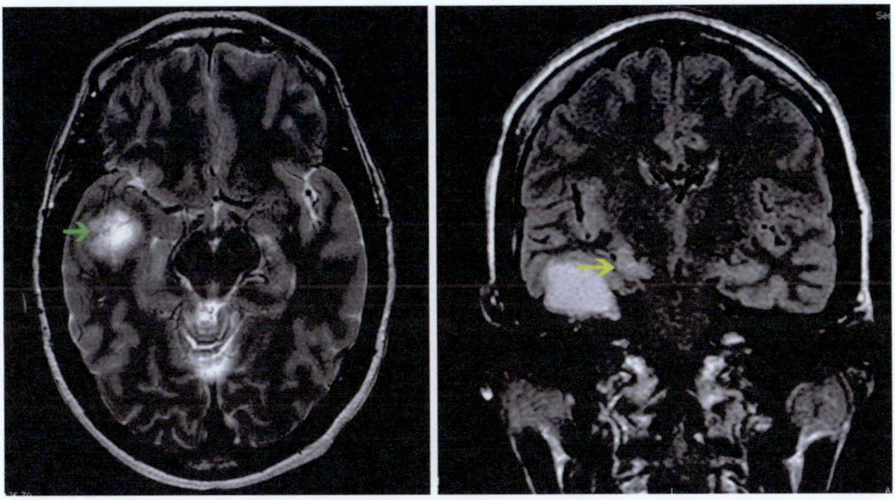

Fig. 16.2 Dual pathology showing a right temporal meningioma (green arrow) and associated right mesial temporal sclerosis (yellow arrow)

tumors, vascular malformations or developmental abnormalities. Low-grade tumors causing epilepsy include dysembryoplastic neuroepithelial tumors, ganglion cell tumors (ganglioglioma, gangliocytoma), and pleomorphic xanthoastrocytoma. These are typically found in the temporal lobe. Vascular malformations causing seizures include cavernomas (best seen as "popcorn" lesions in gradient echo sequences, Fig. 16.3), arteriovenous malformations and Sturge-Weber syndrome (Fig. 16.4). Developmental venous angiomas (DVAs) are often seen in neuroimaging but are thought to be incidental findings that are not epileptogenic. However, they sometimes co-localize with cortical dysplasia or cavernomas, both of which can be epileptogenic.

Tuberous sclerosis often leads to neocortical epilepsy and is associated with widespread hamartomas. This condition can affect multiple internal organs, leading to (1) cortical or subependymal tubers and white matter abnormalities, (2) renal angiomyolipomas, and (3) cardiac rhabdomyomas. The tubers are mostly in the frontal lobe (50%). Subependymal hamartomas (Fig. 16.6) and sub-ependymal giant cell astrocytomas (SEGA) can occur and cannot be distinguished by signal characteristics or enhancement. Evidence of growth in serial scans is the distinguishing feature of SEGA.

16.1.4 Developmental Abnormalities

Neuronal migration abnormalities can include schizencephaly, porencephaly, lissencephaly, agyria, macrogyria, polymicrogyria, pachygyria, microgyria, neuronal heterotopias (including band heterotopia), and agenesis of the corpus callosum.

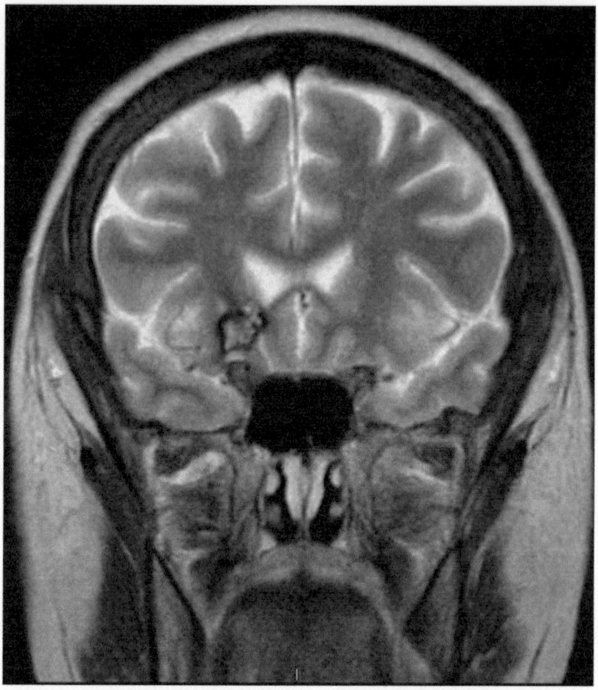

Fig. 16.3 "Popcorn" lesion in cavernous angioma

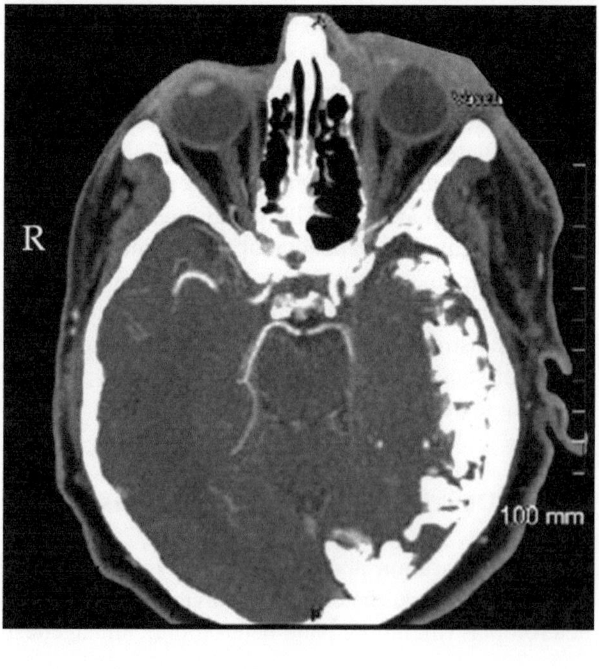

Fig. 16.4 Sturge-Weber syndrome with bilateral calcifications

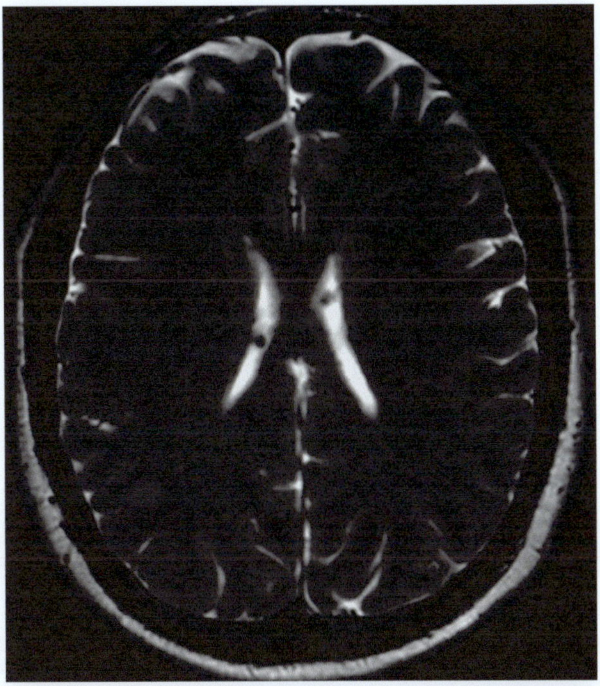

Fig. 16.5 Bilateral subependymal nodules (tuberous sclerosis)

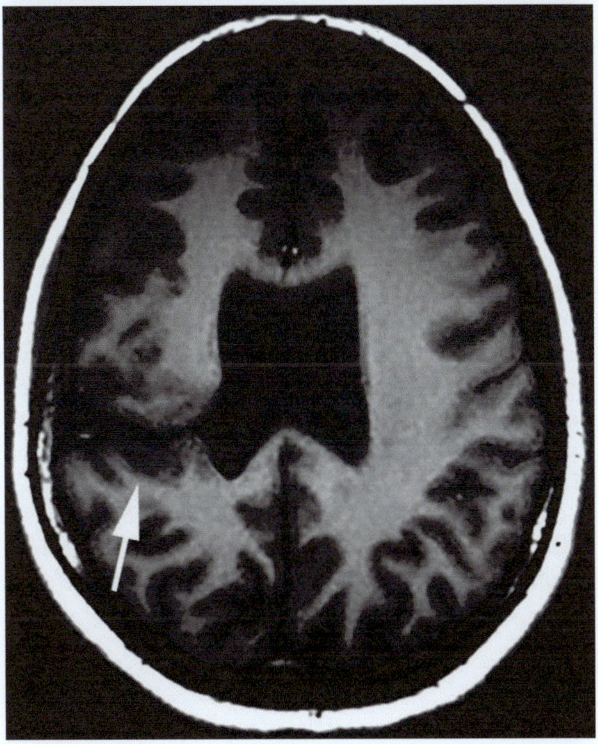

Fig. 16.6 Open lip schizencephaly in a patient with seizures

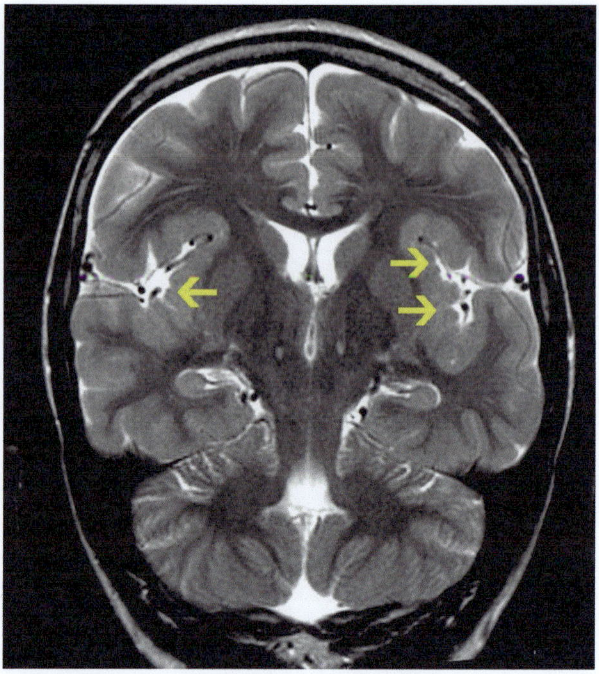

Fig. 16.7 Bilateral perisylvian polymicrogyria. (Image courtesy: Joyce Matsumoto, UCLA, CA)

In **schizencephaly** (Fig. 16.5), there is an abnormal communication between the ventricles and the subarachnoid space, which is lined by gray matter. Structural abnormalities such as polymicrogyria (Fig. 16.8) and heterotopia are best seen in T1 volumetric sequences. **Lissencephaly** can consist of a simplified gyral folding pattern or absence of gyri (**agyria**). This can lead to a thickened appearance of the gyri (**pachygyria**). In **polymicrogyria**, there are multiple small gyri in the gray-white junction appearing as serrations (Fig. 16.8).

Heterotopias (Fig. 16.9) represent normal tissue at an abnormal location, typically gray matter abnormally located within white matter and in the line of radial fiber migration from the ventricle to the superficial cortex. These may be present as nodules around the ventricles (peri-ventricular nodular heterotopia), subjacent to the cortex (subcortical), or as a thick band beneath the gray matter (band heterotopia; double cortex).

Hypothalamic hamartomas are typically seen as projections into the third ventricle and are nearly iso-intense on MRI. **Hemimegalencephaly** (Fig. 16.7) is a special type of hamartomatous overgrowth of one hemisphere of the brain

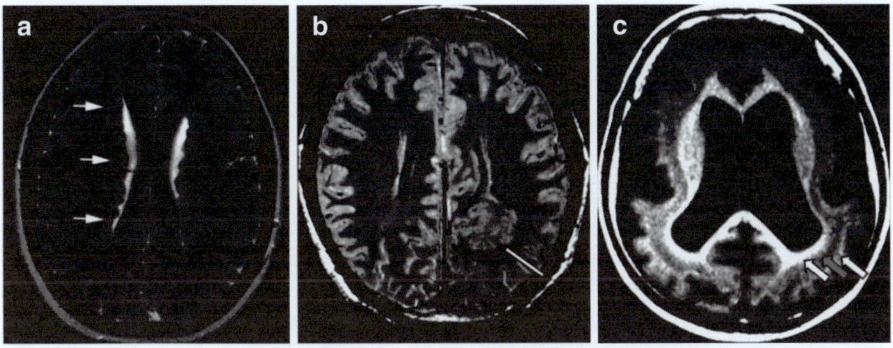

Fig. 16.8 Heterotopia spectrum: (**a**) peri-ventricular (**b**) focal subcortical (**c**) band (double cortex)

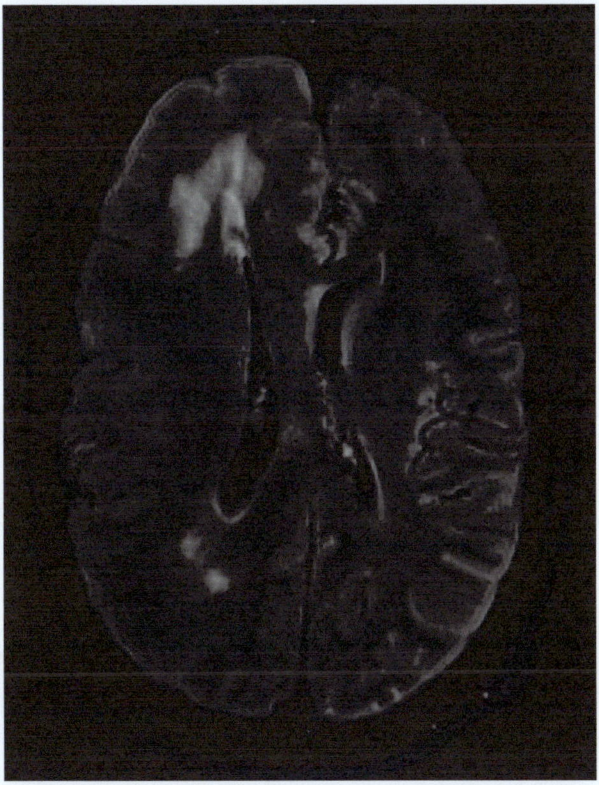

Fig. 16.9 Hemimegalencephaly (right larger than left)

associated with early onset medication refractory seizures, for which hemispherectomy is often required.

Focal cortical dysplasia (FCD) is a common abnormality in patients with TLE [8]. Radiological features can include (1) cortical thickening, (2) subcortical white matter T2/FLAIR signal hyperintensities, which may extend from the ventricle to the cortex termed the "transmantle sign" (Fig. 16.10), and (3) blurred grey-white

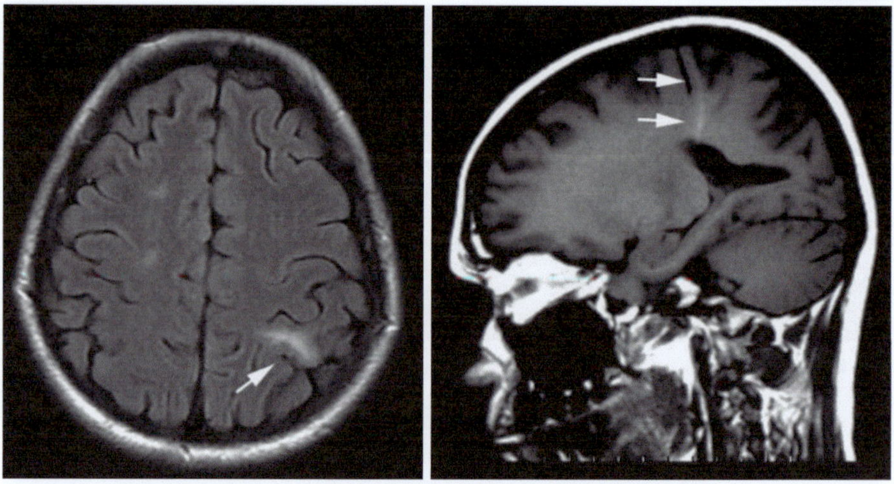

Fig. 16.10 Left parietal cortical dysplasia showing the "transmantle sign"

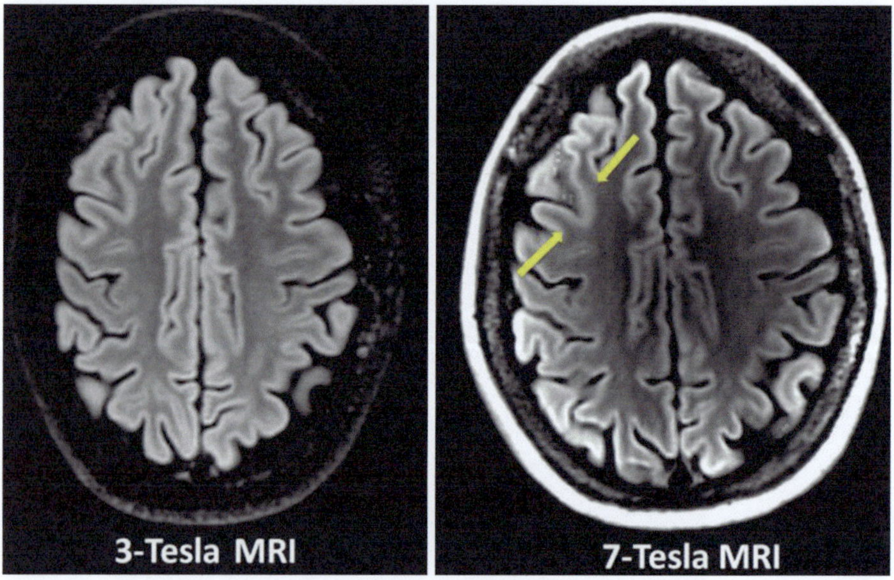

Fig. 16.11 7-Tesla MRI FLAIR sequence shows indistinct gray-white junction (yellow arrows) in a patient with cortical dysplasia, which was not seen in a corresponding 3-Tesla MRI. (Image courtesy: Dr. Renzo Guerrini, University of Florence, Italy)

matter interface, which may be better appreciated at higher magnet strengths (Fig. 16.11) [3]. The transmantle sign is thought to be due to a structural or functional abnormality of radial glial fibers (that later form astrocytes) that provides the scaffolding for neuronal migration from the periventricular germinal matrix to the cortex.

16.2 Positron Emission Tomography (PET)

PET typically studies regional brain glucose utilization as a measure of neuronal activity using radioactive fluorine-18 fluorodeoxy-glucose (^{18}FDG-PET). FDG is transported into tissues and phosphorylated similar to glucose, but accumulates as it cannot undergo further metabolism. As FDG decays, it emits positrons, which provide images indicating glucose utilization within the brain. FDG-PET and MRI images can be co-registered and color-graded with colors reflecting a difference in FDG uptake (Fig. 16.12).

Epileptogenic brain regions are often hypometabolic in inter-ictal FDG-PET imaging and are detected in 60–90% of patients with TLE [9–11], and 67% with extratemporal lobe epilepsy [11]. FDG-PET has an overall diagnostic sensitivity of

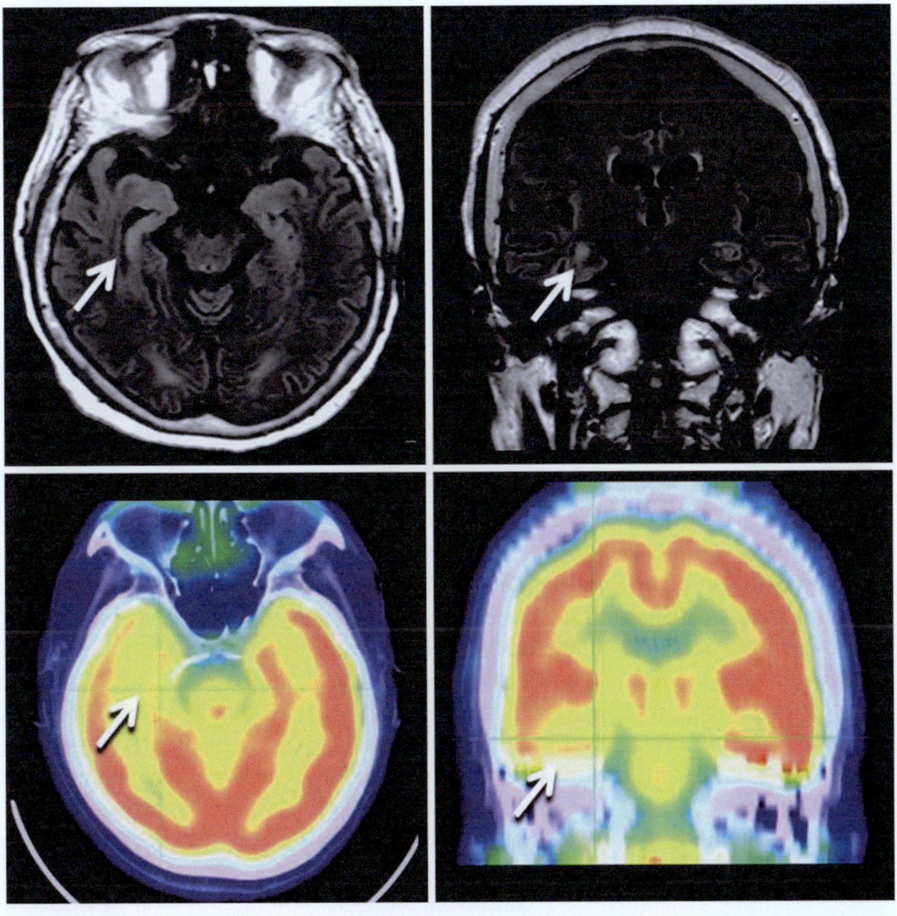

Fig. 16.12 Right mesial temporal lobe epilepsy showing atrophy and hyperintensity in MRI (FLAIR, top) and hypometabolism in PET images (bottom)

44% in detecting "non-lesional" epileptogenic substrates [12]. FDG-PET, when done during or shortly following an epileptic seizure, often reveals complex patterns of both hypometabolism and hypermetabolism.

In addition to FDG-PET, quantification of specific neurotransmitter-receptor relationships is possible using PET using other radioligands (e.g., [11 C] flumazenil with **GABA receptors**, [18 F] 2′-methoxyphenyl-(N-2′-pyridinyl)-p-fluorobenzamidoethyi-piperazine with **serotonin** receptors, [F] Fallypride with **dopamine** receptors, and opioid-based ligands on **opioid** receptors) [1].

16.3 Single Photon Emission Computed Tomography (SPECT)

Brain perfusion SPECT is a functional neuroimaging modality that compares differences in regional cerebral blood flow and is performed in the ictal and the inter-ictal periods during the epilepsy surgery workup. Seizures cause local hyperperfusion of the seizure onset zone, which is detected in ictal SPECT.

During ictal SPECT, injection of a radionuclide tracer (typically 99m Tc-hexa-methyl-propyleneamine-oxime or Tc-ethyl cysteinate dimer) is delivered as soon as possible to the time of ictal onset and preferably within 20 s of the onset. As such, ictal SPECT requires video-EEG monitoring and logistical support to ensure prompt tracer injection. After injection, the tracers are distributed in proportion to cerebral blood flow, and can indicate the region of ictal onset because of the increased perfusion to such regions. After cerebral uptake and conversion to intracellular metabolites, there is minimal redistribution or washout and image acquisition can theoretically be completed up to 4 h after injection. An asymmetry of more than 10% has been considered abnormal (Fig. 16.13). It has been claimed that ictal SPECT correctly identifies the seizure focus in >90% in TLE [13–15], while extra-temporal epilepsy has a lower sensitivity (66%) [16].

In mTLE, ictal SPECT hyperperfusion involves the temporal pole and mesial temporal structures, while the lateral temporal neocortex is more variably involved. In lateral TLE, bilateral temporal lobe hyperperfusion (greater ipsilaterally) or ipsilateral posterolateral temporal hyperperfusion may occur. Due to the normal interval between injection and brain uptake (15–20 s), or due to delayed injection, ictal SPECT image may show areas of seizure propagation in addition to the seizure onset zone. A "postictal switch phenomenon" has also been described with delayed injection, where the seizure onset zone switches to hypoperfusion, while seizure propagation regions become hyperperfused, leading to false localization. Common regions of seizure propagation in TLE include the ipsilateral insula, basal ganglia, frontal lobe, and contralateral temporal lobe. Lower sensitivity in extratemporal epilepsy is possibly related to rapid seizure propagation and secondary generalization. One example of false localization is the hyperperfusion of the temporal lobe that occurs in occipital lobe epilepsy.

The accuracy of ictal SPECT can be enhanced by subtraction ictal SPECT co-registered to MRI (SISCOM) analysis. Here, computer-aided coregistration is used to subtract the inter-ictal SPECT from the ictal SPECT, which is then co-registered

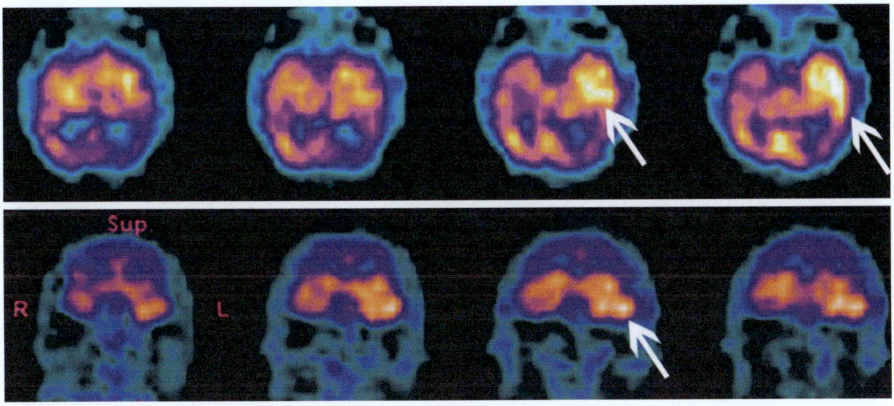

Fig. 16.13 Ictal SPECT showing increased tracer uptake in the left temporal lobe

to the patient's brain MRI. Inter-ictal SPECT by itself has been considered insufficient to confidently localize the seizure focus as it has a sensitivity of only about 50% in identifying the seizure onset zone [13–15].

16.4 Diffusion Tensor Imaging (DTI)

DTI is a structural neuroimaging technique giving detailed information regarding highly organized tissue such as the white matter. DTI evolved from MRI diffusion-weighted imaging (DWI) protocols, which detect the motion of water molecules and is often used clinically to identify cerebral ischemia. While DWI measures water movement in three directions, DTI uses computational methods to integrate diffusion data in more directions (usually 7 or more). These measurements are summarized by a tensor (ellipsoid) model that calculates how far water molecules diffuse in all directions from a point. This "diffusion tensor" is a sensitive detector of tissue microstructural pathology that limits water molecular diffusion.

Common measurements in DTI include fractional anisotropy (FA), which measures the degree of anisotropy (isotropy = equal motion in all directions, anisotropy = directionally constrained motion), and mean diffusivity (MD), which measures the amplitude of diffusion motion. Another method of DTI analysis is fiber tractography (FT, Fig. 16.14), which uses computational algorithms to determine the likely fiber patterns that pass through one or multiple selected regions. Compared to FA and MD, FT is less operator-dependent and more robust.

In TLE, hippocampal cell loss and architectural disruption may cause expanded extracellular space, increasing mean diffusivity. Hippocampal gliosis (sclerosis) may cause architectural distortion and decreased FA. There is a decrease in FA and an increase in MD in multiple white matter regions in TLE in bilateral hemispheres. Tracts closely connected with the affected temporal lobe are most disturbed [17]. DTI may reveal temporal lobe abnormalities that are not detected on conventional

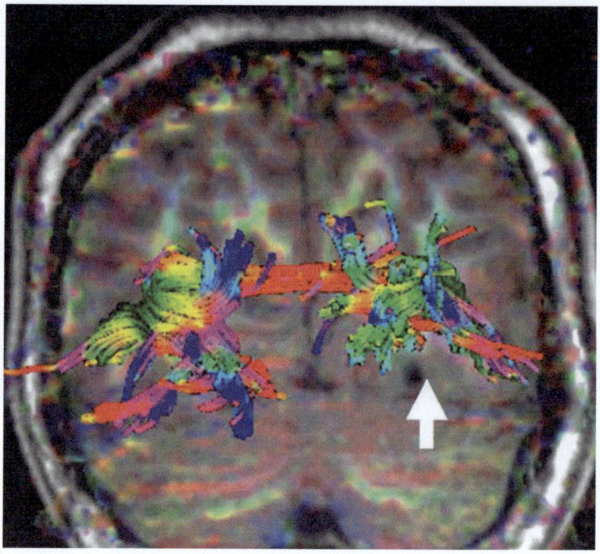

Fig. 16.14 DTI tractography: DTI showing visual fibers superiorly displaced by a dysplastic lesion

MRI. DTI tractography can also assist in surgical decision-making by mapping functionally important fiber tracks. In one study, DTI tractography influenced the surgical procedure in 32% of frontal lobe resections, 25% of occipital resections and 67% of parietal or multilobar resections.

16.5 Functional MRI

Functional MRI exploits the differential T2-weighted magnetic susceptibilities of paramagnetic deoxyhemoglobin and diamagnetic oxyhemoglobin to visualize activity in the cerebral cortex. This difference in magnetic susceptibility can be detected in T2 imaging and is called blood oxygenation level-dependent (BOLD) contrast. Normal brain activity causes reactive vasodilatation that overcompensates the metabolic demand, increasing oxyhemoglobin and the T2 relaxation time in brain areas activated by the task.

Language fMRI is the most common fMRI paradigm used in epilepsy. It has been proposed as a reliable surrogate marker for the more invasive intracarotid amobarbital procedure (Wada test) for memory lateralization. Studies comparing Wada findings to language fMRI have found fMRI to be a viable and non-invasive alternative. Several epilepsy centers use language fMRI in lieu of Wada lateralization, although this approach is disputed by others. Choosing the appropriate fMRI paradigm is important for good results. Verb generation tasks (i.e., think of an action associated with an object. e.g., think of "type" when shown a "keyboard") activate the Broca's area, and semantic/reading tasks activate the posterior language areas. A panel of language tasks is more reliable than a single test. Memory fMRI does not lead to robust activations and is a field of continuing research [1].

16.6 Additional Techniques

Additional techniques are discussed below, and varying degrees of utility have been found in epilepsy surgery workups.

16.6.1 EEG-fMRI

EEG-fMRI is a special application of fMRI that incorporates information from EEG. Here, the patient lies in the MRI scanner with an electrode cap that captures EEG simultaneously while the fMRI is being acquired. It has the advantage of combining the structural resolution of fMRI (~3 mm^3) with the temporal resolution of EEG (milliseconds). fMRI data is modeled based on epileptiform discharges detected in simultaneous EEG in a fashion analogous to fMRI being modeled on verb generation tasks described earlier. EEG-fMRI has been found to be used in some surgical epilepsy programs: it has been reported to be concordant with scalp EEG in 88% of cases and contributes additional information about the epileptic focus in 64% [18]. Simultaneous intracranial EEG recordings and fMRI have also been attempted.

16.6.2 Near-Infrared spectroscopy (NIRS)

NIRS is a low-resolution functional imaging technique using near-infrared spectrum wavelength rays. A transmitter transmits these rays through the cranium to a depth of about 2 cm. Reflected rays following absorption by tissue hemoglobin are detected by a sensor probe and the strength of reflected rays is inversely related to the brain tissue hemoglobin concentration. The signal changes can lateralize and localize the affected hemisphere during seizures. NIRS has the advantage of being portable and easy to use compared to existing techniques. NIRS has also been explored for utility in language lateralization as an additional/alternative procedure to Wada and fMRI [1].

16.6.3 Arterial Spin Labeling (ASL)

ASL uses blood labeling for functional imaging of the brain. Arterial blood is magnetically labeled with a 180° RF pulse prior to imaging. This labeled blood flows into the region imaged with MRI. Subtracting a labeled MRI image from the baseline MRI creates a perfusion image, reflecting the quantity of blood delivered. ASL is non-invasive and repeatable. ASL shows mesial temporal hypometabolism on the side of seizures in TLE, similar to PET hypometabolism [1].

16.6.4 Magnetic Resonance spectroscopy (MRS)

MRS non-invasively measures brain metabolites such as N-acetyl aspartate (NAA), choline (Cho), creatine (Cr), lactate, myo-inositol and GABA. MRS may help localize the epileptogenic zone when conventional MRI is non-revealing. NAA/Cho and NAA/Cr have been reported to be decreased in the lesional temporal lobe in TLE and FLE. MRS changes have also been noted in generalized epilepsy [1].

16.6.5 Magneto Nanoparticles (MNP) Imaging

Magnetonanoparticles, which include iron oxide, are covalently attached to non-radioactive 2-deoxy-glucose. MRI scans are then performed to detect the MNP due to the magnetic properties of iron oxide. Similar to fluoro-deoxy glucose (FDG)-PET scans, there is reduced glucose uptake in abnormal epileptogenic regions in the interictal state and increased uptake with ictal activity. Among the different MNP compounds, 2-deoxy glucose has been approved for human use [1].

Pearls: Neuroimaging

- Routine MRIs are performed in patients with Tuberous Sclerosis to evaluate for Subependymal Giant Cell Astrocytomas (SEGAs).
- CT head in epilepsy related to celiac disease classically shows occipital calcifications.

References

1. Haneef Z, Chen DK. Functional neuro-imaging as a pre-surgical tool in epilepsy. Ann Indian Acad Neurol. 2014;17:S56–64.
2. Kuzniecky R, Burgard S, Faught E, Morawetz R, Bartolucci A. Predictive value of magnetic resonance imaging in temporal lobe epilepsy surgery. Arch Neurol. 1993;50:65–9.
3. De Ciantis A, Barba C, Tassi L, Cosottini M, Tosetti M, Costagli M, Bramerio M, Bartolini E, Biagi L, Cossu M, Pelliccia V, Symms MR, Guerrini R. 7T MRI in focal epilepsy with unrevealing conventional field strength imaging. Epilepsia. 2016;57(3):445–54. https://doi.org/10.1111/epi.13313.
4. Raymond AA, Fish DR, Stevens JM, Cook MJ, Sisodiya SM, Shorvon SD. Association of hippocampal sclerosis with cortical dysgenesis in patients with epilepsy. Neurology. 1994;44:1841–5.
5. Cendes F, Cook MJ, Watson C, Andermann F, Fish DR, Shorvon SD, Bergin P, Free S, Dubeau F, Arnold DL. Frequency and characteristics of dual pathology in patients with lesional epilepsy. Neurology. 1995;45:2058–64.
6. Li LM, Cendes F, Andermann F, Watson C, Fish DR, Cook MJ, Dubeau F, Duncan JS, Shorvon SD, Berkovic SF, Free S, Olivier A, Harkness W, Arnold DL. Surgical outcome in patients with epilepsy and dual pathology. Brain J Neurol. 1999;122(Pt 5):799–805.

7. Mosewich RK, So EL, O'Brien TJ, Cascino GD, Sharbrough FW, Marsh WR, Meyer FB, Jack CR, O'Brien PC. Factors predictive of the outcome of frontal lobe epilepsy surgery. Epilepsia. 2000;41:843–9.
8. Kabat J, Krol P. Focal cortical dysplasia—review. Pol J Radiol. 2012;77:35–43.
9. Drzezga A, Arnold S, Minoshima S, Noachtar S, Szecsi J, Winkler P, Römer W, Tatsch K, Weber W, Bartenstein P. 18F-FDG PET studies in patients with extratemporal and temporal epilepsy: evaluation of an observer-independent analysis. J Nucl Med Off Publ Soc Nucl Med. 1999;40:737–46.
10. Breier JI, Mullani NA, Thomas AB, Wheless JW, Plenger PM, Gould KL, Papanicolaou A, Willmore LJ. Effects of duration of epilepsy on the uncoupling of metabolism and blood flow in complex partial seizures. Neurology. 1997;48:1047–53.
11. Casse R, Rowe CC, Newton M, Berlangieri SU, Scott AM. Positron emission tomography and epilepsy. Mol Imaging Biol MIB Off Publ Acad Mol Imaging. 2002;4:338–51.
12. Lee SK, Lee SY, Kim KK, Hong KS, Lee DS, Chung CK. Surgical outcome and prognostic factors of cryptogenic neocortical epilepsy. Ann Neurol. 2005;58:525–32.
13. Spanaki MV, Spencer SS, Corsi M, MacMullan J, Seibyl J, Zubal IG. Sensitivity and specificity of quantitative difference SPECT analysis in seizure localization. J Nucl Med Off Publ Soc Nucl Med. 1999;40:730–6.
14. Devous MD, Thisted RA, Morgan GF, Leroy RF, Rowe CC. SPECT brain imaging in epilepsy: a meta-analysis. J Nucl Med Off Publ Soc Nucl Med. 1998;39:285–93.
15. Zaknun JJ, Bal C, Maes A, Tepmongkol S, Vazquez S, Dupont P, Dondi M. Comparative analysis of MR imaging, ictal SPECT and EEG in temporal lobe epilepsy: a prospective IAEA multi-center study. Eur J Nucl Med Mol Imaging. 2008;35:107–15.
16. Weil S, Noachtar S, Arnold S, Yousry TA, Winkler PA, Tatsch K. Ictal ECD-SPECT differentiates between temporal and extratemporal epilepsy: confirmation by excellent postoperative seizure control. Nucl Med Commun. 2001;22:233–7.
17. Otte WM, van Eijsden P, Sander JW, Duncan JS, Dijkhuizen RM, Braun KP. A meta-analysis of white matter changes in temporal lobe epilepsy as studied with diffusion tensor imaging. Epilepsia. 2012;53:659–67.
18. Pittau F, Dubeau F, Gotman J. Contribution of EEG/fMRI to the definition of the epileptic focus. Neurology. 2012;78:1479–87.

Anti-Seizure Medications

17

Rohit Marawar and Deepti Zutshi

About 70% of people with new-onset seizures do not have recurrent events. The risk of recurrence is greatest within the first 2 years. Guidelines by the American Academy of Neurology (AAN) and the American Epilepsy Society (AES) state the following variables increase the risk of seizure recurrence—prior brain insult, epileptiform abnormalities on EEG, any significant brain imaging abnormality, or nocturnal seizures. Whereas provoked seizures should not be treated as epilepsy (except reflex seizures), the AAN recommends treatment of any unprovoked seizure with an antiseizure medication (ASM) [1]. The risk of recurrent seizures and their impact and adverse effects of ASM(s) should be considered before starting an ASM.

17.1 Initial Management

Evaluation of a new onset seizure includes screening for metabolic and electrolyte disturbances such as hypoglycemia and hyponatremia, as well as alcohol and drug intoxication. EEG and head imaging (CT, MRI) are routinely performed in the evaluation of all new onset seizures. An AAN practice parameter recommends both EEG and neuroimaging after an *unprovoked* seizure (Level B recommendation) [2].

17.2 ASMs: Mechanism of Action

The vast majority of ASMs act either directly or indirectly on ion channels. The depolarization and excitation of a neuron occur with the inward flow of sodium (Na^+) and calcium (Ca^{++}) ions into the cell. In contrast, the hyperpolarization and inhibition of a neuron occurs with the inward flow of Cl^- and the outward flow of

R. Marawar (✉) · D. Zutshi
Department of Neurology, Wayne State University, Detroit, MI, USA
e-mail: ramarawar@med.wayne.edu

Z. Haneef (ed.), *Epilepsy Fundamentals*,
https://doi.org/10.1007/978-3-031-77741-7_17

K^+. Therefore, shifting the balance of excitation and inhibition toward greater inhibition can be achieved by blocking Na^+/Ca^{++} channels or enhancing Cl^-/K^+ channels (Figs. 17.1 and 17.2).

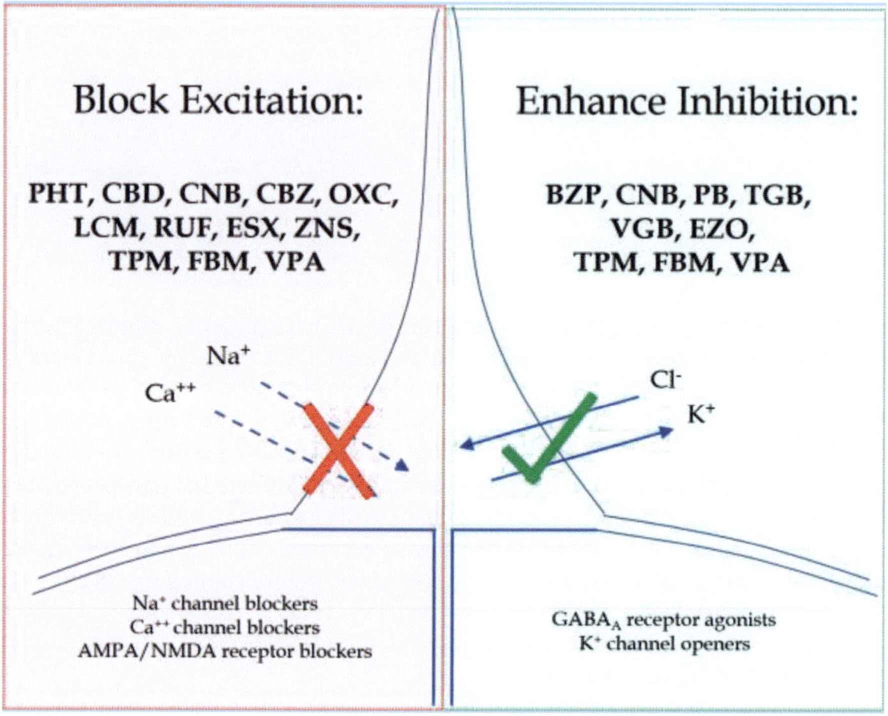

Fig. 17.1 Generic scheme for mechanism of ASM action

Fig. 17.2 Venn Diagram of ASM Mechanisms. Narrow-spectrum underlined, rest broad-spectrum

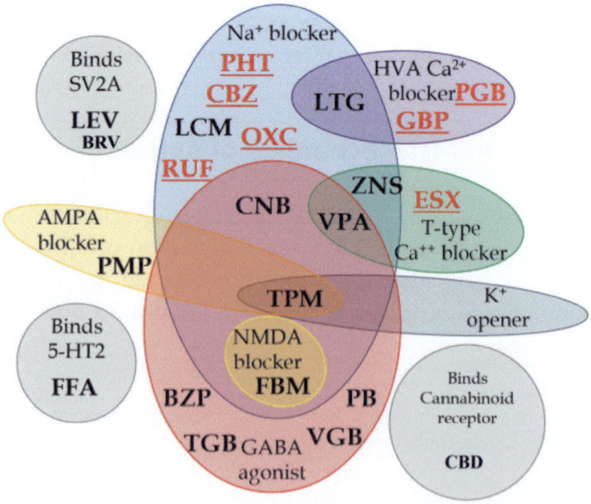

17.3 Choosing an ASM

In a new onset, provoked seizure, treating the underlying etiology should be sufficient. When chronic treatment is warranted, the choice of an ASM depends on multiple factors, starting with the type of seizures and, if known, the epilepsy etiology or syndrome (Table 17.1) [3]. Gender, age, adherence issues, renal and hepatic dysfunction and other comorbidities should also be considered when choosing an ASM. If the patient has evidence of generalized-onset seizures, narrow-spectrum medications should be avoided since they can *exacerbate* generalized seizures (in particular, absence and myoclonic seizures). An exception is a diagnosis of Childhood Absence Epilepsy, for which ESX is the medication of choice for generalized seizures, despite being a narrow-spectrum ASM (Fig. 17.3, Table 17.2)

Table 17.1 Quick reference table for ASM abbreviations

	Generic	Brand		Generic	Brand
BRV	Brivaracetam	Briviact	LTG	Lamotrigine	Lamictal
BZP	Benzodiazepine	N/A	OXC	Oxcarbazepine	Trileptal
CBD	Cannabidiol	Epidiolex	PB	Phenobarbital	Luminal
CBZ	Carbamazepine	Tegretol	PGB	Pregabalin	Lyrica
CLB	Clobazam	Onfi	PMD	Primidone	Mysoline
CLZ	Clonazepam	Klonopin	PHT	Phenytoin	Dilantin
CNB	Cenobamate	Xcopri	PMP	Perampanel	Fycompa
ESL	Eslicarbazepine	Aptiom	RUF	Rufinamide	Banzel
ESX	Ethosuximide	Zarontin	STP	Stiripentol	Diacomit
FBM	Felbamate	Felbatol	TGB	Tiagabine	Gabitril
FFA	Fenfluramine	Fintelpa	TPM	Topiramate	Topamax
GBP	Gabapentin	Neurontin	VGB	Vigabatrin	Sabril
LCM	Lacosamide	Vimpat	VPA	Valproate	Depakote
LEV	Levetiracetam	Keppra	ZNS	Zonisamide	Zonegran

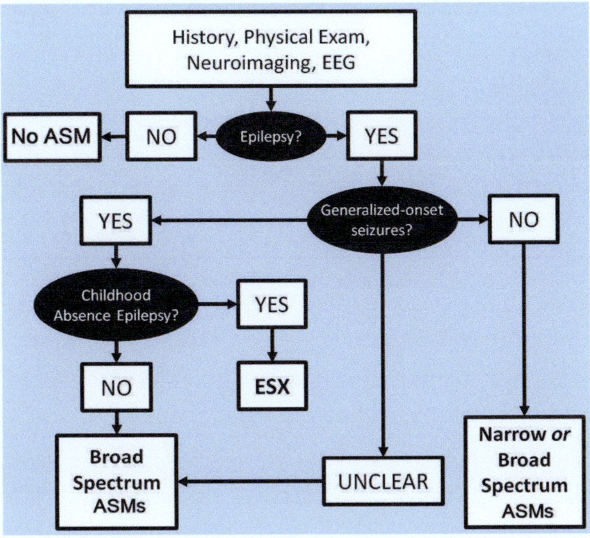

Fig. 17.3 Algorithm for choosing narrow- versus broad-spectrum ASMs

Table 17.2 Syndrome-specific ASM indications

Epileptic syndrome	ASM of choice
Childhood absence epilepsy	ESX > VPA/LTG/LEV
Juvenile myoclonic epilepsy	VPA/LEV/LTG > TPM/ZNS
Dravet syndrome	CLB/CBD/FFA
Lennox-Gastaut syndrome (LGS)	VPA/LTG/FBM/CLB/CBD (CBZ/TGB can lead to status epilepticus)
LGS with drop seizures	RUF
West syndrome	ACTH (VGB if tuberous sclerosis)

[4]. Table 17.3 provides a reference guide for all ASMs. The first ASM controls seizures in approximately half the patients, and the second ASM in about 15%. Approximately one-third of patients remain drug-resistant after multiple ASM trials (Fig. 17.4).

Table 17.3 Quick Reference Guide for ASMs

ASM	Mechanism of action (target/ channel)	Initial dose (mg)	Max dose (mg)	Peak (hrs)	t ½ (hrs)	Metabolism	Notable Side effects[c]
1st Gen							
PHT[a,b]	Na^+	100 tid	200 tid	4–12	22–24	P450 inducer; binds albumin, so the level low in hypoalbuminemia	Hepatotoxicity, marrow suppression, osteomalacia, arrhythmias, gum hypertrophy, hirsutism, neuropathy
CBZ	Na^+	200 bid	400 qid	4–8	5–26	Hepatic inducer, induces its own metabolism.	Agranulocytosis, hyponatremia, osteomalacia, diplopia
VPA[a,b]	GABA/ Na^+/ T-type Ca^{++}	500 bid	1000 tid	16	16	Hepatic, inhibitor of P450.	Tremor, hyperammonemia, hepatic failure, weight gain, hair loss, pancreatitis, thrombocytopenia
PB[a]	GABA	50 bid	100 tid	1–3	75–120	Hepatic, inducer of hepatic P450	Liver toxicity, cognitive slowing, osteomalacia, tolerance/ withdrawal, sedation, respiratory depression
ESX	T-type Ca^{++}	250 bid	750 bid	2–4	60	Hepatic	Blood dyscrasia, depression, GI upset, hiccups
2nd Gen							
LEV[a,b]	SV2A	500 bid	1500 bid	0.6–1.3	6–8	Hepatic (27%), renally excreted	Irritability, anxiety, depression, psychosis
LTG[b]	$Na^{+,}$ HVA Ca^{++}	25 qd-qod	200–350 bid	1–3	24–41	Hepatic; use higher doses with inducers and lower doses with VPA (enzyme inhibitor)	Skin rash (Stevens-Johnson Syndrome), insomnia, aseptic meningitis
OXC[b]	Na^+	300 bid	1200 bid	4	8–10	Hepatic	Hyponatremia (3% symptomatic), diplopia

(continued)

Table 17.3 (continued)

ASM	Mechanism of action (target/channel)	Initial dose (mg)	Max dose (mg)	Peak (hrs)	t ½ (hrs)	Metabolism	Notable Side effects[c]
ZNS	Na^+, T-type Ca^{++}, CA	100 qhs	400 qhs	2–4	60	Hepatic (70%)	Sulfa reaction, psychosis, weight loss, nephrolithiasis
TPM[b]	GABA, Na^+, K^+, AMPA, CA	25 bid	200 bid	2	18–23	Hepatic (P450)	Word finding difficulty, hypohydrosis, paresthesia, weight loss, metabolic acidosis, kidney stones, glaucoma
VGB	GABA transaminase	500 bid	1500 bid	2	4–7	Minimal hepatic metabolism	Permanent visual field deficit, weight gain, depression
GBP	L type Ca^{++}	300 tid	1200 tid	2–4	5–9	No hepatic metabolism, renally excreted	Weight gain, edema
PGB	L type Ca^{++}	75 bid	225 bid	1.5–3	6.3	No hepatic metabolism, renally excreted	Weight gain, edema, euphoria
FBM	Na^+, GABA, NMDA	400 tid	1200 tid	1–4	24	Hepatic	Hepatotoxicity, bone marrow toxicity, weight loss
TGB	GABA reuptake	4 mg qhs	16 tid	0.75	2–9	Hepatic P450 system, not inducer or inhibitor	Rash, seizures
CLB	GABA	5 bid	40 bid	0.5–4	36–42	Hepatic	Sedation (less than other BZP), tolerance/withdrawal
RUF	Na^+	200 bid	1600 bid	4–6	6–10	Non-CYP hydrolysis by carboxylesterases	Headache, short QT
3rd Gen							
LCM[a,b]	Slow Na^+	50 bid	200 bid	1–4	13	Hepatic	PR prolongation, A fib/ flutter, depression
PMP	AMPA	2 qhs	12 qhs	0.5–2.5	105	Hepatic	Irritability, aggression, homicidal intent
ESL	Na^+, T-type Ca^{++}	400 qhs	1200 qhs	1–4	13–20	Hepatic	Diplopia, hyponatremia
BRV[a]	SV2A	50 bid	100 bid	0.5–1	9	Hepatic, renally excreted	Sedation, dizziness, nausea, vomiting, irritability

STP [5]	GABA	50 mg/kg	N/A	1.5	4.5–14	Hepatic, renally excreted	Drowsiness, loss of appetite and weight, hyperexcitability
4th Gen							
CBD [6]	Cannabinoid receptors	2.5 mg/kg bid	10 mg/kg bid	2.5–5	56–61	Hepatic	Sedation, loss of appetite, diarrhea, transaminase elevation
CNB [7]	Na^+, GABA	12.5 mg daily	400 mg daily	1–4	50–60	Hepatic, renally excreted	Ataxia, dizziness, somnolence, diplopia
FFA [8]	5HT-2	0.1 mg/kg bid	0.35 mg/kg bid	4–5	20	Hepatic, renally excreted	Sedation, loss of appetite, diarrhea, serotonin syndrome

[a]Can give IV;

[b]Extended release versions (XR or ER) are available that can be dosed once a day with the same total daily dose, oxcarbazepine extended release has brand name Oxtellar, topiramate extended release has brand name Trokendi

[c]All ASMs have the potential for causing drowsiness, fatigue, dizziness, and rash, CA = carbonic anhydrase, 5HT2 = serotonin receptor

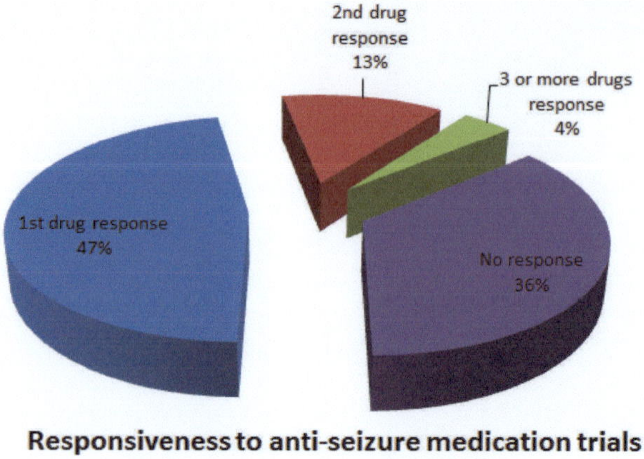

Fig. 17.4 Responsiveness of seizures to trials of one, two, three or more antiseizure medications [9]

17.4 Narrow Versus Broad Spectrum Medications

Narrow-spectrum drugs generally have one mechanism of action and are restricted to certain types of epilepsy. Broad-spectrum drugs can have one or multiple mechanisms of action. Levetiracetam (LEV) is known to bind to synaptic vesicle 2A, but its mechanism of action as an ASM is still unclear.

As opposed to generalized-onset seizures, where treatment options are often limited to broad-spectrum medications and can be dictated by a defined epilepsy syndrome, as seen in Table 17.2, focal seizures do not have well-defined treatment choices. Instead, potential side effects, desired benefits besides seizure control, and comorbidities, *rather than an ASM's particular efficacy*, are the most important considerations. In fact, with focal-onset seizures, there have been very few robust trials comparing ASM efficacies.

Among older ASMs, comparable efficacy was found in epilepsy with focal-onset seizures between CBZ, PHT, PB and PMD, but CBZ was the least well tolerated [10]. ILAE guidelines, based on a review of controlled studies for new-onset epilepsy treatment, found (i) level A evidence (established efficacy) for CBZ and PHT for focal seizures in adults, (ii) OXC for focal seizures in children, (iii) GBP and LTG for focal seizures in the elderly, and (iv) level B evidence (probable efficacy) for VPA for focal seizures in adults [11]. Non-inferiority trials suggest that several newer ASMs are at least as efficacious as older ASMs, with better side effect profiles. However, the general consensus is to choose a medication for a patient on a case-by-case basis depending on the patient's comorbidities and personal preferences. The choice of ASM should also be influenced by its potential utility in comorbid conditions, as noted in Table 17.4.

Polypharmacy The consensus is to treat with a single medication at the lowest possible dose to achieve seizure freedom, without causing side effects. In other words, the ideal treatment goal is **no seizures, no side effects, with the least amount of medication possible**. If a second ASM needs to be added, it is

Table 17.4 Indications for ASMs in addition to seizures

Condition	ASM(s)
Trigeminal neuralgia	CBZ, OXC, LTG
Neuropathic pain	GBP, PGB, CBZ, OXC
Migraine	TPM, VPA > GBP, PGB
Essential tremor	PMD, TPM, ZNS
Anxiety	CLB, BZP
Bipolar disorder	LTG, VPA, CBZ, OXC > GBP

preferable to have a different mechanism of action. Such "rational polypharmacy" is, to a large extent, a theoretical preference as there is a lack of robust studies to allow a mechanistic approach to ASM combinations [12]. An exception is a synergistic effect of lamotrigine and valproic acid, which has been demonstrated [13]. Additionally, additive benefits and side effects can be unpredictable based on enzyme-inducing or enzyme-inhibiting properties of the added ASM.

Systemic Illness Renal dysfunction should entail a decreased dose and reduced dosage frequency of renally excreted ASMs—LEV, LCM, GBP, and PGB. Patients on dialysis should receive an additional post-dialysis dose. Hepatic dysfunction decreases clearance and body protein, leading to elevated free fractions of ASMs. Therefore, dosages of hepatically metabolized ASMs (e.g., PHT, LTG, CBZ) should be appropriately reduced.

Age Neonates have a decreased rate of hepatic metabolism, renal clearance and protein binding, causing elevated free levels in the blood and an increased half-life. During infancy and childhood, the metabolism and volume of distribution increase, leading to a shorter half-life and requiring an increased frequency of dosages. The elderly have physiology similar to neonates, but their care is further complicated by polypharmacy. Thus, newer ASMs with fewer drug interactions are preferred in the elderly [14].

Generic Versus Brand Name The FDA allows 80–125% bioavailability of a generic drug as compared to its brand name counterpart. This difference may be enough to precipitate seizures in some patients. Switching from one generic manufacturer to another can also precipitate seizures. If brand-name medications are not used, then persistence with a single generic manufacturer should generally be attempted. In contrast to a widespread perception of lack of bioequivalence in generic medications, studies such as the EQUIGEN trial, demonstrated strong bioequivalence between generic and brand name LTG [15, 16].

Drug Monitoring Routine drug monitoring in asymptomatic patients is generally not necessary. In fact, it can lead to the inappropriate adjustment of medication, causing breakthrough seizures or side effects. While lab reference ranges are derived from statistical studies, the *therapeutic range* is specific to an individual. Seizure control and adverse effects can occur even when below the laboratory's reference range. In the past, trough levels before the first-morning dose were strongly encouraged. However, serial levels performed at the same time of day can be substituted for trough results [5]. Specific reasons for monitoring include i) assessing adherence, ii) assessing response to physiological changes such as puberty, pregnancy, and elderly age, iii) obtaining a baseline level when seizure and side-effect-free, and iv) evaluating significant or unpredictable interactions between existing and newly added ASMs or other medications [17].

Discontinuing ASMs Seizure freedom is the usual threshold for considering discontinuation of ASMs. However, this approach should be individualized to patient preference, lifestyle and clinical factors. Studies that looked at relapse consisted of mixed populations regarding age and epilepsy etiology, with relapse rates varying from 12–63% over 2–5 years. Favorable prognostic factors for continued seizure freedom include age of onset between 2 and 11 years old, normal mentation and neurological exam, "idiopathic" or genetic etiology, known "benign epilepsies" like childhood absence and childhood centro-temporal, prompt initial response to ASMs, seizure free interval much greater than 2 years, and low therapeutic drug levels. Unfavorable risk factors include age of onset greater than 11 years old, abnormal mentation or neurological exam, presumptive structural etiology, poor initial ASM response, requiring multiple ASMs for seizure freedom, and any EEG abnormalities [18, 19]. The presence of multiple unfavorable risk factors can be additive to the relapse rate [20]. The prognostic value of a pre-discontinuation EEG is probably

ASM Interactions

ASMs often interact with each other and with other medications. This section summarizes the clinically relevant pharmacokinetic and pharmacodynamic interactions of ASMs with other drugs, detailing how ASMs affect and are affected by various medications.

Pharmacokinetic: Induction or inhibition of metabolizing enzymes cytochrome P450 (CYP)

A. ***ASMs*** [6–8, 22].

ASMs *unaffected* by CBZ and PHT (hepatic enzyme inducers): GBP, PGB, LEV, LCM and VGB.

ASMs increased by VPA (hepatic enzyme inhibitor): CBD, CBZ, ESX, LTG, OXC, PB, PHT, RUF, and TGB.

ASMs affecting PHT (increase dosage): CNB

B. Other medications.

(a) Hormonal Contraceptives.

ASMs that increase contraceptives' metabolism—PHT, CNB, CBZ, PB, PMD, OXC, LTG, TPM (>200 mg daily).

ASMs whose clearance is increased by contraceptives—LTG, VPA

(continued)

(b) Warfarin.

ASMS that increase metabolism (decrease levels) of Warfarin—PHB, PMD, CBZ.

ASMs that have an unpredictable response on Warfarin—PHT.

ASMs that decrease metabolism (increase levels) of Warfarin—VPA

(c) ***Antibiotics*:** The following ASM levels are increased.

CBZ—Clarithromycin, Azithromycin, Isoniazid, Ketoconazole, Metronidazole.

ESX—Isoniazid.

PHT—Fluconazole, Isoniazid.

VPA—Isoniazid

(d) ***Psychiatric drugs***: the following ASM levels are increased.

CBZ—Quetiapine.

LTG—Sertraline.

PHT—Fluoxetine, Sertraline, Imipramine, Trazodone.

VPA—Sertraline.

Pharmacodynamic: Two drugs have additive, synergistic or antagonistic effects

A. ASM.

LTG + VPA = synergistic benefit [13].

LTG + CBZ = more side effects.

CBD + CLO = more side effects.

CNB + CLO = more side effects

B. Other medications.

Non-depolarizing neuromuscular blocker (NDNMBs) such as tubocurarine analogs (vecuronium, rocuronium etc.)—Best studied with PHT, PB and CBZ [23].

Acute administration of ASMs potentiates the effect of NDNMBs

Chronic administration of ASMs increases resistance to NDNMBs.

low when compared to other risk factors. Regardless of the rate of withdrawal of ASMs (many weeks versus months), the eventual relapse rate is similar [21].

With adults, the psychosocial impact of recurrence of seizures is much greater and should be taken into consideration. Twenty to thirty percent of patients who were previously seizure free and have a recurrence of seizures after ASM withdrawal fail to achieve seizure freedom again despite restarting the same ASM(s) at the pre-withdrawal dose.

Rescue Therapies

Seizure clusters are defined as having either 2 to 3 or more seizures within a 6- to 24-hour period and are associated with a negative impact on quality of life, daily function, and increased utilization of emergency rooms [24]. Rescue medications can provide a fast and effective way to reduce the risk of recurrent seizures and progression into status epilepticus. Currently available rescue medications are benzodiazepines. Currently, there are three FDA-approved rescue therapies.

A. Diazepam rectal gel (Diastat®)
   - For patients ≥2 years of age, weight-based dosing from 5–20 mg is available.
   - Age 2–5 years: 0.5 mg/kg; Age 6–11 years (0.3 mg/kg); Age 12 years and older (0.2 mg/kg).
   - Repeat dose can be given 4–12 hours after the first dose.
   - Maximum dose recommendation is no more than five episodes in a month and no more than one episode every five days.
   - Contraindications: hypersensitivity to diazepam, acute narrow angle-glaucoma.
   - Most common adverse events: somnolence without respiratory depression.

B. Midazolam nasal spray (Nayzilam®) [25].
   - For patients ≥12 years or older; 5 mg per dose/spray.
   - Administer 1 spray in one nostril, can repeat one additional spray (5 mg) into the other nostril after 10 minutes if a patient has not responded to the initial dose.
   - Maximum doses recommended up to five times per month, no more than one episode every three days and no more than two doses within 24 hours.
   - Contraindications: Hypersensitivity to midazolam, acute narrow-angle glaucoma.
   - Most common side effects: somnolence, headache, nasal discomfort, throat irritation and rhinorrhea.

C. Diazepam nasal spray (Valtoco®) [26]
   - For patients ≥6 years or older; 5 mg, 7.5 mg and 10 mg doses are available.
   - Age 6–11 years (0.3 mg/kg) up to 20 mg total; Age 12 years and older (0.2 mg/kg) up to 20 mg total. Limitation of 10 mg total per nostril; if going above, administer excess into another nostril by the second sprayer in a blister pack.
   - Second dose can be administered at least 4 hours after the initial dose with a new blister pack.

(continued)

- Maximum doses are no more than five per month and no more than one episode every five days.
- Contraindications: Hypersensitivity to diazepam, acute narrow-angle glaucoma.
- The most common adverse effects include somnolence, headache, and nasal discomfort.

Other benzodiazepines may also be used off-label such as disintegrating tablet formulations, which can be placed in the oral cavity such as in the cheek or under the tongue. This route may affect the absorption and bioavailability of the active agent and there is an increased risk of aspiration and biting injury. Future therapies such as oral inhalation of benzodiazepines are currently under investigation.

Pearls: Antiseizure Medications

Benzodiazepines and barbiturates both act on $GABA_A$ receptors to increase inhibition, but benzodiazepines increase the duration of open $GABA_A$ receptors, while barbiturates increase the frequency of open $GABA_A$ receptors. Both can lead to dependency, with tolerance and withdrawal symptoms.

Bone health: Enzyme-inducing ASMs (e.g., CBZ and PHT) can cause osteoporosis over the long term. DEXA (bone density) scans should be completed every 2 years, and treatment with Calcium (500 mg bid) and Vitamin D (400 IU bid) should be instituted in all patients on these medications. If osteoporosis results, bisphosphonates should be instituted.

Anecdotal evidence supports supplementation with 50 mg bid of pyridoxine (Vitamin B6) if patients develop irritability after starting LEV.

As a general rule of thumb, ASM dosage can be increased weekly until the maximum dose is reached, intolerable side effects develop, or seizure freedom results.

If changing a medication due to side effects, either stop the offending drug immediately (life-threatening/severe side effect) or taper as a second medication is titrated up (non-life-threatening side effect).

If changing a medication due to suboptimal efficacy, add the second medication until the highest tolerable dose is achieved, and then consider tapering off the first ASM if the second medication brings seizure freedom.

Erythromycin leads to CBZ toxicity by directly inhibiting its metabolism [27].

ASMs associated with weight gain: VPA, PGB, GBP, VGB.

ASMs associated with weight loss: TPM, ZNS, FBM, FFA.

FFA treatment requires cardiac screening with echocardiogram [8].

"Forced normalization" refers to the development of psychotic behavior on suppression of seizures and "normalization" (improvement of interictal activity) of EEG (Table 17.5).

Table 17.5 Cross-sensitivity of skin rashes with ASM use: the likelihood of patients having a skin rash if previously experienced a rash with another antiseizure medication [28]

CBZ ⊛ PHT: 57.6%	OXC ⊛ CBZ: 71.4%
CBZ ⊛ LTG: 20%	OXC ⊛ LTG: 37.5%
CBZ ⊛ OXC: 33%	PB ⊛ CBZ: 66.7%
CBZ ⊛ PB: 26.7%	PB ⊛ PHT: 53.3%
LTG ⊛ CBZ: 26.3%	PHT ⊛ CBZ: 42%
LTG ⊛ OXC: 20%	PHT ⊛ PB: 19.5%
LTG ⊛ PHT: 38.9%	PHT ⊛ LTG: 18.9%

Acknowledgments Atul Maheshwari MD, Baylor College of Medicine, Houston, Texas for contribution to previous versions of this chapter.

References

1. Krumholz A, Wiebe S, Gronseth GS, Gloss DS, Sanchez AM, Kabir AA, Liferidge AT, Martello JP, Kanner AM, Shinnar S, Hopp JL, French JA. Evidence-based guideline: management of an unprovoked first seizure in adults: report of the guideline development Subcommittee of the American Academy of neurology and the American Epilepsy Society. Neurology. 2015;84(16):1705–13. https://doi.org/10.1212/WNL.0000000000001487.
2. Krumholz A, Wiebe S, Gronseth G, Shinnar S, Levisohn P, Ting T, Hopp J, Shafer P, Morris H, Seiden L, Barkley G, French J, Quality Standards Subcommittee of the American Academy of Neurology; American Epilepsy Society. Practice parameter: evaluating an apparent unprovoked first seizure in adults (an evidence-based review): report of the quality standards Subcommittee of the American Academy of neurology and the American Epilepsy Society. Neurology. 2007;69(21):1996–2007. https://doi.org/10.1212/01.wnl.0000285084.93652.43.
3. Glauser T, Ben-Menachem E, Bourgeois B, Cnaan A, Guerreiro C, Kälviäinen R, Mattson R, French JA, Perucca E, Tomson T, ILAE Subcommission on AED Guidelines. Updated ILAE evidence review of antiepileptic drug efficacy and effectiveness as initial monotherapy for epileptic seizures and syndromes. Epilepsia. 2013;54(3):551–63. https://doi.org/10.1111/epi.12074.
4. Glauser TA, Cnaan A, Shinnar S, Hirtz DG, Dlugos D, Masur D, Clark PO, Capparelli EV, Adamson PC, Childhood Absence Epilepsy Study Group. Ethosuximide, valproic acid, and lamotrigine in childhood absence epilepsy. N Engl J Med. 2010;362(9):790–9. https://doi.org/10.1056/NEJMoa0902014.
5. Chiron C, Marchand MC, Tran A, Rey E, d'Athis P, Vincent J, Dulac O, Pons G. Stiripentol in severe myoclonic epilepsy in infancy: a randomised placebo-controlled syndrome-dedicated trial. STICLO study group. Lancet. 2000;356(9242):1638–42. https://doi.org/10.1016/s0140-6736(00)03157-3.
6. Sekar K, Pack A. Epidiolex as adjunct therapy for treatment of refractory epilepsy: a comprehensive review with a focus on adverse effects. F1000Research. 2019;8:F1000 Faculty Rev-234.
7. Roberti R, De Caro C, Iannone LF, Zaccara G, Lattanzi S, Russo E. Pharmacology of cenobamate: mechanism of action, pharmacokinetics, drug-drug interactions and tolerability. CNS Drugs. 2021;35(6):609–18. https://doi.org/10.1007/s40263-021-00819-8.
8. Simon K, Sheckley H, Anderson CL, Liu Z, Carney PR. A review of fenfluramine for the treatment of Dravet syndrome patients. Curr Res Pharmacol Drug Discov. 2022;3:100078.
9. Kwan P, Brodie MJ. Early identification of refractory epilepsy. N Engl J Med. 2000;342:314–9.
10. Mattson RH, Cramer JA, Collins JF, Smith DB, Delgado-Escueta AV, Browne TR, Williamson PD, Treiman DM, McNamara JO, McCutchen CB, et al. Comparison of carbamazepine, phenobarbital, phenytoin, and primidone in partial and secondarily generalized tonic-clonic seizures. N Engl J Med. 1985;313(3):145–51. https://doi.org/10.1056/NEJM198507183130303.

11. Glauser T, Ben-Menachem E, Bourgeois B, Cnaan A, Chadwick D, Guerreiro C, Kalviainen R, Mattson R, Perucca E, Tomson T. ILAE treatment guidelines: evidence-based analysis of anti-epileptic drug efficacy and effectiveness as initial monotherapy for epileptic seizures and syndromes. Epilepsia. 2006;47(7):1094–120. https://doi.org/10.1111/j.1528-1167.2006.00585.x.
12. Brodie MJ, Covanis A, Gil-Nagel A, Lerche H, Perucca E, Sills GJ, White HS. Antiepileptic drug therapy: does mechanism of action matter? Epilepsy Behav. 2011;21(4):331–41. https://doi.org/10.1016/j.yebeh.2011.05.025.
13. Brodie MJ, Yuen AW. Lamotrigine substitution study: evidence for synergism with sodium valproate? 105 study group. Epilepsy Res. 1997;26:423–32.
14. Engel J Jr, Moshe S. Epilepsy: a comprehensive textbook. Philadelphia, PA: Wolters Kluwer Health/Lippincott Williams & Wilkins; 2008.
15. Ting TY, Jiang W, Lionberger R, Wong J, Jones JW, Kane MA, Krumholz A, Temple R, Polli JE. Generic lamotrigine versus brand-name Lamictal bioequivalence in patients with epilepsy: a field test of the FDA bioequivalence standard. Epilepsia. 2015;56(9):1415–24. https://doi.org/10.1111/epi.13095.
16. Privitera MD, Welty TE, Gidal BE, Diaz FJ, Krebill R, Szaflarski JP, Dworetzky BA, Pollard JR, Elder EJ Jr, Jiang W, Jiang X, Berg M. Generic-to-generic lamotrigine switches in people with epilepsy: the randomised controlled EQUIGEN trial. Lancet Neurol. 2016;15(4):365–72. https://doi.org/10.1016/S1474-4422(16)00014-4.
17. Patsalos PN, Berry DJ, Bourgeois BF, Cloyd JC, Glauser TA, Johannessen SI, Leppik IE, Tomson T, Perucca E. Antiepileptic drugs—best practice guidelines for therapeutic drug monitoring: a position paper by the subcommission on therapeutic drug monitoring, ILAE commission on therapeutic strategies. Epilepsia. 2008;49(7):1239–76. https://doi.org/10.1111/j.1528-1167.2008.01561.x.
18. Britton JW. Antiepileptic drug withdrawal: literature review. Mayo Clin Proc. 2002;77:1378–88.
19. Lamberink HJ, Otte WM, Geerts AT, Pavlovic M, Ramos-Lizana J, Marson AG, Overweg J, Sauma L, Specchio LM, Tennison M, Cardoso TMO, Shinnar S, Schmidt D, Geleijns K, Braun KPJ. Individualised prediction model of seizure recurrence and long-term outcomes after withdrawal of antiepileptic drugs in seizure-free patients: a systematic review and individual participant data meta-analysis. Lancet Neurol. 2017;16(7):523–31. https://doi.org/10.1016/S1474-4422(17)30114-X. Epub 2017 May 5. Erratum in: Lancet Neurol. 2017 Aug;16(8):584. https://doi.org/10.1016/S1474-4422(17)30221-1
20. Prognostic index for recurrence of seizures after remission of epilepsy. Medical Research Council antiepileptic drug withdrawal study group. BMJ. 1993;306:1374–8.
21. Berg AT, Shinnar S. Relapse following discontinuation of antiepileptic drugs: a meta-analysis. Neurology. 1994;44:601–8.
22. Perucca E. Clinically relevant drug interactions with antiepileptic drugs. Br J Clin Pharmacol. 2006;61:246–55.
23. Soriano SG, Martyn JAJ. Antiepileptic-induced resistance to neuromuscular blockers: mechanisms and clinical significance. Clin Pharmacokinet. 2004;43:71–81.
24. Jafarpour S, Hirsch LJ, Gaínza-Lein M, Kellinghaus C, Detyniecki K. Seizure cluster: definition, prevalence, consequences, and management. Seizure. 2019;68:9–15.
25. Detyniecki K, Van Ess PJ, Sequeira DJ, Wheless JW, Meng TC, Pullman WE. Safety and efficacy of midazolam nasal spray in the outpatient treatment of patients with seizure clusters-a randomized, double-blind, placebo-controlled trial. Epilepsia. 2019;60:1797–808.
26. Wheless JW, Miller I, Hogan RE, Dlugos D, Biton V, Cascino GD, Sperling MR, Liow K, Vazquez B, Segal EB, Tarquinio D, Mauney W, Desai J, Rabinowicz AL, Carrazana E, DIAZ.001.05 Study Group. Final results from a phase 3, long-term, open-label, repeat-dose safety study of diazepam nasal spray for seizure clusters in patients with epilepsy. Epilepsia. 2021;62:2485–95.
27. Turner PV, Renton KW. The interaction between carbamazepine and erythromycin. Can J Physiol Pharmacol. 1989;67:582–6.
28. Hirsch LJ, Arif H, Nahm EA, Buchsbaum R, Resor SR Jr, Bazil CW. Cross-sensitivity of skin rashes with antiepileptic drug use. Neurology. 2008;71:1527–34.

18 Neurostimulation

Manan Nath, Zulfi Haneef, and Irfan Ali

18.1 Introduction

The evaluation of epilepsy surgery involves intricate decision-making, with neuromodulation recommended when resection faces challenges due to a broad seizure network, multifocal seizure onsets or functional eloquence of the seizure onset zone (Fig. 18.1). Electrical neurostimulation to disrupt epileptic networks was pioneered by Penfield and Jasper in 1954 [1]. The current neurostimulation modalities approved by the US Food and Drug Administration (FDA) for focal epilepsy are vagus nerve stimulation (VNS), deep brain stimulation (DBS) of anterior nucleus of thalamus (ANT) and responsive neurostimulation (RNS). Please refer to Table 18.1 for main features differentiating the three neurostimulation modalities. Intracranial monitoring helps identify potential neurostimulation sites within the epileptogenic network. Neurostimulation may be an open loop providing continuous stimulation without any feedback or a closed loop providing stimulation in response to ictal activity on electrocorticogram (ECoG), electrothalamogram (EThG) or other biosignals such as cardiac activity.

M. Nath (✉)
Clinical Neurophysiology and Epilepsy Fellow, Texas Children's Hospital, Baylor College of Medicine, Houston, TX, USA
e-mail: mnath@uams.edu

Z. Haneef
Neurology, Kellaway Section of Neurophysiology, Baylor College of Medicine, Epilepsy Center of Excellence, DeBakey VA Medical Center, Houston, TX, USA

I. Ali
Neurology and Neurophysiology, Texas Children's Hospital, Baylor College of Medicine, Houston, TX, USA

Z. Haneef (ed.), *Epilepsy Fundamentals*,
https://doi.org/10.1007/978-3-031-77741-7_18

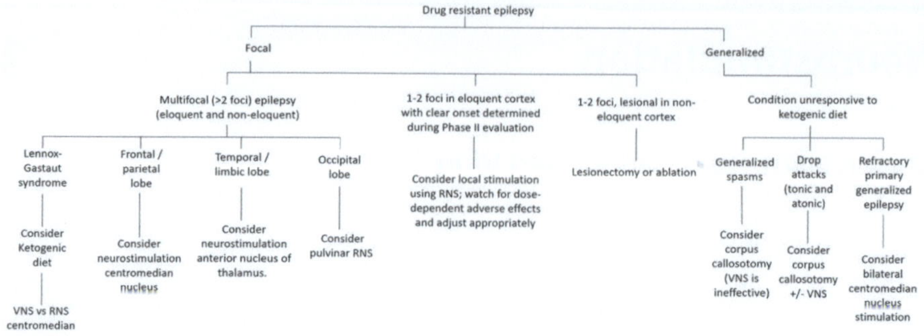

Fig. 18.1 General approach for management of drug-resistant epilepsy

18.2 Vagus Nerve Stimulation (VNS)

VNS involves electrical stimulation of the vagus nerve below its cervical and cardiac branches. VNS was developed as an open-loop system. Following the development of an ictal tachycardia detection feature with responsive vagus nerve stimulation, it can also serve as a closed loop system [2]. Additionally a "magnet" is provided to the patient/caregiver to be swiped on the VNS to interrupt an ongoing seizure. The mechanism of VNS involves stimulation of afferent fibers to the nucleus tractus solitarius, which leads to norepinephrine release from the locus coeruleus and serotonin from the raphe nucleus as well as increased inhibitory neurotransmitter release from projections to the subcortical structures (particularly the thalamus) to disrupt the epileptic circuit [3]. This alters cortical synchronization over time or acutely to even abort the seizure [2].

Prior to VNS FDA approval, several double-blind, parallel, randomized controlled trials by the VNS group were conducted (E03 trial, 1995; E05 trial, 1998) [4, 5]. Comparable pivotal trials with long-term follow-up similar to RNS and DBS are not available. VNS was approved by the FDA in 1997 for use as adjunctive therapy for reducing seizure frequency in medically refractory "partial onset" seizures in patients older than 12 years. In 2017, the FDA approved VNS for adjunctive therapy in patients $\geq$4 years of age with "partial onset seizures that are refractory." The Aspire SR model (approved 2015) used a customizable cardiac-based algorithm to detect relative heart rate increases during seizures and deliver responsive vagal stimulation to abort them, thus allowing the detection and treatment of seizures in real time for selected patients. Studies show that seizure detection based on heart rate changes 20% above baseline had a high sensitivity [6].

Table 18.1 gives indications for the approved neurostimulation devices used in epilepsy. There is a progressive decrease in seizures with a 51%–63% responder rate over 2–3 years and an approximate 63%–78% responder rate over an extended 10-year period (Fig. 18.7) [7, 8]. VNS implantation is associated with use of fewer antiseizure medications compared to patients on conventional medical therapy. VNS can decrease emergency room visits, hospital stay duration, costs, and leads to

Table 18.1 Comparison between VNS, DBS, and RNS devices

	VNS	DBS	RNS
System type	Open-loop in "Normal mode" (closed loop with ictal tachycardia)	Open-loop	Closed-loop
Stimulation type	Extracranial	Intracranial	Intracranial
Mechanism	Increased inhibitory neurotransmission via projections to the thalamus. Long-term neuroplasticity	Direct stimulation causing increased inhibitory neurotransmission and long-term neuroplasticity	
Seizure detection	Indirect (tachycardia)	None	Yes
FDA approval (year)	Adjunctive therapy in patients ≥4 years with medically refractory focal seizures. Approved in Europe for generalized epilepsy	DBS-ANT for adjunctive therapy in patients ≥18 years with medically refractory focal seizures	Adjunctive therapy in patients ≥18 years with medically refractory focal seizures with ≤2 epileptogenic foci
Conditions treated (includes off-label use reported in the literature)	Epileptic encephalopathies such (e.g., LGS, LKS, PME, absence epilepsy and JME) [12]	Thalamic stimulation for neuromodulation of broad epileptic networks	Focal seizures in non-resectable locations and thalamic stimulation for broad epileptic networks
Implantable pulse generator (IPG) location[b]	Left upper chest	Left upper chest (subclavicular)	Placed in a ferrule located within the cranium
Electrode location	Wrapping vagus nerve with rostral negative electrode	Thalamic nuclei	Strip or depth electrodes in cortex/ thalamus
MRI compatibility	Needs to be switched off and then re-started following scan	Conditionally safe if certain conditions are met	Conditionally safe; needs to be placed in MRI mode
Impact on SUDEP[a]	Progressive decrease over 3–10 year period [10]	2 per 1000 patient-years (SANTÉ 10 year follow up) [13]	2 per 1000 patient years [14]
Lateralizes seizures	No	No	Yes
Adverse effects (all can cause local infection & hemorrhage)	Cough, voice change, neck muscle spasm, throat paresthesias, worsening COPD and sleep apnea	Chest paresthesias, hemi-body paresthesia, sleep disruption, depression, memory and mood problems	Paresthesias, visual perceptual phenomena with thalamic implants

Abbreviations: ANT—Anterior nucleus of thalamus; LGS—Lennox-Gastaut Syndrome; LKS—Landau-Kleffner syndrome; PME—Progressive myoclonic epilepsy; JME—Juvenile myoclonic epilepsy; SUDEP—Sudden unexpected death in epilepsy.

a For comparison, the SUDEP rate is 11.4 per 1000 patient-years in patients who have failed epilepsy surgery

b IPG can serve as an anode in all three neurostimulation modalities an overall improvement in quality of life [9]. VNS can reduce SUDEP risk over 3–10 year period after implantation [10] and may improve alertness and cognition. While formal studies on pregnancy are lacking, there are reports of favorable seizure and pregnancy outcomes [11].

VNS implantation can be performed under general, local or regional anesthesia using a small chest incision. Lead wires are wrapped around the left vagus nerve, and the generator is subcutaneously implanted in the left upper chest (Fig. 18.2). Principally, a rostral negative electrode emits electrical impulses rostro-caudally through the vagus nerve, which has afferent connections with the brainstem nuclei and the central nervous system. The electrons traveling caudally are prevented from impacting organs such as the heart and gastrointestinal tract by the positive caudally placed electrode along the vagus nerve. During implantation, system integrity is assessed by analyzing impedance. The output current is initially set to 0 mA and then titrated to target settings (Table 18.2) [15]. The target settings determine the lifespan of the VNS battery with lower stimulation settings and higher tachycardia threshold settings leading to longer battery life (medically refractory "partial onset" seizures). Table 18.3 [16] gives information regarding manufacturers and models of VNS device, the battery life, and features of current devices in use.

Complications The pre-FDA approval E03 and E05 studies revealed a small possibility of incision site infection (0.8%–1.5%), at times needing device explantation [4, 5]. VNS stimulation can be associated with dose-dependent complications such as hoarseness, cough, paresthesia and shortness of breath, and muscle spasms, which tend to improve with prolonged use (Table 18.1) [17]. Patients with comorbid COPD and sleep apnea may experience exacerbated respiratory issues with VNS [18]. Although uncommon, VNS lead fracture can occur, causing improper delivery of impulses. This can be remedied by lead re-positioning, although device removal may be necessary.

18.3 Deep Brain Stimulation (DBS)

DBS has a long history of being used in movement disorders such as Parkinson's disease, essential tremor and dystonia. In epilepsy, DBS operates as an open-loop system to modulate thalamic nuclei for modulation of multifocal/regional seizures not amenable to surgical resection. Individuals with refractory temporal lobe or other focal seizures without underlying psychiatric illness may be suitable candidates for DBS. The target nuclei for thalamic DBS are the anterior nucleus for

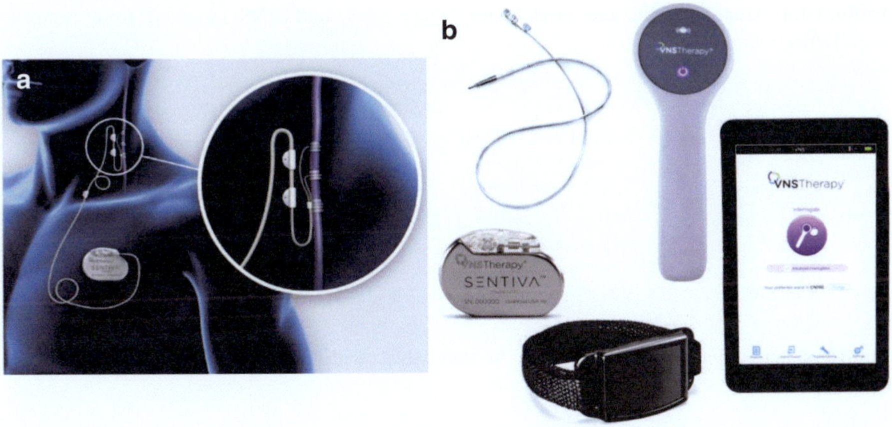

Fig. 18.2 (**a**) The implanted pulse generator in the upper left chest with lead wire and stimulating wires around the vagus nerve. (**b**) VNS device, wand, programming tablet, magnet and lead wire (Images used with permission from LivaNova, Inc.)

Table 18.2 Typical neurostimulation settings used for VNS, DBS, RNS

	VNS [16]				RNS [20, 21]
	Normal	Autostim	Magnet	DBS [19]	
Charge density	0.125–0.25 mA with increments of 0.125–0.25 mA to target 1.5–2 mA	0.25–0.375 mA with increments of 0.125–0.25 mA to max 1.5–2 mA	0.5 mA with increments of 0.25 mA to target 2 mA	2–3 mA (or V) per cathode with an increment of 1–2 mA (or V) to target 5 mA (or V) or effect	0.5μC/cm^2 with increments of 0.5μC/cm^2 Target charge density[a]: 2–4 μC/cm^2 Maximum charge density: 25μC/cm^2 (for safety)
Pulse width	250μSec	250μSec	500μSec	90μSec	160μSec (40–1000)
Frequency	<50 Hz	20–30 Hz	20–30 Hz	145 Hz	200 Hz (local) and 150 Hz (thalamic) (1–333 Hz)
Duty cycle	10% to max 16–58%				

mA—milliampere; V—Volt; Hz—Hertz; μsec—microsecond; μC—microcoulombs.

[a] Charge density (μC/cm^2): is the amount of electric current introduced per stimulation phase. Charge density = (Electric charge in μC per second × Pulse width in second) / Surface area of electrode contact in cm^2

Table 18.3 Manufacturers and models for VNS, DBS, and RNS (adapted from Simpson et al. [16])

	Manufacturer	Device	Battery life	Features
VNS	LivaNova (formerly Cyberonics)	AspireHC (M105), AspireSR (M106), SenTiva (M1000)	3–8 y	Guided programming, autostim (Aspire, SenTiva), day/night programming (SenTiva)
DBS	Medtronic	Activa PC, Activa RC, percept PC	3–15 y	Limited sensing, rechargeable, handheld device to switch between treatment parameters
RNS	Neuropace	300 M, 320 M	3.5–12.5 y	EEG recording

Company and device names are used in this table, with permission from the companies. Abb: y-years

temporal lobe epilepsy, centromedian nucleus, dorso-medial nucleus and reticular nucleus for frontal lobe and generalized seizures. DBS is thought to inhibit neuronal excitation or disrupt neuronal synchronicity [22]. Animal studies suggest that DBS may modulate neurogenesis, network connectivity, neuronal transmission and synaptic function [23]. Optimal DBS efficacy (Fig. 18.7) in temporal lobe seizures is achieved through precise electrode placement within the anterior nucleus of the thalamus at the termination of the mammillothalamic tract [24]. Refractory generalized and frontal lobe seizures are treated by stimulating the centromedian nucleus [25]. Pulvinar stimulation has been used for parieto-occipital seizures as well as for posterior temporal lobe seizures [26]. Other potential nuclei for consideration are dorsalis medialis for frontal lobe seizures, given its close relationship with centromedian nucleus [27] as well as nucleus reticularis for generalized epilepsy [28]. The sentinel trial for DBS in epilepsy was the "stimulation of anterior nucleus of the thalamus in epilepsy" trial (SANTE, Fig. 18.7) [19], multicenter, randomized control trials launched following initial pilot studies in 1988 and between 2000 and 2008 [29, 30]. DBS-ANT received FDA approval in 2018 for adjunctive use in patients ≥18 years with refractory focal epilepsy with demonstrated safety and effectiveness of ≥6 months. Table 18.1 gives some common off-label indications for DBS use in epilepsy DBS can be performed either using a single cathode within the desired thalamic nucleus and the pulse generator serving as the anode or bipolar using a cathode and anode in the same electrode or 2 electrodes both serving as cathodes (Fig. 18.3). Stimulation parameters in clinical practice are generally lower than in the SANTÉ trial and are increased every 1–3 months to target amplitude or desired efficacy. Table 18.3 [16] gives information regarding manufacturers and models of DBS device, the battery life as well as features of current devices in use. Utilizing a handheld device, individuals can seamlessly transition between two stimulation parameters in the home setting, facilitating the effective management of distinct day and night-time stimulation parameters (Fig. 18.4).

Complications DBS implantation may be associated with adverse effects such as infection or hemorrhage. Additionally, transient adverse effects such as paresthesias

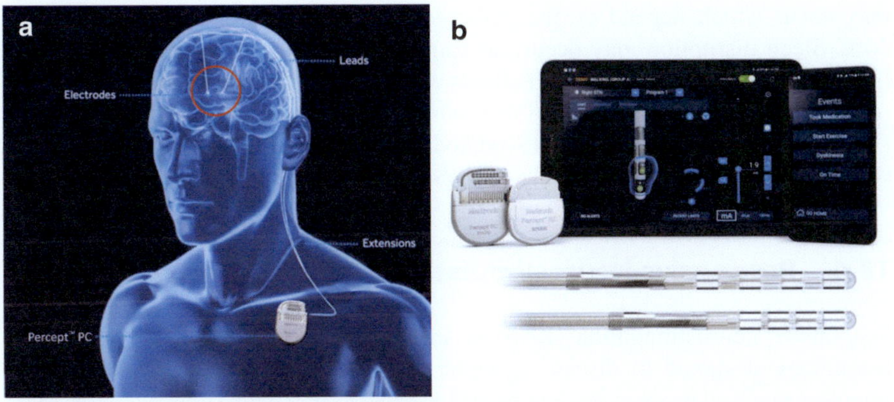

Fig. 18.3 (**a**) Deep brain stimulation intermittent pulse generator (DBS IPG) device placed subclavicularly on the left with lead wire travelling up the neck to the brain and electrodes in the thalamic subnuclei. (**b**) DBS IPG device, programming tablet, patient handset and electrodes (Image permissions from Medtronic, Inc.)

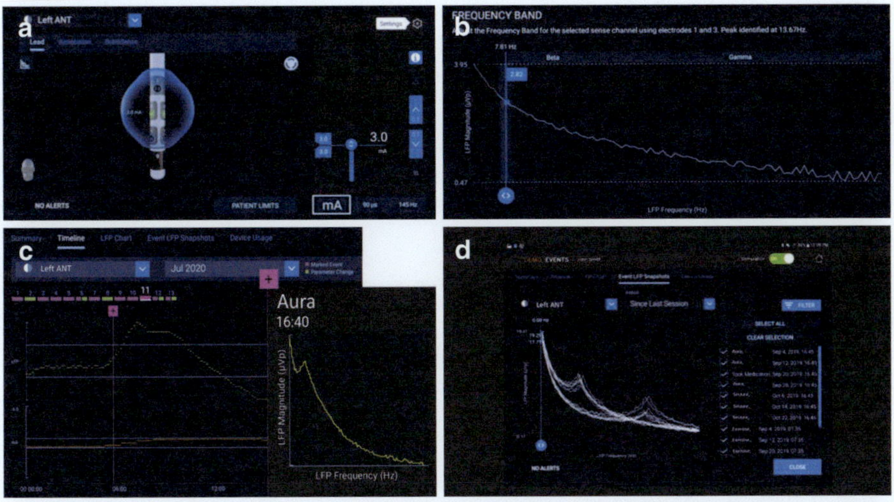

Fig. 18.4 (**a**) DBS interface with the volume of neural activation (VNA) surrounding the configured electrode. VNA may be symmetric or asymmetric based on active electrode contacts. Charge density is set in milliamperes (mA) with depicted pulse duration and frequency. (**b**) Configuration screen for Local Field Potential (LFP) where a center point for a 5 Hz recording window is chosen either during in-clinic "Streaming" or out-of-clinic "Timeline." Following this, patient events such as "seizure," "aura," "medication intake" are created with the capability of LFP capture, which in turn generates (**c**) a timeline with plotted LFP (averaged over 10 min). A specific time point with a marked event can be chosen with LFP recording for detailed review. (**d**) LFP trends of multiple selected events can be analyzed to generate an event diary with corresponding LFP changes. (Image permissions from Medtronic, Inc)

may occur which depend on the thalamic nuclei stimulated and the stimulation dose. Sleep disruption may occur but can be ameliorated by programming lower frequencies during bedtime [31]. Patients with DBS may exhibit memory and mood issues, with instances of reported new-onset or worsening pre-existing depression [32]. Of particular concern is DBS-related suicidality (0.3–0.7%) noted in patients with Parkinson's disease treated with DBS despite good surgical outcomes [33].

18.4 Responsive Neurostimulation (RNS)

Responsive neurostimulation (RNS) is an adaptive closed-loop neurostimulation technology designed to discern epileptic activity and deliver targeted electrical impulses via 2 electrodes (depth or strip electrodes), each with 4 contacts (Fig. 18.5), which interrupt the propagation of seizures. The RNS system can store electrocorticographic or electrothalamographic data and provides responsive stimulation for the detection of incipient seizures. The life of RNS battery depends on the stimulation and detection settings set by the physician (Table 18.3) [16]. Notably, the latest version of the stimulator is compatible with MRI using an "MRI mode." The benefit of RNS is thought to be due to the responsive component and plasticity that results from chronic stimulation. Over an extended duration, it aims to moderate and attenuate the epileptic network, thereby diminishing the frequency of seizures.

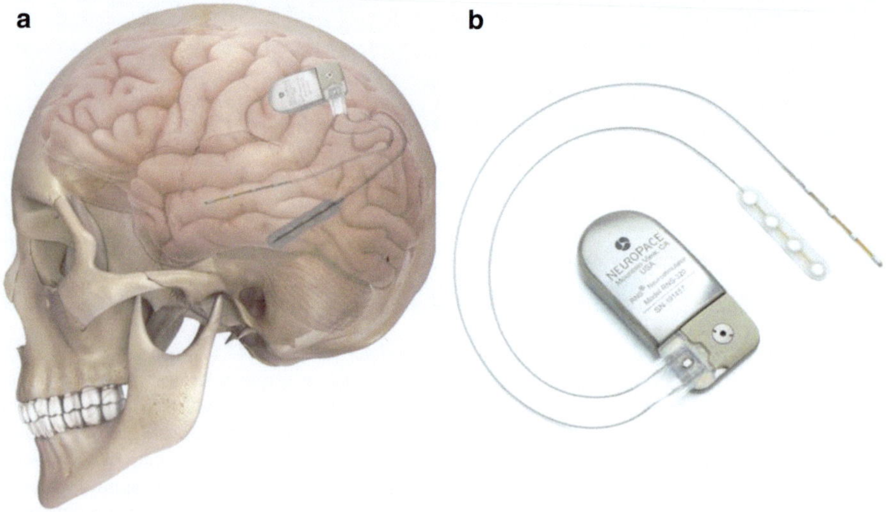

Fig. 18.5 (**a**) Illustration of RNS device implanted in a ferrule in craniotomy cavity with active depth electrodes and/or strips. 2 backup electrodes are implanted that serve as alternate targets. (Image(s) downloaded and used under license from Shutterstock.com. RNS device image used with permission from NeuroPace, Inc). (**b**) Picture of RNS device with depth and strip electrode (Image used with permission from NeuroPace, Inc.; © 2021 NeuroPace, Inc.)

The RNS neurostimulator is placed in a ferrule that is attached via screws to a craniotomy in the skull. In this procedure, a section of bone, which is the size of the neurostimulator, is removed, and the neurostimulator is affixed inside the skull defect to flush with the rest of the skull.

RNS stimulation can be provided either as monopolar or bipolar or in a lead-to-lead configuration. The guiding principle behind RNS in disrupting epileptic networks lies in activating inhibitory neurons through the application of high-frequency electrical currents [34] when neurons within the epileptic network are selectively activated during the onset of seizures. While RNS can specifically target the seizure onset zone or cortical irritability, it also demonstrates broader application by targeting the thalamic nuclei such as the centromedian nucleus for frontal lobe and generalized seizures, the anterior nucleus for temporal lobe and pulvinar for parieto-occipital epilepsy [19, 35]. The effects of thalamic stimulation are broader, possibly due to extensive thalamic connectivity.

Once RNS is implanted, it is programmed to the "line length detection" mode, which has high sensitivity to detect seizures but low specificity for cortical detection and area under cover mode for thalamic detection of seizures [36]. Eventually this mode is modified to a more specific bandpass mode. Once programmed, the best practice is to allow a minimum of 2–3 months to assess the effect of programming change. During each office visit, the epileptologist checks the RNS impedance, which may be elevated if there is a break in the RNS electrode or its connection with the neurostimulator device.

Patients are provided with a magnet, which, when swiped over the device, marks and stores seizures or clinical episodes of concern for later review. Unlike the VNS, the magnet swipe does NOT invoke stimulation, and patients do not have the capability to provide voluntary neurostimulation. Patients are required to download daily data and upload it on a weekly basis to the platform accessible by the physician for review and adjustment of detection and stimulation parameters. The recorded electrocorticogram (ECoG) and electrothalamogram (EThG) are securely stored in an online database called Patient Data Management System (PDMS) (See Fig. 18.6). ECoG/EThG can be of variable lengths ranging from 30 to 240 s with a maximum storage capacity of 12 min at a time. However, the potential for data overwrite exists if the patient fails to download it daily or in case of elevated seizure burden, resulting in expedited filling of the device storage.

The RNS device was approved by the FDA in 2013 as an adjunctive therapy for patients ≥18 years with focal refractory seizures localized to no more than 2 epileptogenic foci. A pivotal trial involved 191 adult patients with medically refractory focal epilepsy [20] on stable ASMs with seizures localized to maximum 2 epileptogenic regions as determined on intracranial EEG monitoring (Fig. 18.7). An ongoing trial (NAUTILUS, NCT05147571) seeks to assess the efficacy of RNS in idiopathic generalized epilepsy. Table 18.1 gives some common indications for RNS use.

RNS data trends find application in seizure predictions by identifying ultradian, circadian and multidien seizure patterns. Response to antiseizure medication intervention can also be noted on RNS trending but keeping in mind the limited spatial

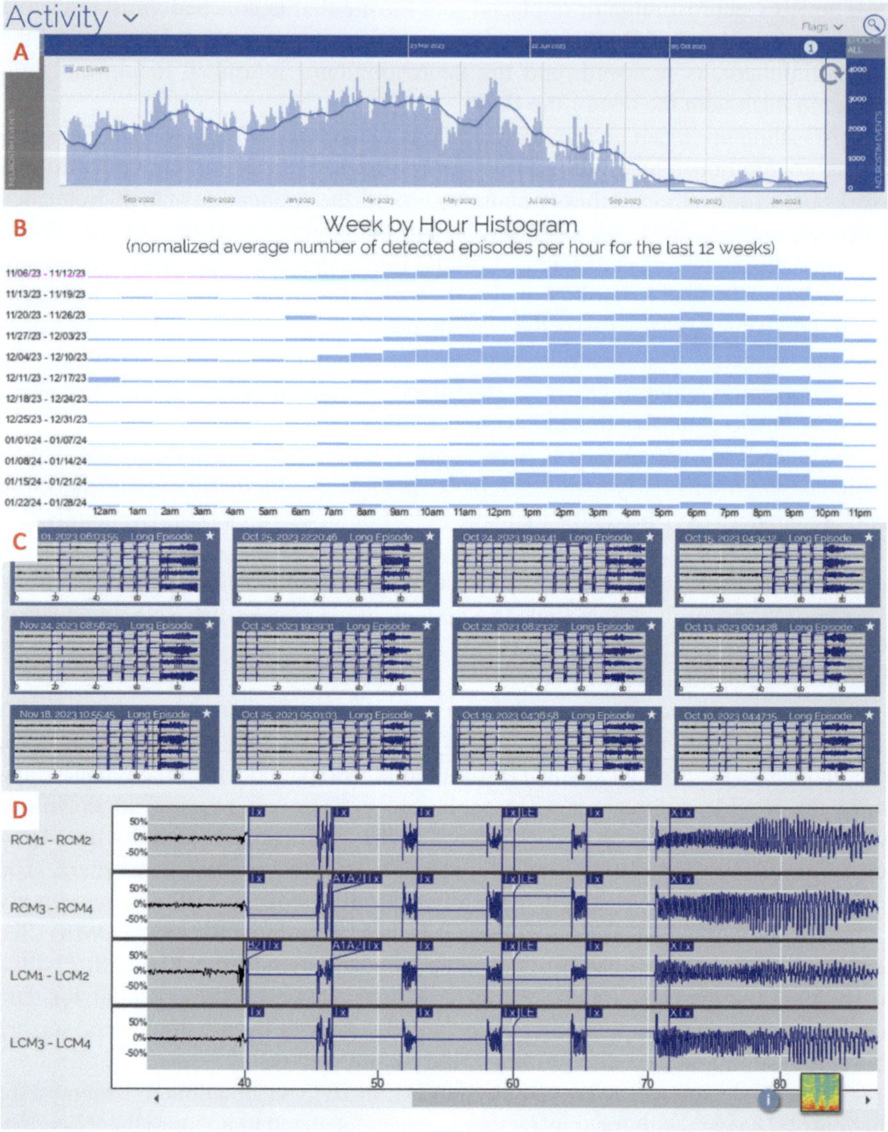

Fig. 18.6 (**a**) Patient Data Management System (PDMS)) interface with daily histograms of all events with initial increasing frequency of "all events," however, subsequent reduction, following initiation of effective antiseizure medication. (**b**) Week-by-hour histogram depicting detections of epileptiform activity over a 12-week period. (**c**) Electrocorticogram (EcoG) library depicting individual EcoGs without any filter applied. A total of twelve, 60 s EcoGs can be stored; however, EcoGs may be programmer-specified with magnet-activated EcoGs up to 240 s and long episodes up to 120 s. Up to 4 "scheduled" EcoGs can be stored per day. (**d**) EcoG with widened timebase showing a 10 s epoch with ictal detection and back-to-back responsive stimulation, following which baseline electrical activity is observed. (Image permissions NeuroPace, Inc.; © 2021 NeuroPace, Inc.)

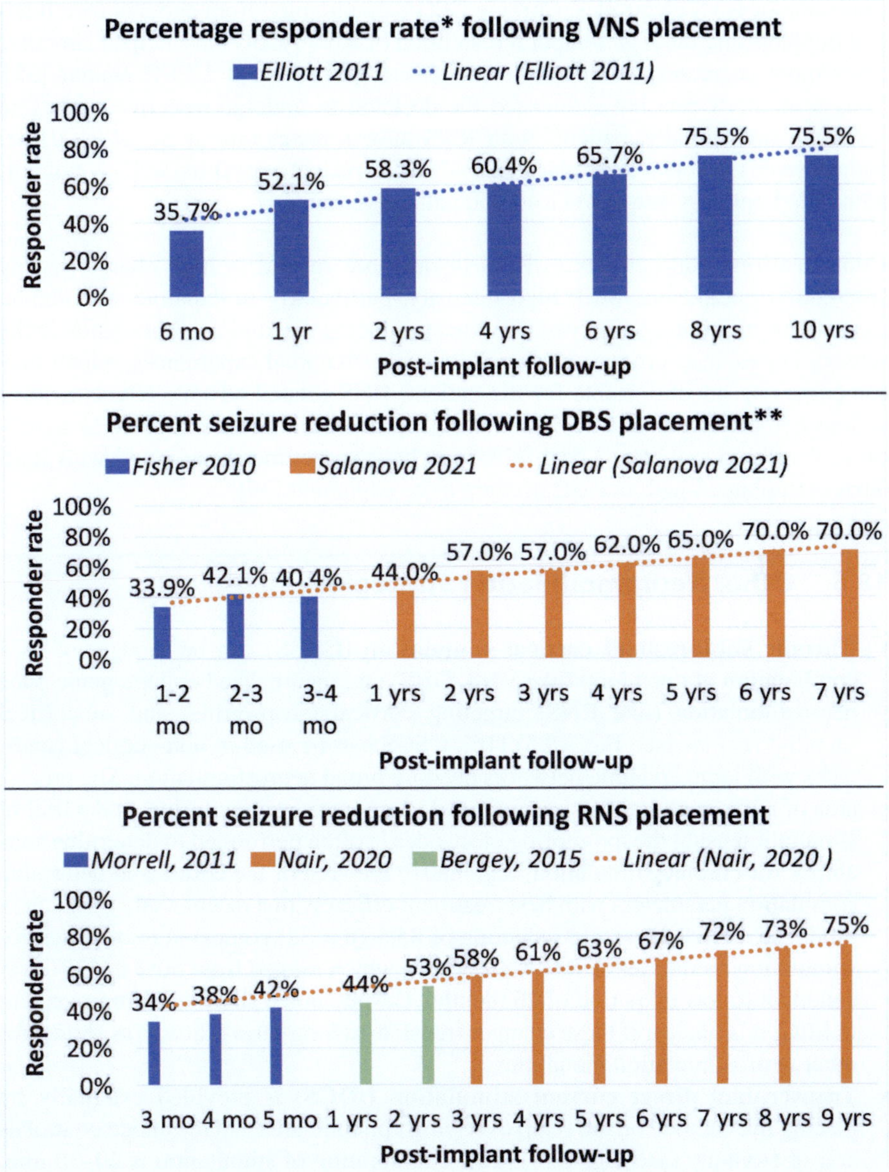

Fig. 18.7 Graphical representation of long term data from VNS [8], DBS [1, 13] [9], and RNS [20, 21, 37] neurostimulation studies* Responder rate is the percentage of patients with >50% seizure reduction** Initial 4 months were part of a double-blind placebo-controlled period subsequently with all patients under treatment arm- Median seizure frequency change includes all randomized patients between active and placebo as well as last observation carried forward (LOCF) patients. The seizure frequency at the last observation was carried forward for drop-outs and used to compute the missing values

resolution of RNS electrodes. Chronic EEG data obtained from patients with bilateral implants and the high temporal resolution of RNS ECoG have helped lateralize cases more accurately compared to the shorter phase 1 and 2 EEG studies [38]. Long-term RNS data has influenced the decision to undergo resective surgery in some patients. Finally, patients with RNS have a lower rate of SUDEP (2/1000 patient-years) compared to placebo (6–7/1000 patient-years) as well as patients with failed epilepsy surgery (11.4/1000 patient-years) [14].

Complications Side effects to RNS devices are related to high charge density stimulation parameters. Such high intensity, particularly in thalamic stimulation, can activate adjoining sensory nuclei, thus producing discomfort. Stimulation in the sensory cortex may produce tactile and visual perceptual experiences, which may sometimes be uncomfortable for the patient. RNS-related adverse effects involve serious hemorrhage in 2.1–2.7% patients [20]; however, this did not lead to significant neurologic sequelae. Local infections have sometimes been noted with RNS devices implanted with devices needing to be explanted [39].

18.5 Other Neuromodulation Strategies

- **Chronic subthreshold cortical stimulation (CSCS)** can be thought of as a combination of open loop (like VNS, DBS) and regionalized epileptogenic zone neurostimulation (like RNS) targeting cortical (strips/grids) and subcortical (depths) regions (see Fig. 18.8) [16]. CSCS can be used in non-surgical candidates with large epileptic networks needing broad neurostimulation. The precise area of neurostimulation is regionalized using invasive monitoring, and a trial of stimulation using the monitoring electrodes is often performed to determine suitability for chronic stimulation, regionalize the area of the cortex and determine stimulation parameters with best treatment efficacy. In a recent study [40] CSCS produced a median seizure reduction of 85% (n = 32) compared to other neurostimulation modalities (DBS, ANS, VNS), which ranged from 50% to 63% over a median follow up period of 26 months. Despite noted short-term improvement in EEG, it is unclear if these changes translate to long-term efficacy, as shown for other neurostimulation modalities.
- **Transcranial direct current stimulation (tDCS)** is provided externally by placing the cathode on the scalp overlying epileptic area and the anode contralateral or frontally (see Fig. 18.8) [16]. The duration of stimulation is 20–30 min. The methodology and protocols are variable, and there are no clear recommendations. T-DCS involves subthreshold stimulation of the cortex, producing depolarization/hyperpolarization of pyramidal neurons, which alters resting membrane potential by influencing ion concentration, transmembrane proteins and synaptic function and decreases cortical excitability. A double-blind, randomized, placebo-controlled trial in patients with mesial temporal lobe epilepsy and hippocampal sclerosis in 2017 [41] demonstrated a significant reduction in mean

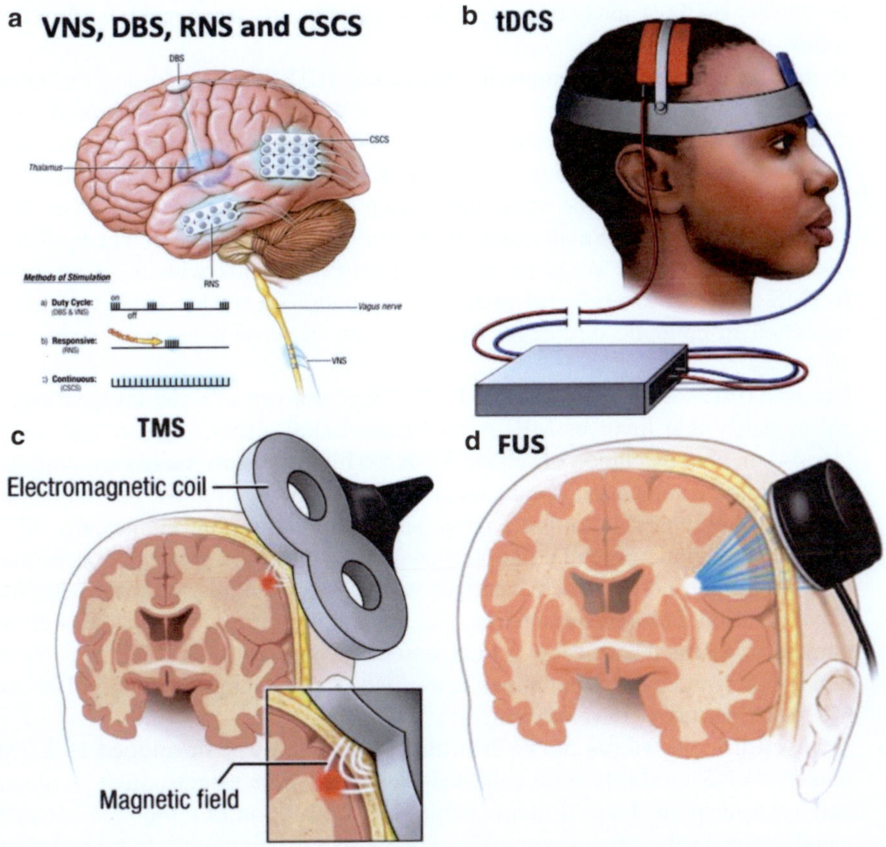

Fig. 18.8 (**a**) Standard neurostimulation strategies such as VNS, DBS, RNS and chronic subthreshold cortical stimulation (CSCS).[a] (**b**) transcranial direct current stimulation (tDCS) with external cathode on scalp overlying region of epileptogenicity (red band) and frontally placed anode (blue band).[a, b] (**c**) Transcranial magnetic stimulation (TMS) producing a focal magnetic field underneath the coil over the epileptogenic region.[a, b] (**d**) Focused ultrasound (FUS) for transcranial seizure focus ablation or neuromodulation.[a, b, c] See text for details. Images used with permission of [a]Mayo Foundation for Medical Education and Research, all rights reserved © Mayo Clinic, [b]Simpson et al. [16], and [c]Starnes et al. [44]

seizure frequency (43.4%–54.6%) in the treatment group versus 6.25% in the placebo group.

- **Trigeminal nerve stimulation (TNS)** is a non-invasive treatment modality that uses adhesive skin electrodes to stimulate bilateral trigeminal nerves at high frequency using an external pulse generator. This neurostimulation modality is approved in Europe, Canada, and Australia, for the treatment of epilepsy, depression, and attention deficit hyperactivity disorder. One double-blind, randomized control trial in 2013 [42] demonstrated a responder rate of 40.5% in the treatment group versus 15.6% in controls, with the benefits mostly sustained in a 12-month follow-up.

- **Repetitive transcranial magnetic stimulation (rTMS)** alters cortical excitability via magnetic pulses, which produce a focal magnetic field in the brain region underneath the coil, depolarizing a small area of cortex and hyperpolarizing of nearby axons thus decreasing cortical excitability (see Fig. 18.8) [16]. R-TMS has FDA approval for the treatment of depression, headaches and obsessive-compulsive disorders. Small randomized controlled trials have shown equivocal results. In one larger study in 2012 [43] on 60 patients with focal seizures, R-TMS daily for 2 weeks led to an 80% reduction in seizure frequency and a reduction in epileptic discharges in EEG, with these benefits maintained over 10 weeks.
- **Focused ultrasound (FUS):** While FUS is better known as a non-invasive surgical modality that involves MRI-guided transcranial ultrasound (see Fig. 18.8) [16, 44] mediated ablation of seizure focus via high-intensity sonication [45], the low-frequency sonication parameter is used for non-destructive neuromodulation that is comparable to direct intracranial stimulation. Mechanistically, FUS can affect synaptic potential and neuronal conduction via membrane deformation that, when coupled with endogenous action potentials, disrupts neuronal depolarization and synchronization. Studies are ongoing in animals as well as humans to assess the efficacy of FUS, and data comparing FUS against conventional neuromodulation modalities is awaited. Device integration and FUS and MR machines are in development.
- **Transcutaneous vagus nerve stimulation (tVNS)** was developed as a less invasive VNS modality with relatively fewer adverse effects, such as apnea and hypopneas in sleep. It involves bipolar stimulation of cutaneous sensory vagal fibers in the medial auricular concha using parameters that are below patient's perception threshold. Reductions in seizure frequency have been reported in small studies ranging from a modest 23% to 54% over a 2–6 month period [46, 47]. While tVNS is well tolerated, evidence for efficacy remains incomplete.
- **Epicranial Application of Stimulation Electrodes for Epilepsy (EASEE)** is a long-term neurostimulation modality for an epileptogenic focus on the cortex where high frequency, low frequency and direct current (DC) stimulation is used in a closed loop setting. EASEE comprises an implantable pulse generator, a five-contact electrode, a programming device for the physician to adjust stimulation parameters and a handheld device for the patient. The large central electrode and four surrounding electrodes are placed in a sub-scalp (subgaleal) location to record and deliver neurostimulation in a closed-loop setting. In a 2023 study involving pooled analyses of 2 non-randomized uncontrolled trials involving 32 adult patients [48], there was a 52% median seizure reduction compared to baseline over six months.

Acknowledgements Mr. Sanjay Anand, LivaNova Inc.
Mr. Todd Hewgley, Medtronic Inc.
Mr. AJ Cuadra, Neuropace Inc.
Ms. Julie Park, Neuropace Inc.
Ms. Kaytlyn Thomas, Neuropace Inc.

References

1. Penfield W, Jasper HH. Epilepsy and the functional anatomy of the human brain. Boston, MA: Little, Brown & Co; 1954.
2. Eggleston KS, Olin BD, Fisher RS. Ictal tachycardia: the head–heart connection. Seizure. 2014;23:496–505.
3. Ben-Menachem E, Hamberger A, Hedner T, Hammond EJ, Uthman BM, Slater J, Treig T, Stefan H, Ramsay RE, Wernicke JF, et al. Effects of vagus nerve stimulation on amino acids and other metabolites in the CSF of patients with partial seizures. Epilepsy Res. 1995;20:221–7.
4. Handforth A, DeGiorgio CM, Schachter SC, Uthman BM, Naritoku DK, Tecoma ES, Henry TR, Collins SD, Vaughn BV, Gilmartin RC, Labar DR, Morris GL 3rd, Salinsky MC, Osorio I, Ristanovic RK, Labiner DM, Jones JC, Murphy JV, Ney GC, Wheless JW. Vagus nerve stimulation therapy for partial-onset seizures: a randomized active-control trial. Neurology. 1998;51:48–55.
5. A randomized controlled trial of chronic vagus nerve stimulation for treatment of medically intractable seizures. Neurology. 1995;45:224–30.
6. Boon P, Vonck K, van Rijckevorsel K, El Tahry R, Elger CE, Mullatti N, Schulze-Bonhage A, Wagner L, Diehl B, Hamer H, Reuber M, Kostov H, Legros B, Noachtar S, Weber YG, Coenen VA, Rooijakkers H, Schijns OE, Selway R, Van Roost D, Eggleston KS, Van Grunderbeek W, Jayewardene AK, RM MG. A prospective, multicenter study of cardiac-based seizure detection to activate vagus nerve stimulation. Seizure. 2015;32:52–61.
7. Chrastina J, Novák Z, Zeman T, Kočvarová J, Pail M, Doležalová I, Jarkovský J, Brázdil M. Single-center long-term results of vagus nerve stimulation for epilepsy: a 10–17 year follow-up study. Seizure. 2018;59:41–7.
8. Elliott RE, Morsi A, Tanweer O, Grobelny B, Geller E, Carlson C, Devinsky O, Doyle WK. Efficacy of vagus nerve stimulation over time: review of 65 consecutive patients with treatment-resistant epilepsy treated with VNS > 10years. Epilepsy Behav. 2011;20:478–83.
9. Ryvlin P, Gilliam FG, Nguyen DK, Colicchio G, Iudice A, Tinuper P, Zamponi N, Aguglia U, Wagner L, Minotti L, Stefan H, Boon P, Sadler M, Benna P, Raman P, Perucca E. The long-term effect of vagus nerve stimulation on quality of life in patients with pharmacoresistant focal epilepsy: the PuLsE (open prospective randomized long-term effectiveness) trial. Epilepsia. 2014;55:893–900.
10. Ryvlin P, So EL, Gordon CM, Hesdorffer DC, Sperling MR, Devinsky O, Bunker MT, Olin B, Friedman D. Long-term surveillance of SUDEP in drug-resistant epilepsy patients treated with VNS therapy. Epilepsia. 2018;59:562–72.
11. Husain MM, Stegman D, Trevino K. Pregnancy and delivery while receiving vagus nerve stimulation for the treatment of major depression: a case report. Ann General Psychiatry. 2005;4:16.
12. Zamponi N, Rychlicki F, Corpaci L, Cesaroni E, Trignani R. Vagus nerve stimulation (VNS) is effective in treating catastrophic 1 epilepsy in very young children. Neurosurg Rev. 2008;31:291–7.
13. Salanova V, Sperling MR, Gross RE, Irwin CP, Vollhaber JA, Giftakis JE, Fisher RS, SANTÉ Study Group. The SANTÉ study at 10 years of follow-up: effectiveness, safety, and sudden unexpected death in epilepsy. Epilepsia. 2021;62:1306–17.

14. Devinsky O, Friedman D, Duckrow RB, Fountain NB, Gwinn RP, Leiphart JW, Murro AM, Van Ness PC. Sudden unexpected death in epilepsy in patients treated with brain-responsive neurostimulation. Epilepsia. 2018;59:555–61.
15. Agnew WF, McCreery DB. Considerations for safety with chronically implanted nerve electrodes. Epilepsia. 1990;31 Suppl 2:S27–32.
16. Simpson HD, Schulze-Bonhage A, Cascino GD, Fisher RS, Jobst BC, Sperling MR, Lundstrom BN. Practical considerations in epilepsy neurostimulation. Epilepsia. 2022;63:2445–60.
17. Ramsay RE, Uthman BM, Augustinsson LE, Upton AR, Naritoku D, Willis J, Treig T, Barolat G, Wernicke JF. Vagus nerve stimulation for treatment of partial seizures: 2. Safety, side effects, and tolerability. First international Vagus nerve stimulation study group. Epilepsia. 1994;35:627–36.
18. Hsieh T, Chen M, McAfee A, Kifle Y. Sleep-related breathing disorder in children with vagal nerve stimulators. Pediatr Neurol. 2008;38:99–103.
19. Fisher R, Salanova V, Witt T, Worth R, Henry T, Gross R, Oommen K, Osorio I, Nazzaro J, Labar D, Kaplitt M, Sperling M, Sandok E, Neal J, Handforth A, Stern J, DeSalles A, Chung S, Shetter A, Bergen D, Bakay R, Henderson J, French J, Baltuch G, Rosenfeld W, Youkilis A, Marks W, Garcia P, Barbaro N, Fountain N, Bazil C, Goodman R, McKhann G, Babu Krishnamurthy K, Papavassiliou S, Epstein C, Pollard J, Tonder L, Grebin J, Coffey R, Graves N, SANTE Study Group. Electrical stimulation of the anterior nucleus of thalamus for treatment of refractory epilepsy. Epilepsia. 2010;51:899–908.
20. Morrell MJ. Responsive cortical stimulation for the treatment of medically intractable partial epilepsy. Neurology. 2011;77:1295–304.
21. Nair DR, Laxer KD, Weber PB, Murro AM, Park YD, Barkley GL, Smith BJ, Gwinn RP, Doherty MJ, Noe KH, Zimmerman RS, Bergey GK, Anderson WS, Heck C, Liu CY, Lee RW, Sadler T, Duckrow RB, Hirsch LJ, Wharen RE Jr, Tatum W, Srinivasan S, GM MK, Agostini MA, Alexopoulos AV, Jobst BC, Roberts DW, Salanova V, Witt TC, Cash SS, Cole AJ, Worrell GA, Lundstrom BN, Edwards JC, Halford JJ, Spencer DC, Ernst L, Skidmore CT, Sperling MR, Miller I, Geller EB, Berg MJ, Fessler AJ, Rutecki P, Goldman AM, Mizrahi EM, Gross RE, Shields DC, Schwartz TH, Labar DR, Fountain NB, Elias WJ, Olejniczak PW, Villemarette-Pittman NR, Eisenschenk S, Roper SN, Boggs JG, Courtney TA, Sun FT, Seale CG, Miller KL, Skarpaas TL, Morrell MJ, RNS System LTT Study. Nine-year prospective efficacy and safety of brain-responsive neurostimulation for focal epilepsy. Neurology. 2020;95:e1244–56.
22. Wu C, Sharan AD. Neurostimulation for the treatment of epilepsy: a review of current surgical interventions. Neuromodulation. 2013;16:10–24.
23. Fisher RS. Deep brain stimulation of thalamus for epilepsy. Neurobiol Dis. 2023;179:106045.
24. Jiltsova E, Möttönen T, Fahlström M, Haapasalo J, Tähtinen T, Peltola J, Öhman J, Larsson EM, Kiekara T, Lehtimäki K. Imaging of anterior nucleus of thalamus using 1.5T MRI for deep brain stimulation targeting in refractory epilepsy. Neuromodulation. 2016;19(8):812–7.
25. Velasco F, Velasco M, Ogarrio C, Fanghanel G. Electrical stimulation of the Centromedian thalamic nucleus in the treatment of convulsive seizures: a preliminary report. Epilepsia. 1987;28:421–30.
26. Benarroch EE. Pulvinar: associative role in cortical function and clinical correlations. Neurology. 2015;84:738–47.
27. Encabo H, Bekerman AJ. Responses evoked in nucleus medialis dorsalis of the thalamus by subcortical stimulation. A microelectrode study. Brain Res. 1971;28:35–46.
28. Cox CL, Huguenard JR, Prince DA. Nucleus reticularis neurons mediate diverse inhibitory effects in thalamus. Proc Natl Acad Sci. 1997;94:8854–9.
29. Hodaie M, Wennberg RA, Dostrovsky JO, Lozano AM. Chronic anterior thalamus stimulation for intractable epilepsy. Epilepsia. 2002;43:603–8.
30. Osorio I, Overman J, Giftakis J, Wilkinson SB. High frequency thalamic stimulation for inoperable mesial temporal epilepsy. Epilepsia. 2007;48:1561–71.

31. Drane DL, Pedersen NP. Finding the sweet spot: fine-tuning DBS parameters to cure seizures while avoiding psychiatric complications. Epilepsy Curr. 2019;19:174–6.
32. Järvenpää S, Peltola J, Rainesalo S, Leinonen E, Lehtimäki K, Järventausta K. Reversible psychiatric adverse effects related to deep brain stimulation of the anterior thalamus in patients with refractory epilepsy. Epilepsy Behav. 2018;88:373–9.
33. Appleby BS, Duggan PS, Regenberg A, Rabins PV. Psychiatric and neuropsychiatric adverse events associated with deep brain stimulation: a meta-analysis of ten years' experience. Mov Disord. 2007;22:1722–8.
34. Rich S, Hutt A, Skinner FK, Valiante TA, Lefebvre J. Neurostimulation stabilizes spiking neural networks by disrupting seizure-like oscillatory transitions. Sci Rep. 2020;10:15408.
35. Elder C, Friedman D, Devinsky O, Doyle W, Dugan P. Responsive neurostimulation targeting the anterior nucleus of the thalamus in 3 patients with treatment-resistant multifocal epilepsy. Epilepsia Open. 2019;4:187–92.
36. Gotman J. Automatic recognition of epileptic seizures in the EEG. Electroencephalogr Clin Neurophysiol. 1982;54:530–40.
37. Bergey GK, Morrell MJ, Mizrahi EM, Goldman A, King-Stephens D, Nair D, Srinivasan S, Jobst B, Gross RE, Shields DC, Barkley G, Salanova V, Olejniczak P, Cole A, Cash SS, Noe K, Wharen R, Worrell G, Murro AM, Edwards J, Duchowny M, Spencer D, Smith M, Geller E, Gwinn R, Skidmore C, Eisenschenk S, Berg M, Heck C, Van Ness P, Fountain N, Rutecki P, Massey A, O'Donovan C, Labar D, Duckrow RB, Hirsch LJ, Courtney T, Sun FT, Seale CG. Long-term treatment with responsive brain stimulation in adults with refractory partial seizures. Neurology. 2015;84:810–7.
38. Chiang S, Fan JM, Rao VR. Bilateral temporal lobe epilepsy: how many seizures are required in chronic ambulatory electrocorticography to estimate the laterality ratio? Epilepsia. 2022;63:199–208.
39. Heck CN, King-Stephens D, Massey AD, Nair DR, Jobst BC, Barkley GL, Salanova V, Cole AJ, Smith MC, Gwinn RP, Skidmore C, Van Ness PC, Bergey GK, Park YD, Miller I, Geller E, Rutecki PA, Zimmerman R, Spencer DC, Goldman A, Edwards JC, Leiphart JW, Wharen RE, Fessler J, Fountain NB, Worrell GA, Gross RE, Eisenschenk S, Duckrow RB, Hirsch LJ, Bazil C, O'Donovan CA, Sun FT, Courtney TA, Seale CG, Morrell MJ. Two-year seizure reduction in adults with medically intractable partial onset epilepsy treated with responsive neurostimulation: final results of the RNS system pivotal trial. Epilepsia. 2014;55:432–41.
40. Alcala-Zermeno JL, Gregg NM, Starnes K, Mandrekar JN, Van Gompel JJ, Miller K, Worrell G, Lundstrom BN. Invasive neuromodulation for epilepsy: comparison of multiple approaches from a single center. Epilepsy Behav. 2022;137:108951.
41. San-Juan D, Espinoza López DA, Vázquez Gregorio R, Trenado C, Fernández-González Aragón M, Morales-Quezada L, Hernandez Ruiz A, Hernandez-González F, Alcaraz-Guzmán A, Anschel DJ, Fregni F. Transcranial direct current stimulation in mesial temporal lobe epilepsy and hippocampal sclerosis. Brain Stimul. 2017;10:28–35.
42. DeGiorgio CM, Soss J, Cook IA, Markovic D, Gornbein J, Murray D, Oviedo S, Gordon S, Corralle-Leyva G, Kealey CP, Heck CN. Randomized controlled trial of trigeminal nerve stimulation for drug-resistant epilepsy. Neurology. 2013;80:786–91.
43. Sun W, Mao W, Meng X, Wang D, Qiao L, Tao W, Li L, Jia X, Han C, Fu M, Tong X, Wu X, Wang Y. Low-frequency repetitive transcranial magnetic stimulation for the treatment of refractory partial epilepsy: a controlled clinical study. Epilepsia. 2012;53:1782–9.
44. Starnes K, Schulze-Bonhage A, Lundstrom B. Neurostimulation for epilepsy: advances, applications and opportunities. Amsterdam: Elsevier Science; 2023.
45. Labate A, Bertino S, Morabito R, Smorto C, Militi A, Cammaroto S, Anfuso C, Tomaiuolo F, Tonin P, Marino S, Cerasa A, Quartarone A. MR-guided focused ultrasound for refractory epilepsy: where are we now? J Clin Med. 2023;12:7070.
46. He W, Jing X, Wang X, Rong P, Li L, Shi H, Shang H, Wang Y, Zhang J, Zhu B. Transcutaneous auricular vagus nerve stimulation as a complementary therapy for pediatric epilepsy: a pilot trial. Epilepsy Behav. 2013;28:343–6.

47. Bauer S, Baier H, Baumgartner C, Bohlmann K, Fauser S, Graf W, Hillenbrand B, Hirsch M, Last C, Lerche H, Mayer T, Schulze-Bonhage A, Steinhoff BJ, Weber Y, Hartlep A, Rosenow F, Hamer HM. Transcutaneous Vagus nerve stimulation (tVNS) for treatment of drug-resistant epilepsy: a randomized, double-blind clinical trial (cMPsE02). Brain Stimul. 2016;9:356–63.
48. Schulze-Bonhage A, Hirsch M, Knake S, Kaufmann E, Kegele J, Rademacher M, Vonck K, Coenen VA, Glaser M, Jenkner C, Winter Y, Groppa S, EASEE Study Group. Focal cortex stimulation with a novel implantable device and Antiseizure outcomes in 2 prospective multicenter single-arm trials. JAMA Neurol. 2023;80:588.

19 Thalamic Neuromodulation

Debopam Samanta, Fuad Aloor, and Zulfi Haneef

Thalamic neuromodulation has emerged as a crucial treatment option for drug-resistant epilepsy (DRE) with widespread and/or undefined epileptic networks (generalized, bilateral multifocal, and epilepsy with unclear seizure-onset zones). In thalamic neuromodulation, thalamic nuclei are electrically stimulated through deep brain stimulation (DBS) and responsive neurostimulation (RNS) depth electrodes, either at predetermined intervals (open loop) or in response to impending seizures (closed loop).

19.1 Thalamic Nuclei and Connectivity

The thalamus, a bilateral paired diencephalic structure divided into approximately 60 distinct nuclei, has long been recognized for its role in epilepsy. Among thalamic nuclei, anterior nucleus (ANT), centromedian nucleus (CMN), and pulvinar neuromodulation are particularly useful for limbic or temporal lobe epilepsies, generalized epilepsies, and posterior temporal and posterior quadrant epilepsies respectively, due to their respective connections and roles within the thalamus [1, 2]. (Fig. 19.1).

D. Samanta (✉)
Child Neurology section, Department of Pediatrics, University of Arkansas for Medical Sciences, Little Rock, AR, USA
e-mail: dsamanta@uams.edu

F. Aloor
Department of Psychiatry and Behavioral Science, Galveston, TX, USA

Z. Haneef
Kellaway Section of Neurophysiology, Baylor College of Medicine, Epilepsy Center of Excellence, DeBakey VA Medical Center, Houston, TX, USA

Z. Haneef (ed.), *Epilepsy Fundamentals*,
https://doi.org/10.1007/978-3-031-77741-7_19

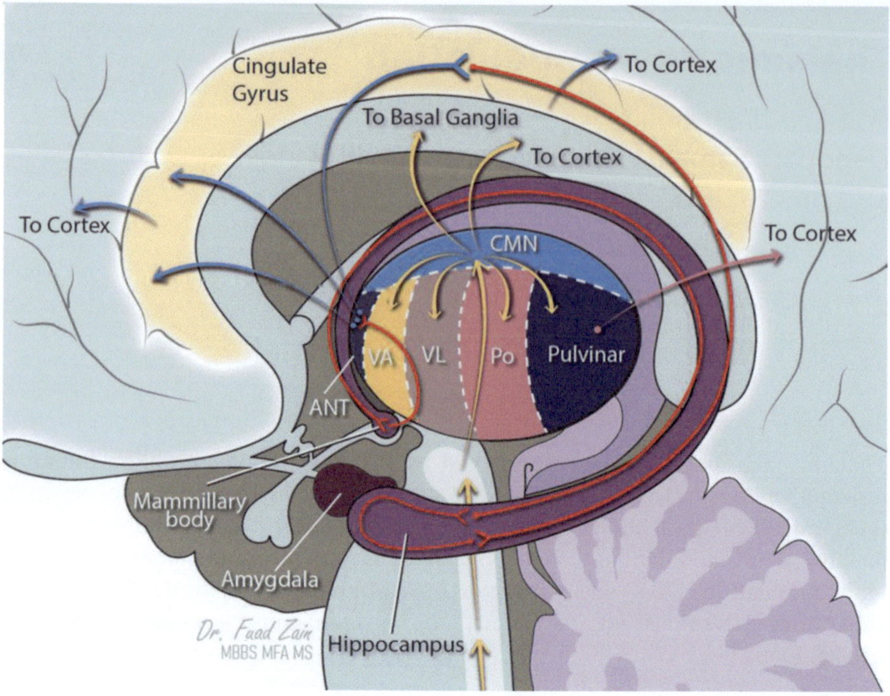

Fig. 19.1 The anterior nuclei (ANT) receive afferents from the mammillary bodies via the mammillothalamic tract and the subiculum via the fornix. In turn, they project to the cingulate gyrus. The centromedian nucleus (CMN) receives diffuse subcortical, intrathalamic, and cortical connectivity, with inputs from the cerebral cortex, vestibular nuclei, globus pallidus, superior colliculus, reticular formation, and spinothalamic tract, while projecting to various regions including the striatum, globus pallidus, substantia nigra, subthalamic nucleus, and somatomotor cortices. The medial pulvinar has reciprocal connections with the amygdala, hippocampus, temporal neocortex, cingulate cortex, orbitofrontal cortex, posterior parietal, premotor, and prefrontal cortical areas

19.1.1 Anterior Nucleus of Thalamus (ANT)

The pivotal SANTE (Stimulation of the Anterior Nucleus of Thalamus for Epilepsy) study and subsequent open-label follow-up studies demonstrated progressive benefits of ANT-DBS in seizure control in multifocal DRE [3, 4, 5]. A wide range of seizure responsiveness has been observed with ANT stimulation in various clinical trials, including randomized controlled trials (RCTs), open-label extensions, and retrospective or prospective registries (Table 19.1). This variability in response may be attributed to patient and epilepsy characteristics and differences in surgical techniques (ability to successfully target ANT) or stimulation parameters.

Table 19.1 RCTs and selective observational DBS studies for ANT stimulation

Study type (author)	Follow-up (n)	Seizure reduction	Additional comments
RCT of ANT DBS (*Fisher 2010* [3])	3 months (110)	40.4% in stim. vs 14.5% in control group (p = 0.002)	Stimulation group reported more depression or memory problems
Open label extension of SANTE trial (ANT DBS) (*Salanova 2015, 2021* [4, 5])	Up to 10 years (103 at 1 yr.; 83 at 5 yr.; 62 at 10 yr)	41% at 1 yr., 69% at 5 yrs., 75% at 7 yrs.	Responder rate: 43% at 1 yr., 68% at 5 yrs. improvement in seizure severity and quality of life
RCT of ANT DBS (*Herrman 2018* [6])	6 months (18)	No statistical significance; 23% ON vs 0% OFF	Significant reduction in the frequency of all seizures (22%) at 6 months (P = 0.009), and small but significant reduction in seizure severity
Observation registry for ANT DBS (*Peltola 2023* [7])	2 and 5 years (170)	33.1% at 2 yrs. 55.1% at 5 yrs	Less improvement in patients with cognitive impairment

RCT randomized controlled trial, *ANT* anterior nucleus of thalamus, *DBS* deep brain stimulation, *Yrs* years

19.1.2 Centromedian Nucleus (CMN) of Thalamus

For the past several decades, the CMN has been recognized as a potential neuromodulation target. In 1987, Velasco et al. made pioneering observations on the benefits of electrical stimulation of the CMN in treating primary generalized or multifocal uncontrollable seizures, which also led to an improvement in psychological performance [8]. Subsequent studies indicated potentially greater benefits in generalized epilepsy, including Lennox-Gastaut syndrome (LGS), compared to focal epilepsy, such as frontal lobe seizures [9, 10]. Several RCTs and observational studies have reported seizure reductions ranging from 30% to 90%, with greater benefits observed over longer follow-up periods (Table 19.2). Interestingly, CM stimulation performed in conjunction with ANT stimulation did not yield additional benefits [11].

Other than the CMN-DBS approach described above, RNS depth electrodes have also been utilized to stimulate thalamic nuclei with or without cortical strips. RNS in CMN showed positive responses in both generalized (e.g., bilateral tonic-clonic seizures (BTCS), tonic seizures, myoclonic seizures, absence seizures, drop seizures) and focal epilepsy (Table 19.3).

19.1.3 Pulvinar

Recent investigations have explored stimulation of the medial pulvinar for posterior temporal and posterior quadrant epilepsies. This approach may facilitate seizure termination, prevent focal-to-bilateral tonic-clonic propagation, decrease impaired consciousness, and promote faster recovery following seizures [21]. RNS has also been employed to detect ictal discharge from the pulvinar or posterior cortex, with successful stimulation of the pulvinar observed in these cases [22].

Table 19.2 RCTs and selective observational DBS studies for centromedian nucleus (CMN) stimulation

Study type (author)	Epilepsy(n)	Mean or median follow-up	Mean or median seizure reduction and other outcomes
RCT of CMN DBS (*Fisher 1992* [12])	DRE (7)	3 months	30% ON vs 8% OFF (not statistically significant between ON and OFF period)
RCT of CMN DBS (*Velasco* et al. 2000 [13])	DRE (13)	3 months	No statistical significance between on and off stimulation
RCT of CMN DBS (*Dalic* et al, *2022* [14])	LGS (20)	6 months	Not statistically significant for clinical seizures but significant for electrographic seizures (59% vs 0% had ≥50% reduction)
Prospective open-label study Bilateral CMN-DBS (*Cukiert 2020* [15])	DEE/LGS (20)	2.55 year	74.5 ± 17.8%; ninety percent of the patients were considered responders (>50% seizure frequency reduction). All patients showed some improvement in attention.
Retrospective chart review of bilateral CM-DBS (*Yang 2022* [16])	DEE, IGE, bifrontal focal (14)	19 ± 5 months	91%; 86% were considered responders (≥ 50% decrease in seizure frequency
Retrospective single-center CM with or without ANT DBS (*Alcala-Zermeno 2021* [11])	DEE, focal (16)	80 months	58%; ten patients (63%) reported ≥50% seizure frequency reduction. Median seizure frequency reduction and responder rate were not significantly different for CM + ANT versus CM only.
Retrospective two-center CM DBS (*Valentin 2013* [10])	DEE, IGE, frontal lobe epilepsy (11)	DEE or IGE—12–60 months, frontal lobe 5–36 months	During the blinded phase, DEE-IGE: All patients had >50% seizure reduction. Frontal lobe epilepsy: Only 1 had >50% seizure reduction; all patients reported an increment in quality of life that was statistically significant

DRE Drug resistant epilepsy, *RCT* randomized clinical trial, *CMN* centromedian nucleus, *ANT* anterior nucleus of thalamus, *LGS* Lennox Gastaut syndrome, *DEE* Developmental and epileptic encephalopathy, *IGE* Idiopathic generalized epilepsy

Table 19.3 Selective RNS studies targeting thalamic nuclei with or without cortical strips

Author	Epilepsy (n), follow-up	Seizure reduction	Additional comments
Burdette 2020 [17]	Focal (7), 17 months	88% (mean: 80%, range: 55–100%)	Four patients had simultaneous or near-simultaneous seizure onsets in the neocortex and CMT, and three had onsets in the neocortex that spread to the CMT.
Beaudreault 2022 [18]	DEE, IGE, focal (14), 35 months	0–24% (n = 3), 25–49% (n = 4), 50–74% (n = 1), 75–99% (n = 4), not available (n = 2)	Ictal activity in neocortical leads preceded thalamic involvement in all patients with combined neocortical and thalamic lead sets.
Roa 2022 [19]	IGE, DEE, focal (23), 22.3 months	0–24% (8.7%), 25–49% (26.1%), 50–99% (60.9%), seizure-free (4.3%)	All patients reported significant improvement in seizure duration and severity, and 17 patients (74%) reported improved post-ictal state.
Fields 2023 [20]	IGE, DEE, focal (25), varies	33% at 6 months, 55% at 1 year, 65% at 2 years, and 74% at >2 years	Global clinical impression—Nine patients were very much improved, six patients were much improved, seven patients were minimally improved, and three patients had no change.

RNS responsive neurostimulation, *CMN* centromedian nuclei, *ANT* anterior nucleus of the thalamus, *DEE* developmental and epileptic encephalopathy, *IGE* idiopathic generalized epilepsy

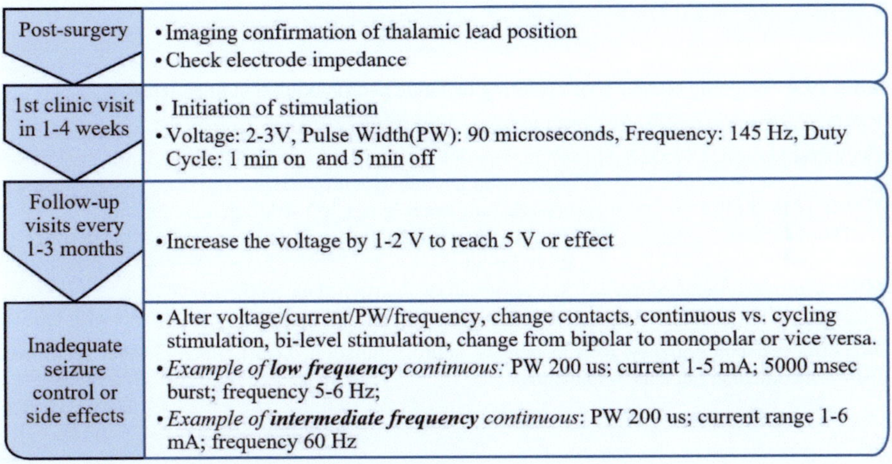

Fig. 19.2 Optimizing thalamic neuromodulation

19.2 Optimizing Thalamic Neuromodulation Parameters

Thalamic neuromodulation involves several factors: target location, voltage/current choice, stimulation frequency (low, intermediate, or high), pulse width, bipolar/referential polarity, and stimulus pattern- (continuous, intermittent, or responsive, especially for RNS). Optimizing parameter settings often requires experimentation, as shown in Fig. 19.2.

19.3 Complications

In the SANTE study, patients reported experiencing various adverse effects, including paresthesia (18%), pain at the implant site (10.9%), and infection at the implant site (9.1%) [3, 4, 5]. Furthermore, subjective symptoms of depression such as dysphoria, anhedonia, pessimism, hopelessness, and sleep disturbances were observed in 37.3% of patients during the long-term follow-up of the SANTE trial [4, 5]. Notably, among these patients, 66% had a pre-existing history of depression. Additionally, suicidal ideation was reported by 11.8% of patients, with one subject committing suicide; however, the suicide was not believed to be related to the device. Memory impairment was noted in approximately 27.3% of patients, with half of these individuals having a history of memory impairment before the study commenced [4, 5].

19.4 Future Directions

Thalamic stimulation represents a cutting-edge approach in epilepsy treatment, and the future holds promise for clinical and research advancements in this area, including expanded applications to target complex epileptic networks. Several clinical trials are currently underway, and patients are actively recruited to explore these potentials further (Table 19.4).

Table 19.4 Ongoing clinical trial exploring thalamic neuromodulation in epilepsy

Clinical trial	Study focus
PULSE study (NCT04692701)	Evaluating duty-cycle of bilateral pulvinar thalamic nucleus stimulation in patients with DRE
NAUTILUS study (NCT05147571)	Investigating the role of CMN-RNS in patients with drug-resistant idiopathic generalized epilepsy who are over 12 years old
RNS system LGS feasibility study (NCT05339126)	Assessing safety and preliminary effectiveness of corticothalamic CMN-RNS stimulation in patients with LGS

DRE drug-resistant epilepsy, *RNS* responsive neurostimulation, *LGS* Lennox-Gastaut syndrome, *CMN* centromedian nucleus of thalamus

References

1. Warsi NM, Yan H, Suresh H, Wong SM, Arski ON, Gorodetsky C, et al. The anterior and centromedian thalamus: anatomy, function, and dysfunction in epilepsy. Epilepsy Res. 2022;182:106913.
2. Rosenberg DS, Mauguière F, Demarquay G, Ryvlin P, Isnard J, Fischer C, et al. Involvement of medial pulvinar thalamic nucleus in human temporal lobe seizures. Epilepsia. 2006;47(1):98–107.

3. Fisher R, Salanova V, Witt T, Worth R, Henry T, Gross R, et al. Electrical stimulation of the anterior nucleus of thalamus for treatment of refractory epilepsy. Epilepsia. 2010;51(5):899–908.
4. Salanova V, Witt T, Worth R, Henry TR, Gross RE, Nazzaro JM, et al. Long-term efficacy and safety of thalamic stimulation for drug-resistant partial epilepsy. Neurology. 2015;84(10):1017–25.
5. Salanova V, Sperling MR, Gross RE, Irwin CP, Vollhaber JA, Giftakis JE, et al. The SANTÉ study at 10 years of follow-up: effectiveness, safety, and sudden unexpected death in epilepsy. Epilepsia. 2021;62(6):1306–17.
6. Herrman H, Egge A, Konglund AE, Ramm-Pettersen J, Dietrichs E, Taubøll E. Anterior thalamic deep brain stimulation in refractory epilepsy: a randomized, double-blinded study. Acta Neurol Scand. 2019;139(3):294–304.
7. Peltola J, Colon AJ, Pimentel J, Coenen VA, Gil-Nagel A, Gonçalves Ferreira A, et al. Deep brain stimulation of the anterior nucleus of the thalamus in drug-resistant epilepsy in the MORE multicenter patient registry. Neurology. 2023;100(18):e1852–e65.
8. Velasco F, Velasco M, Ogarrio C, Fanghanel G. Electrical stimulation of the centromedian thalamic nucleus in the treatment of convulsive seizures: a preliminary report. Epilepsia. 1987;28(4):421–30.
9. Velasco F, Velasco M, Jimenez F, Velasco AL, Marquez I. Stimulation of the central median thalamic nucleus for epilepsy. Stereotact Funct Neurosurg. 2001;77(1–4):228–32.
10. Valentín A, García Navarrete E, Chelvarajah R, Torres C, Navas M, Vico L, et al. Deep brain stimulation of the centromedian thalamic nucleus for the treatment of generalized and frontal epilepsies. Epilepsia. 2013;54(10):1823–33.
11. Alcala-Zermeno JL, Gregg NM, Wirrell EC, Stead M, Worrell GA, Van Gompel JJ, et al. Centromedian thalamic nucleus with or without anterior thalamic nucleus deep brain stimulation for epilepsy in children and adults: a retrospective case series. Seizure. 2021;84:101–7.
12. Fisher RS, Uematsu S, Krauss GL, Cysyk BJ, McPherson R, Lesser RP, et al. Placebo-controlled pilot study of centromedian thalamic stimulation in treatment of intractable seizures. Epilepsia. 1992;33(5):841–51.
13. Velasco F, Velasco M, Jiménez F, Velasco AL, Brito F, Rise M, et al. Predictors in the treatment of difficult-to-control seizures by electrical stimulation of the centromedian thalamic nucleus. Neurosurgery. 2000;47(2):295–304; discussion -5.
14. Dalic LJ, Warren AEL, Bulluss KJ, Thevathasan W, Roten A, Churilov L, et al. DBS of thalamic Centromedian nucleus for Lennox-Gastaut syndrome (ESTEL trial). Ann Neurol. 2022;91(2):253–67.
15. Cukiert A, Cukiert CM, Burattini JA, Mariani PP. Seizure outcome during bilateral, continuous, thalamic centromedian nuclei deep brain stimulation in patients with generalized epilepsy: a prospective, open-label study. Seizure. 2020;81:304–9.
16. Yang JC, Bullinger KL, Isbaine F, Alwaki A, Opri E, Willie JT, et al. Centromedian thalamic deep brain stimulation for drug-resistant epilepsy: single-center experience. J Neurosurg. 2022;137(6):1591–600.
17. Burdette DE, Haykal MA, Jarosiewicz B, Fabris RR, Heredia G, Elisevich K, et al. Brain-responsive corticothalamic stimulation in the centromedian nucleus for the treatment of regional neocortical epilepsy. Epilepsy Behav. 2020;112:107354.
18. Beaudreault CP, Muh CR, Naftchi A, Spirollari E, Das A, Vazquez S, et al. Responsive Neurostimulation targeting the anterior, Centromedian and Pulvinar thalamic nuclei and the detection of electrographic seizures in pediatric and young adult patients. Front Hum Neurosci. 2022;16:876204.
19. Roa JA, Abramova M, Fields M, Vega-Talbott M, Yoo J, Marcuse L, et al. Responsive Neurostimulation of the thalamus for the treatment of refractory epilepsy. Front Hum Neurosci. 2022;16:926337.
20. Fields MC, Eka O, Schreckinger C, Dugan P, Asaad WF, Blum AS, et al. A multicenter retrospective study of patients treated in the thalamus with responsive neurostimulation. Front Neurol. 2023;14:1202631.

21. Filipescu C, Lagarde S, Lambert I, Pizzo F, Trébuchon A, McGonigal A, et al. The effect of medial pulvinar stimulation on temporal lobe seizures. Epilepsia. 2019;60(4):e25–30.
22. Burdette D, Mirro EA, Lawrence M, Patra SE. Brain-responsive corticothalamic stimulation in the pulvinar nucleus for the treatment of regional neocortical epilepsy: a case series. Epilepsia Open. 2021;6(3):611–7.

Status Epilepticus

20

Sally Mathias and Jyoti Pillai

Status epilepticus (SE) is an emergency associated with increased morbidity, mortality and significant healthcare costs, estimated at $4 billion in the US [1]. The incidence of SE increased from 3.5 to 12.5/100,000 between 1979 and 2010, possibly due to increasing recognition by continuous EEG (cEEG) monitoring and evolving SE definitions. SE incidence has a bimodal peak in the first and after the fifth decades of life [2]. The average cumulative incidence of SE is 36.1/ 100,000 [3].

20.1 Definition

There has been a gradual shortening in the time needed to diagnose SE. It is known that a seizure lasting for >5–10 min is unlikely to spontaneously remit and requires treatment [5, 6]. In 2015, the International League Against Epilepsy (ILAE) and the Commission on Epidemiology proposed a definition that helps in standardizing the diagnosis (Table 20.1) [4]. The time point t1 refers to the time when treatment should be initiated, and t2 determines the intensity of treatment to prevent irreversible brain damage. These time points for convulsive SE are based on animal/clinical research suggesting that most convulsive seizures last <5 min and irreversible brain damage usually occurs after 30 min of seizures. There is limited information on t1 and t2 in focal SE and absence SE [4].

S. Mathias (✉)
Associate Professor of Neurology, University of Kentucky College of Medicine, Chief of Neurology Service, Lexington Veterans Affairs (VA) Health Care System, Lexington, Kentucky, USA
e-mail: sally.mathias@va.gov; sally.mathias@uky.edu

J. Pillai
Program Director, Neurology Residency, Medical Director, Epilepsy, Tower Health, Reading Hospital, Philadelphia, Pennsylvania, USA

Associate Professor of Neurology, Drexel University College of Medicine, Philadelphia, Pennsylvania, USA

Z. Haneef (ed.), *Epilepsy Fundamentals*,
https://doi.org/10.1007/978-3-031-77741-7_20

Table 20.1 ILAE definition of status epilepticus. Operational dimensions (ILAE) indicate times t1 (when seizure is likely to get prolonged and emergency treatment should be started) and t2 (when long-term neuronal consequences are expected) for different kinds of SE [4]

	t1	t2
Tonic-clonic SE	5 min	30 min
Focal SE with impaired consciousness	10 min	>60 min
Absence SE	10–15 min	Unknown

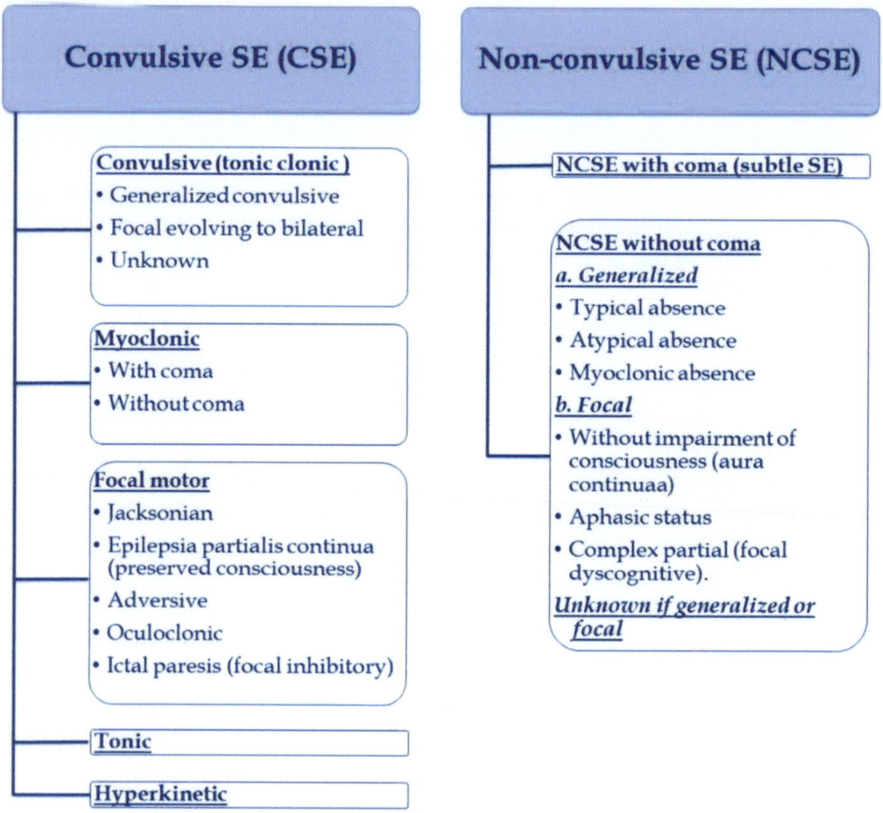

Fig. 20.1 Classification of status epilepticus

20.2 Classification

The ILAE task force proposed classifying SE on the following 4 axes: (1) semiology, (2) etiology, (3) EEG correlates, and (4) age. SE can also be classified based on semiology into convulsive (with prominent motor symptoms) and nonconvulsive (Fig. 20.1).

Convulsive SE (CSE) (SE with Prominent Motor Phenomena) CSE manifests as self-perpetuating seizures featuring a series of sustained or interrupted generalized tonic-clonic seizures without return of consciousness to baseline in between [7]. CSE has the highest morbidity among all types of SE. Clonic movements diminish as seizures progress and ultimately cease, even with ongoing electrographic SE.

Nonconvulsive SE (NCSE) NCSE is defined as seizure activity on EEG without prominent motor phenomena. The absence of significant improvement in consciousness 20–30 min after the cessation of convulsions should raise the concern of NCSE, which is seen in almost half the patients, and an EEG should be performed urgently to confirm this [8]. Patients may exhibit subtle symptoms such as confusion, lethargy, staring, mutism, catatonia, agitation, psychosis, delirium, blinking, automatisms, facial twitching, tremulousness, rhythmic muscle twitches, tonic eye deviation or nystagmus. NCSE affects up to 10% of critically ill patients with altered consciousness, especially after GCSE (48%), intracranial hemorrhage, or CNS infection. About 75% of seizures in the critically ill are nonconvulsive. Unlike CSE, NCSE relies heavily on expert EEG interpretation.

20.3 Etio-pathogenesis

The etiology of SE (Table 20.2) can vary by age and geographic region. Animal and human studies indicate that permanent cerebral damage may occur in SE due to systemic metabolic disturbances (e.g., hypoxia, hypoglycemia) or as the direct consequence of the electrical discharges in the brain [9]. Initially, compensatory mechanisms including increases in cerebral blood flow can meet the metabolic demand of the intense motor and cerebral activity. After 5–30 min of GCSE, compensatory mechanisms progressively fail, and cerebral blood flow becomes dependent on systemic blood pressure. Ongoing seizure activity may result in rhabdomyolysis, hyperthermia, and multiple organ failure. Intracranial hypertension and cerebral edema may ensue [8].

In SE, events at the receptor level involve a decrease in inhibitory GABA A, beta 2/beta 3 and gamma 2 receptor subunits through endocytosis, and increased internalization of GABA-A receptors, reducing the effect of GABAergic stimulation. This process may result in reduced clinical effectiveness of GABA antagonists such as benzodiazepines as SE persists over time. In addition, with increasing duration of SE, there is an increase in excitatory NMDA receptors. Further changes include changes in excitatory and inhibitory neuropeptide expression, maintaining the hyperexcitable state, changes in genetic and epigenetics, including altered regulation of microRNA in days and weeks after SE [1].

Table 20.2 Causes of status epilepticus

Subtherapeutic ASM
Breakthrough seizures
Remote symptomatic
Prior stroke
Traumatic brain injury (TBI)
Post-encephalitis
Acute stroke
Acute trauma
Metabolic
Hypo- and hyperglycemia
Electrolyte imbalance (Mg, Ca, Na)
Hepatic/renal failure
Infectious
Sepsis
CNS infections
Anoxic brain injury
Drug intoxications
Alcohol withdrawal
Hypertensive encephalopathy
Neoplasms, primary or metastatic.
Congenital malformations
Cryptogenic
Immune-mediated
Autoimmune encephalitis (anti-NMDAR encephalitis, VGKC-related proteins such as LGI1 antibody-mediated limbic encephalitis, AMPA receptor, GABA receptor encephalitis)
Paraneoplastic syndromes

20.4 Management

Prompt initiation of management is crucial in SE, given its life-threatening nature. Management should begin with early recognition and a detailed and relevant history from family, caregivers or emergency personnel.

20.4.1 Treatment of Convulsive SE (Fig. 20.2)

Initial management includes maintenance of the airway, breathing, and circulation. CSE should be treated emergently to enhance responsiveness and outcome. In addition to the rapid administration of a first-line ASM, metabolic factors causing or contributing to SE should be corrected, including administration of thiamine followed by glucose when glucose levels are low or unknown and correction of electrolyte imbalances. Aggressive control of hyperthermia is essential and definitive control of CSE should be established within 30 min of onset.

First-line therapy: Benzodiazepines are the preferred agents for emergent initial treatment. Multiple studies demonstrate the higher efficacy of IV lorazepam in aborting SE [10, 11]. IM midazolam is a safe and effective alternative [12]. Intranasal or buccal midazolam, as well as rectal or intranasal diazepam are

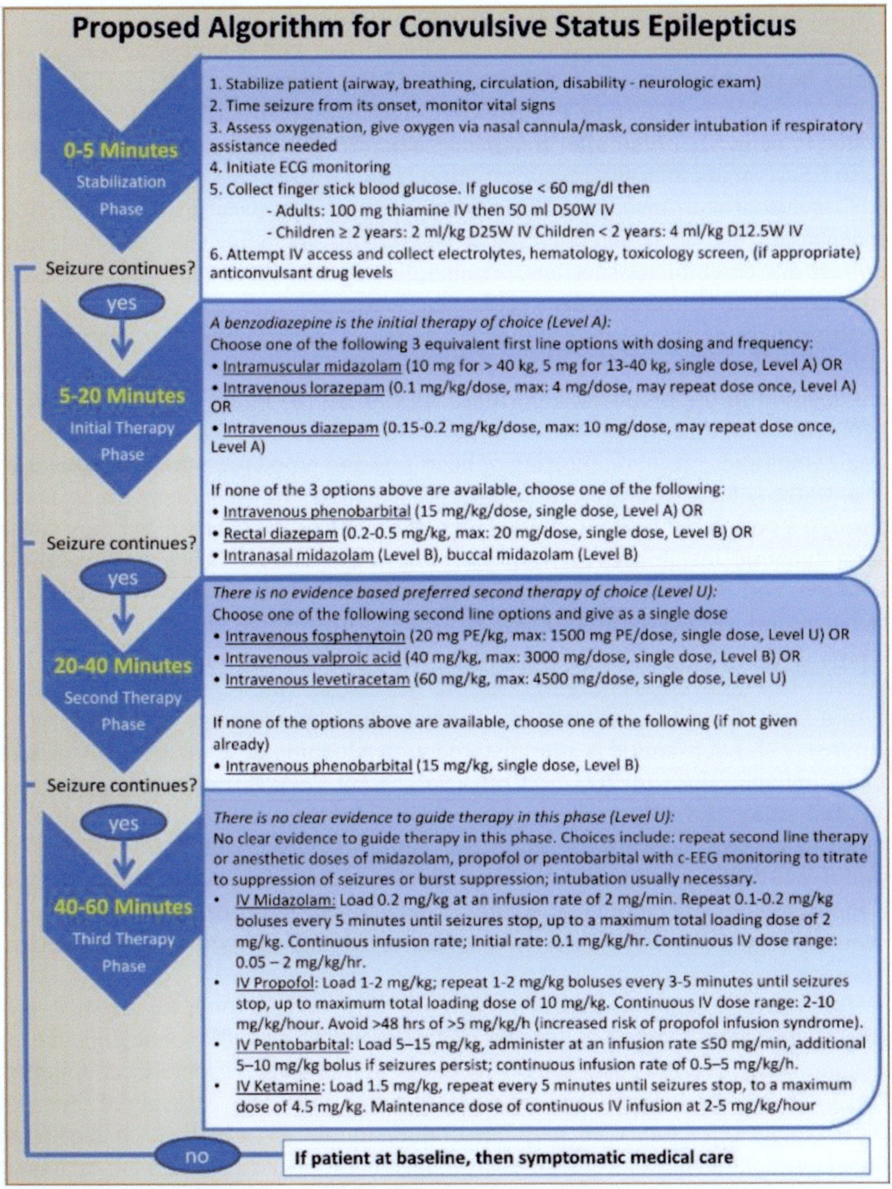

Fig. 20.2 Managementalgorithm for status epilepticus adapted from the proposal by the Guideline Committee of the American Epilepsy Society (CBC: complete blood count, PE: phenytoin equivalents)

considered safe options [13, 14]. Although respiratory compromise is a concern with benzodiazepines, it happens at a rate lower than with placebo, suggesting that it may be the consequence of untreated CSE [15]. Underdosing of benzodiazepines is a real problem, as seen in a prospective study where 70% of CSE patients continued to be in SE 1 hour after treatment, and 58% of NCSE patients continued to be in SE 12 hours after treatment initiation [16].

Concurrent with managing SE, workup for underlying causes like meningitis, encephalitis, and acute cerebrovascular pathologies, including intra-cerebral hemorrhage and cerebral mass lesions, is imperative. A third of the patients with SE are unresponsive to benzodiazepines [17].

Regardless of the response to benzodiazepines, second-line ASMs are administered to all patients to prevent recurrence. ASM options available in IV form are detailed in the proposed algorithm Fig. 20.2. [18] Phenytoin poses risks of hypotension and arrhythmias- requiring slow infusion, avoidance of peripheral small veins and careful monitoring of heart rate and blood pressure. Fosphenytoin is a water-soluble pro-drug alternative that can be infused three times faster through a peripheral line and carries no risk of soft-tissue necrosis but may cause hemodynamic instability. IV valproate offers the same efficacy with minimal sedation and without cardiorespiratory suppression. Levetiracetam [19] and lacosamide [20, 21] show promise in small retrospective series. A systematic review found equal efficacy and safety between lacosamide (57.3%) and phenytoin (45.7%) in controlling SE. Serious adverse events were higher in the phenytoin arm (cardiac arrhythmias, bradycardia, hypotension and respiratory failure) [22]. Lacosamide is administered with a loading dose of 200–400 mg and a maintenance dose of 200–300 mg every 12 h [23]. A randomized, double-blinded study (ESETT trial) comparing IV fosphenytoin (20 mg/kg), valproic acid (40 mg/kg), and levetiracetam (60 mg/kg) infused over 10 min found no significant difference in efficacy between them for SE cessation and improved responsiveness at 60 min [17]. Another study found no difference in respiratory depression and hypotension between phenobarbital, phenytoin, levetiracetam, lacosamide, valproate and diazepam [24].

Treatment choices may depend on comorbidities and etiology; for instance, valproate should be avoided in patients with hepatic failure, thrombocytopenia, or coagulopathy, but it is the drug of choice in idiopathic generalized epilepsy. Phenytoin is contraindicated in such patients due to potential worsening of seizures. Patients with refractory CSE may need intubation for IV anesthetic medications. Rapid termination of SE is crucial as prolonged duration of SE can lead to treatment refractoriness. The management algorithm (Fig. 20.2) outlines the necessary steps in management, including laboratory tests [18].

20.4.2 Treatment of Nonconvulsive SE

Treatment of refractory NCSE differs from refractory CSE, due to the lower risk of severe acute systemic complications. The need for aggressive treatment in NCSE is debated. Proponents suggest that aggressive treatment prevents refractoriness and adverse outcomes [25, 26], while opponents question the unknown impact on outcomes, suggesting a cautious approach to avoid cardiorespiratory compromise and associated complications, especially outside of the ICU. It may be prudent to delay anesthetic agents in NCSE, avoiding unnecessary intubation and exploring other non-sedating ASMs. The Salzburg diagnostic criteria for NCSE (Table 20.3) help prevent overdiagnosis in EEGs affected solely by the underlying condition [27]. If absence SE is confirmed on EEG, intravenous diazepam or derivatives, rather than phenytoin or the barbiturates, are recommended, with intravenous valproate also being considered effective.

20.4.3 Treatment of Focal-Aware SE

Treatment of focal aware SE (simple partial SE/ epilepsia partialis continua, see Chapter 2) should be less aggressive, often requiring several hours to days to control. The risk with intubation/sedation is often greater than the risk of continued focal seizures.

Refractory SE (RSE) As per the international Consensus definitions for SE subtypes [28], RSE is SE persisting despite administration of at least two appropriately selected and dosed parenteral medications, including a benzodiazepine, with no specific seizure duration required. It involves aggressive treatment, such as

Table 20.3 Salzburg Criteria proposed for diagnosing NCSE [37] [40]

	EEG pattern	And either of the following
Patients without known epileptic encephalopathy	• EDs >2.5 Hz.	
	• EDs ≤2.5 Hz **OR** rhythmic delta/theta activity (>0.5 Hz).	• EEG and clinical improvement after IV ASM. • Subtle clinical ictal phenomena in EEG patterns. • Typical spatiotemporal evolution.
Patients with known epileptic encephalopathy.	• Increased epileptiform discharges compared to baseline **OR** increased rhythmic delta/theta activity compared to baseline.	• Observable change in clinical state. • Improvement in clinical and EEG with IV ASM.

Abbreviations: EDs—epileptiform discharges (spikes, poly-spikes, sharp waves, sharp and slow wave complexes), IV ASM—intravenous anti-seizure medications therapeutic coma, with an anesthetic dose (initial bolus/es followed by infusion) of an ASM [29] like midazolam, propofol, and pentobarbital. There is little evidence to choose between these agents. Midazolam may cause tachyphylaxis and breakthrough seizures, pentobarbital can lead to hypotension and immunosuppression, while propofol poses a risk of "propofol infusion syndrome" (metabolic acidosis, renal failure, cardiac dysfunction, and rhabdomyolysis, especially with prolonged administration >48 h at a high dose >5 mg/kg/h). Assisted ventilation, cardiovascular monitoring, and potential vasopressor use to treat hypotension are required during their use.

Super-Refractory SE (SRSE) SE persisting at least 24 hours after onset of anesthesia, either without interruption, recurring while on appropriate anesthetic treatment, or recurring after withdrawal of anesthesia and requiring anesthetic reintroduction is termed SRSE. It represents up to 20% of cases of RSE [30]. It is more common in younger patients with encephalitis. Therapeutic options include a second trial of the same anesthetic drug; switching to another anesthetic drug; combining multiple anesthetic drugs and ASMs, including ketamine; immune therapies such as steroids, plasma exchange, or IVIg (particularly in patients suspected of immune mediated SE such as Rasmussen's encephalitis, Hashimoto's encephalopathy or related to a paraneoplastic process), ketogenic diet, hypothermia, electroconvulsive therapy, surgical management, and VNS [30]. RNS, direct cortical stimulation, and electroconvulsive therapy have also been described. Brivaracetam, pregabalin, topiramate, perampanel and clobazam are additional add-on options based on small studies in patients with refractory and super refractory SE [31–33].

Prolonged SRSE (PSRSE) SRSE that persists for at least 7 days, including ongoing need for anesthetics.

NORSE and FIRES New-onset refractory SE (NORSE) is a clinical presentation in patients without a history of active epilepsy, preexisting neurological disorders or clear acute or active structural, metabolic, or toxic cause. It can occur in patients with remote brain injuries or resolved epilepsy. Febrile infection–related epilepsy syndrome (FIRES) is a subset of NORSE, requiring a prior febrile infection between 2 weeks and 24 h before the onset of RSE whereas the presence of fever at the onset of SE is not a defining criterion [28]. Infections are the most prevalent etiology of pediatric NORSE [34], while adults often have unidentified etiologies in 50% patients with autoimmune encephalitis (either paraneoplastic or non-paraneoplastic) [35] being a frequent culprit, followed by viral infections including HSV1. A fulminant CNS inflammatory response is the proposed mechanism [30]. NORSE is highly resistant to conventional treatment with ASMs and often requires immunotherapy. NORSE does not encompass refractory SE with fully retained consciousness such as epilepsies partialis continua [28].

EEG EEG is not required for the diagnosis and the early management of GCSE but is valuable in identifying electrographic SE in patients who fail to regain consciousness after a generalized tonic-clonic seizure, in patients with NCSE, and to guide management of SE with anesthetics.

The 2013 Salzburg criteria proposal definition of NCSE, as outlined in Table 20.3 [36, 37] provides a high sensitivity (97.2%) and specificity (95.9%) for diagnosing NCSE [38]. The American Clinical Neurophysiology Society (ACNS) adopted the Salzburg Consensus Criteria in their 2021 Standard Critical Care EEG Terminology. Electroclinical SE/nonconvulsive SE is defined as an electroclinical seizure lasting for ≥10 continuous minutes or a total duration of ≥20% of any 60-minute recording period [39]. However, diagnosing NCSE can be challenging, requiring careful interpretation due to overlapping patterns with ictal, interictal or metabolic encephalopathy-related activity.

A 30- to 60-minute EEG detects seizures in only 45–58% of patients. The ACNS recommends cEEG monitoring for critically ill adults and children. The monitoring duration depends on the pre-test probability for seizures, with a minimum of *24 hours* recommended. In specific cases like comatose patients, those under sedation, or with periodic discharges on the EEG, prolonged monitoring of *48 hours* or more is recommended [41].cEEG should be continued for at least 24 h after controlling SE to exclude recurrence of nonconvulsive seizures or NCSE [42].

Neuroimaging MRI can reveal peri-ictal abnormalities in SE, with cerebral hyperperfusion indicating increased metabolic demand due to ongoing ictal activity and cytotoxic edema, suggesting neuronal damage. Increased cortical/subcortical signals on T2/DWI are the most common manifestations. In most patients, MRI changes match the EEG focus, particularly LPDs. In most cases (80%), MRI changes improve within 1–6 weeks, but persistent alterations may lead to local atrophy. Temporal lobe SE is associated with thalamic DWI changes [13, 43]. Head CT is indicated once convulsions are controlled and lumbar puncture is recommended if an infectious, inflammatory or autoimmune cause of SE is suspected after seizure control.

Ictal-Interictal Continuum (IIC) IIC, synonymous with "possible electrographic seizure" or "possible electrographic SE," suggests a reasonable chance of EEG pattern contributing to impaired alertness, clinical symptoms, and/or contributing to neuronal injury. It often warrants a diagnostic treatment trial, typically with a parenteral ASM.

Patterns on the IIC include **(a)** any periodic discharge (PD) or spike and wave (SW) pattern that averages >1.0 and ≤ 2.5 Hz over 10 s (>10 and ≤ 25 discharges in 10 s), **(b)** any PD or SW pattern that averages ≥0.5 Hz and ≤ 1.0 Hz over 10 s (≥5 and ≤ 10 discharges in 10 s) and has a plus modifier or fluctuation, or **(c)** any lateralized rhythmic delta activity (LRDA) averaging >1 Hz for at least 10 s (at least 10 waves in 10 s) with a plus modifier or fluctuation. This includes any LRDA, BIRDA, UIRDA, and MfRDA, but not GRDA and **(d)** It does not qualify as an electrographic seizure or electrographic SE.

If treatment of a pattern on the IIC with a parenteral ASM leads to improvement in the EEG and definite clinical improvement, this would meet the secondary criterion of an electrographic seizure or electroclinical SE. If treatment of an IIC pattern with a parenteral ASM leads to improvement in the EEG without clinical improvement, this would be possible electroclinical SE [39]. If there is evidence of clinical seizures, these are considered ictal. The potential for neuronal injury (y-axis, Fig. 20.3) is a helpful guide in deciding how aggressively to treat.

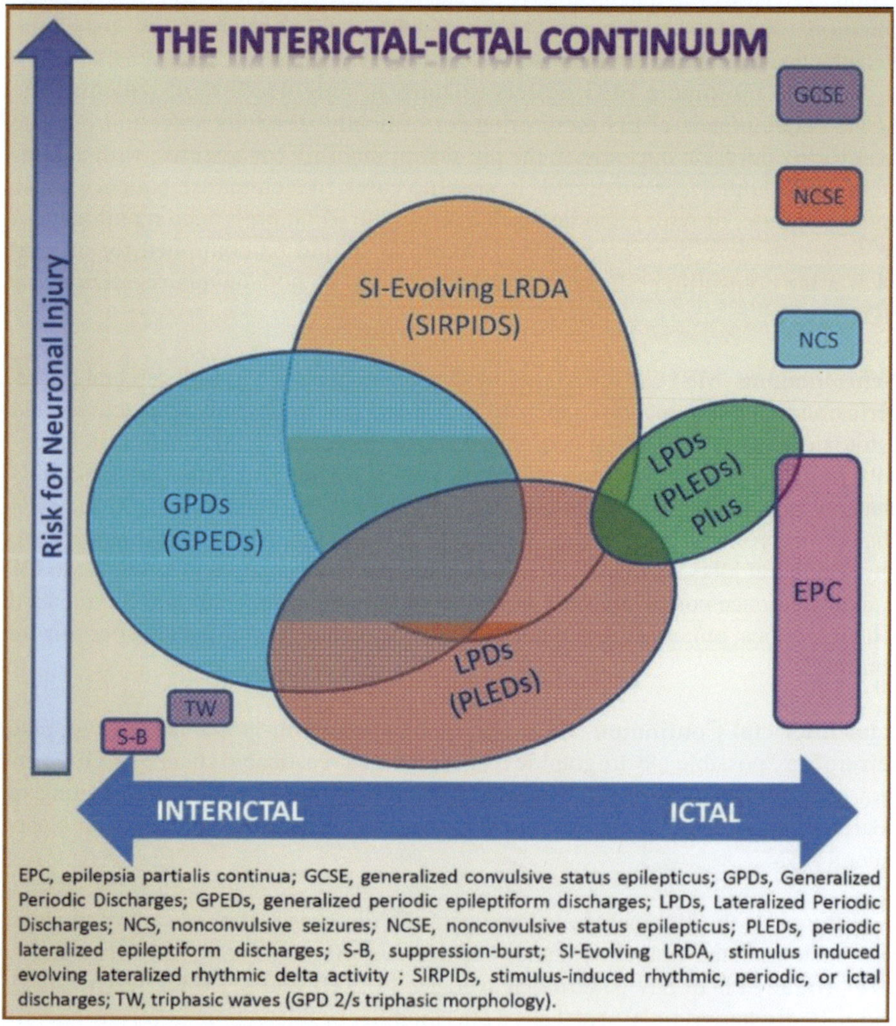

Fig. 20.3 The schematic of an ictal-interictal continuum as adapted from Chong and Hirsch. [40] Treatment should be more aggressive with a greater risk of neuronal injury (y-axis)

20.5 Outcomes

A systematic review reported SE mortality rates of 15.9% in adults, 3.6% in children, and 17.3% in RSE [44].Poor outcomes are associated with anoxia, hypoxia, or multiple medical problems, while low mortality is linked to a known epilepsy diagnosis and low serum ASMs. Underlying structural cause predicts recurrent epilepsy, as does age less than 4 years. Long-term cognitive sequelae occur in 30% of children, and functional sequelae in 20–60% of adults [34].

Age is a significant predictor, with the youngest and oldest (>65 years) having the worst outcomes. SE is associated with high recurrence, development of subsequent epilepsy and worsening of previous epilepsy. The recurrence rates of SE ranged from 10 to 56% in children and 13–37% in a mixed population of adults and children. [45, 48]

In RSE, 75% have a poor outcome, with a 32% in-hospital mortality. Poor predictors include duration of drug-induced coma, arrhythmias requiring intervention, pneumonia, and prolonged mechanical ventilation. Achieving seizure control without an isoelectric EEG or burst suppression pattern is associated with a good functional recovery [46].

A prognostic score, SE Severity Score (STESS) (Table 20.4) relies on four outcome predictors. A score of 0–2 is favorable, indicating a low risk of death, with an excellent negative predictive value (0.97). A favorable STESS was highly related to survival and a return to baseline clinical condition in survivors. Early aggressive treatment may not be routinely warranted in patients with a favorable STESS, who are likely to survive their SE episode [47].

Table 20.4 Status Epilepticus Severity Score (STESS)

	Features	Score
Consciousness	Alert or somnolent/ confused	0
	Stuporous or comatose	1
Worst seizure type	Simple-partial, absence, myoclonic	0
	GCSE	1
	NCSE in coma	2
Age	< 65 years	0
	≥ 65 years	2
Previous seizures	Yes	0
	No or unknown	1
Total:		**0–6**

Pearls: Status Epilepticus

- The time duration at which treatment for convulsive SE should be started is 5 min, and 30 min is when long-term neuronal consequences are expected if a patient continues to be in convulsive SE.
- With increasing duration in SE, there is a reduction of GABA-A receptors and an increase in excitatory NMDA receptors, thus making GABAergic anti-seizure medications less effective.
- Regardless of the response to initial intravenous benzodiazepines, second-line anti-seizure medications are administered to all patients in SE to prevent recurrence.
- Topiramate and Zonisamide should be avoided if a patient is on propofol, is acidotic or has renal stones as they can cause metabolic acidosis.
- Phenytoin and lacosamide should be avoided if patients have preexisting cardiac conduction delay.

Acknowledgements Saba Jafarpour and Tobias Loddenkemper for authoring a previous version of this chapter.

References

1. Betjemann JP, Lowenstein DH. Status epilepticus in adults. Lancet Neurol. 2015;14:615–24.
2. Dham BS, Hunter K, Rincon F. The epidemiology of status epilepticus in the United States. Neurocrit Care. 2014;20:476–83.
3. Leitinger M, Trinka E, Giovannini G, et al. Epidemiology of status epilepticus in adults: a population-based study on incidence, causes, and outcomes. Epilepsia. 2019;60:53–62.
4. Trinka E, Cock H, Hesdorffer D, et al. A definition and classification of status epilepticus—report of the ILAE task force on classification of status epilepticus. Epilepsia. 2015;56:1515–23.
5. Shinnar S, Berg AT, Moshe SL, Shinnar R. How long do new-onset seizures in children last? Ann Neurol. 2001;49:659–64.
6. Jenssen S, Gracely EJ, Sperling MR. How long do most seizures last? A systematic comparison of seizures recorded in the epilepsy monitoring unit. Epilepsia. 2006;47:1499–503.
7. Blume WT, Lüders HO, Mizrahi E, Tassinari C, van Emde BW, Engel J. Glossary of descriptive terminology for ictal semiology: report of the ILAE task force on classification and terminology. Epilepsia. 2001;42:1212–8.
8. Hirsch LJ, Gaspard N. Status epilepticus Contin Minneap Minn. 2013;19:767–94.
9. Chen JWY, Wasterlain CG. Status epilepticus: pathophysiology and management in adults. Lancet Neurol. 2006;5:246–56.
10. Alldredge BK, Gelb AM, Isaacs SM, et al. A comparison of lorazepam, diazepam, and placebo for the treatment of out-of-hospital status epilepticus. N Engl J Med. 2001;345:631–7.
11. Treiman DM, Meyers PD, Walton NY, et al. A comparison of four treatments for generalized convulsive status epilepticus. Veterans affairs status epilepticus cooperative study group. N Engl J Med. 1998;339:792–8.

12. Silbergleit R, Lowenstein D, Durkalski V, Conwit R, NETT Investigators. Lessons from the RAMPART study—and which is the best route of administration of benzodiazepines in status epilepticus. Epilepsia. 2013;54(Suppl 6):74–7.
13. Trinka E, Leitinger M. Management of Status Epilepticus, refractory status epilepticus, and super-refractory status epilepticus. Contin Minneap Minn. 2022;28:559–602.
14. Brigo F, Nardone R, Tezzon F, Trinka E. Nonintravenous midazolam versus intravenous or rectal diazepam for the treatment of early status epilepticus: a systematic review with meta-analysis. Epilepsy Behav EB. 2015;49:325–36.
15. Glauser T, Shinnar S, Gloss D, et al. Evidence-based guideline: treatment of convulsive status epilepticus in children and adults: report of the guideline Committee of the American Epilepsy Society. Epilepsy Curr. 2016;16:48–61.
16. Kellinghaus C, Rossetti AO, Trinka E, et al. Factors predicting cessation of status epilepticus in clinical practice: data from a prospective observational registry (SENSE). Ann Neurol. 2019;85:421–32.
17. Kapur J, Elm J, Chamberlain JM, et al. Randomized trial of three anticonvulsant medications for status epilepticus. N Engl J Med. 2019;381:2103–13.
18. Glauser T, Shinnar S, Gloss D, et al. Evidence-based guideline: treatment of convulsive status epilepticus in children and adults: report of the guideline Committee of the American Epilepsy Society. Epilepsy Curr Am Epilepsy Soc. 2016;16:48–61.
19. Mundlamuri RC, Sinha S, Subbakrishna DK, et al. Management of generalised convulsive status epilepticus (SE): a prospective randomised controlled study of combined treatment with intravenous lorazepam with either phenytoin, sodium valproate or levetiracetam—pilot study. Epilepsy Res. 2015;114:52–8.
20. Höfler J, Trinka E. Lacosamide as a new treatment option in status epilepticus. Epilepsia. 2013;54:393–404.
21. Kellinghaus C, Berning S, Immisch I, et al. Intravenous lacosamide for treatment of status epilepticus. Acta Neurol Scand. 2011;123:137–41.
22. Panda PK, Panda P, Dawman L, Sharawat IK. Efficacy of lacosamide and phenytoin in status epilepticus: a systematic review. Acta Neurol Scand. 2021;144:366–74.
23. Hantus S. Epilepsy Emergencies Contin Minneap Minn. 2016;22:173–90.
24. Brigo F, Del Giovane C, Nardone R, Trinka E, Lattanzi S. Intravenous antiepileptic drugs in adults with benzodiazepine-resistant convulsive status epilepticus: a systematic review and network meta-analysis. Epilepsy Behav EB. 2019;101:106466.
25. Topjian AA, Gutierrez-Colina AM, Sanchez SM, et al. Electrographic status epilepticus is associated with mortality and worse short-term outcome in critically ill children. Crit Care Med. 2013;41:215–23.
26. Abend NS, Arndt DH, Carpenter JL, et al. Electrographic seizures in pediatric ICU patients: cohort study of risk factors and mortality. Neurology. 2013;81:383–91.
27. Leitinger M, Beniczky S, Rohracher A, et al. Salzburg consensus criteria for nonconvulsive status epilepticus—approach to clinical application. Epilepsy Behav EB. 2015;49:158–63.
28. Hirsch LJ, Gaspard N, van Baalen A, et al. Proposed consensus definitions for new-onset refractory status epilepticus (NORSE), febrile infection-related epilepsy syndrome (FIRES), and related conditions. Epilepsia. 2018;59:739–44.
29. Gaínza-Lein M, Fernández IS, Ulate-Campos A, Loddenkemper T, Ostendorf AP. Timing in the treatment of status epilepticus: from basics to the clinic. Seizure. 2019;68:22–30.
30. VanHaerents S, Gerard EE. Epilepsy emergencies: status epilepticus, acute repetitive seizures, and autoimmune encephalitis. Contin Minneap Minn. 2019;25:454–76.
31. Ochoa JG, Dougherty M, Papanastassiou A, Gidal B, Mohamed I, Vossler DG. Treatment of super-refractory status epilepticus: a review. Epilepsy Curr. 2021;21:1535759721999670.
32. Vossler DG, Bainbridge JL, Boggs JG, et al. Treatment of refractory convulsive status epilepticus: a comprehensive review by the American Epilepsy Society treatments committee. Epilepsy Curr. 2020;20:245–64.
33. Mahmoud SH, Rans C. Systematic review of clobazam use in patients with status epilepticus. Epilepsia Open. 2018;3:323–30.

34. Sculier C, Barcia Aguilar C, Gaspard N, et al. Clinical presentation of new onset refractory status epilepticus in children (the pSERG cohort). Epilepsia. 2021;62:1629–42.
35. Gofton TE, Gaspard N, Hocker SE, Loddenkemper T, Hirsch LJ. New onset refractory status epilepticus research: what is on the horizon? Neurology. 2019;92:802–10.
36. Johnson EL, Kaplan PW. Status epilepticus: definition, classification, pathophysiology, and epidemiology. Semin Neurol. 2020;40:647–51.
37. Beniczky S, Hirsch LJ, Kaplan PW, et al. Unified EEG terminology and criteria for nonconvulsive status epilepticus. Epilepsia. 2013;54(Suppl 6):28–9.
38. Trinka E, Leitinger M. Which EEG patterns in coma are nonconvulsive status epilepticus? Epilepsy Behav EB. 2015;49:203–22.
39. Hirsch LJ, Fong MWK, Leitinger M, et al. American clinical neurophysiology Society's standardized critical care EEG terminology: 2021 version. J Clin Neurophysiol Off Publ Am Electroencephalogr Soc. 2021;38:1–29.
40. Chong DJ, Hirsch LJ. Which EEG patterns warrant treatment in the critically ill? Reviewing the evidence for treatment of periodic epileptiform discharges and related patterns. J Clin Neurophysiol Off Publ Am Electroencephalogr Soc. 2005;22:79–91.
41. Herman ST, Abend NS, Bleck TP, et al. Consensus statement on continuous EEG in critically ill adults and children, part I: indications. J Clin Neurophysiol Off Publ Am Electroencephalogr Soc. 2015;32:87–95.
42. Brophy GM, Bell R, Claassen J, et al. Guidelines for the evaluation and management of status epilepticus. Neurocrit Care. 2012;17:3–23.
43. Cornwall CD, Dahl SM, Nguyen N, et al. Association of ictal imaging changes in status epilepticus and neurological deterioration. Epilepsia. 2022;63:2970–80.
44. Giovannini G, Monti G, Tondelli M, et al. Mortality, morbidity and refractoriness prediction in status epilepticus: comparison of STESS and EMSE scores. Seizure. 2017;46:31–7.
45. Sculier C, Gaínza-Lein M, Sánchez Fernández I, Loddenkemper T. Long-term outcomes of status epilepticus: a critical assessment. Epilepsia. 2018;59(Suppl 2):155–69.
46. Hocker SE, Britton JW, Mandrekar JN, Wijdicks EFM, Rabinstein AA. Predictors of outcome in refractory status epilepticus. JAMA Neurol. 2013;70:72–7.
47. Rossetti AO, Logroscino G, Milligan TA, Michaelides C, Ruffieux C, Bromfield EB. Status epilepticus severity score (STESS): a tool to orient early treatment strategy. J Neurol. 2008;255:1561–6.
48. Pujar SS, Neville BGR, Scott RC, Chin RFM. North London epilepsy research network. Death within 8 years after childhood convulsive status epilepticus: a population-based study. Brain. J Neurol. 2011;134:2819–27.

21 Surgical Evaluation

Francis G. Tirol

Epilepsy surgery is the most effective method to treat drug-resistant focal epilepsy. Accurate localization of the "epileptogenic zone" (EZ), defined as the brain region necessary and sufficient to generate seizures, is critical in the surgical workup of epilepsy. Presurgical testing to identify the EZ includes video EEG monitoring, brain MRI, neuropsychology testing and optional functional imaging methods such as PET, SPECT, MEG/MSI, and others. If more precise delineation of the EZ is deemed necessary after these tests, invasive intracranial EEG recording may be attempted. These diagnostic techniques are also utilized to assess if the EZ may overlap with the eloquent cortex that should be spared during surgical resection to avoid postoperative deficits.

21.1 Drug-Resistant Epilepsy

The ILAE defines drug-resistant (synonyms: drug-resistant, intractable, refractory) epilepsy as a failure of the trial of at least two tolerated, appropriately chosen ASMs to achieve seizure freedom. Although several definitions exist for "seizure freedom," one definition describes it as a seizure-free interval of at least three times the longest seizure-free period prior to intervention, or a period of 1 year, whichever is greater [1].

21.2 Appropriate Patient Selection

Worldwide, epilepsy affects about 50 million people. Approximately 1–4% of the US population has epilepsy. About 60% of newly diagnosed patients will have a good response to ASMs. Among this 60%, the first ASM controls seizures in 47% of patients in a drug-naïve population. Seizure control can be achieved in 13% of patients with the second ASM, and in 4% of patients with the third ASM/

F. G. Tirol (✉)
Comprehensive Epilepsy Center, Department of Neurology, MedStar Georgetown University Hospital, Washington, DC, USA

Z. Haneef (ed.), *Epilepsy Fundamentals*,
https://doi.org/10.1007/978-3-031-77741-7_21

combination ASM [2]. This finding leads to the common definition of drug-resistant epilepsy as "the failure of two ASMs" to control epilepsy. Apparent drug resistance (pseudoresistance) can be secondary to drug noncompliance, or wrong diagnosis (e.g., non-epileptic seizures). The appropriate choice of ASM based on the epilepsy syndrome being treated should also be ensured.

Patients initially well controlled on ASMs may later become refractory to drug therapy. Temporal lobe epilepsy (TLE) due to mesial temporal sclerosis (MTS), for example, is characteristically a progressive epilepsy syndrome. In one series, the average time to failure of the first two ASMs in MTS patients was 9.1 years.

Epilepsy surgery is effective and safe in selected patients with intractable focal epilepsy. Seizure freedom following surgical resection is achieved in 80% of patients with a structural lesion such as a cavernous angioma or low-grade neoplasm, in approximately 60–70% of all patients with TLE, and in 50% of patients with focal, neocortical epilepsy.

Although the epilepsy surgery workup has some variability between centers, there are some common themes, which have been simplified in Fig. 21.1.

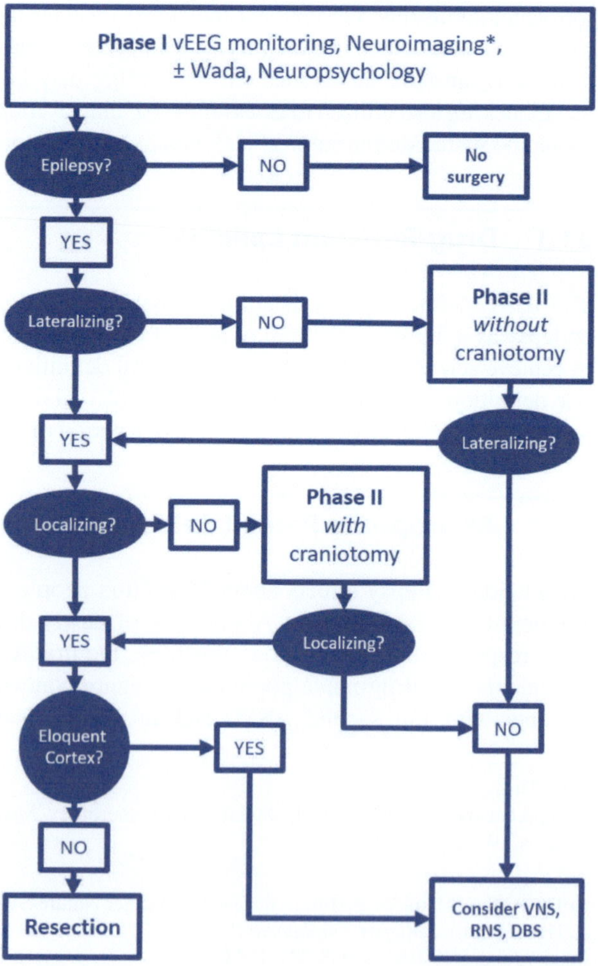

Fig. 21.1 Flow diagram suggesting a simplified algorithm of a typical epilepsy surgery workup. *Neuroimaging can include MRI, PET, SPECT, MEG, etc. as indicated

21.3 Video EEG

The goals of video EEG monitoring are: (1) **characterization**: to differentiate non-epileptic events from epilepsy; (2) **presurgical**: to help localize the **epileptogenic zone (EZ)** for epilepsy surgery; and (3) **quantification** of the seizure burden.

The **"epileptogenic zone" (EZ)** can be defined as the brain region necessary and sufficient to generate seizures, the resection of which will lead to seizure freedom. Interictal EEG findings help localize the **"irritative zone,"** while the ictal EEG often reveals the **"seizure onset zone"** (Fig. 21.2). Symptoms reported by the patient help to delineate the **"symptomatogenic zone,"** which occurs when the ictal activity reaches a brain region that produces symptoms. The **"epileptogenic lesion"** is the radiologically visible lesion.

Ictal EEG findings include a variety of changes, including localized EEG attenuation, low voltage fast activity, rhythmic spikes/spike-waves, and rhythmic slow waves. Some seizures, particularly focal aware seizures and seizures from deeper foci, may have no visible changes on scalp EEG. Careful synthesis of semiology and the interictal/ictal EEG findings during video EEG monitoring is considered the gold standard in the presurgical localization of the epileptogenic zone.

Video EEG monitoring can be divided into two phases. "Phase I" implies non-invasive, scalp video EEG monitoring. Certain centers also consider the use of sphenoidal electrodes for improved temporal coverage with "phase I." If more precise delineation of the EZ is necessary, the evaluation may proceed to "phase II" monitoring with intracranial EEG. "Phase III" refers to the actual surgical resection, which may include the utilization of intraoperative electrocorticography monitoring.

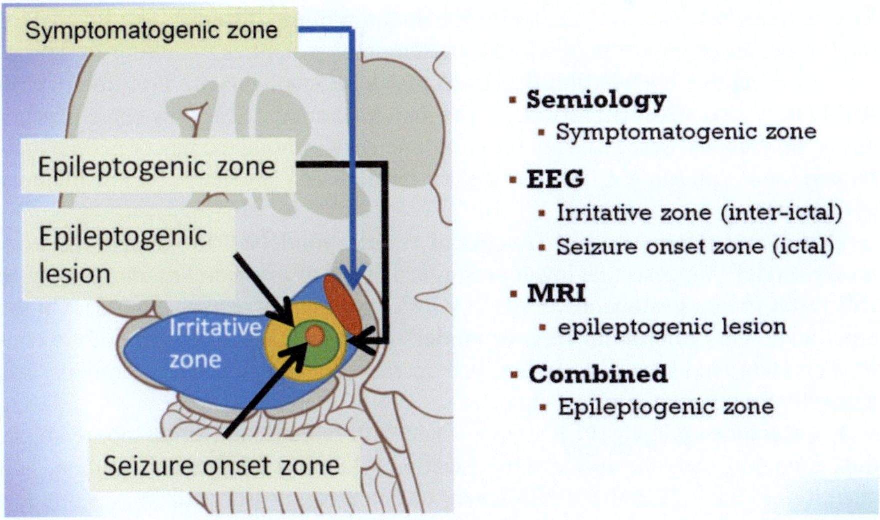

Fig. 21.2 "Zones" relevant to the epilepsy surgery workup

21.4 MRI and Functional Neuroimaging

Magnetic Resonance Imaging (MRI) is an essential tool utilized in the presurgical workup of drug-resistant epilepsy. If there is discordance in the data derived from video EEG monitoring and the MRI, additional functional neuroimaging techniques may aid the localization of the epileptogenic zone, described in more detail in Chap. 10 (Neuroimaging).

21.5 Neuropsychology

The primary purpose of neuropsychological testing in the presurgical workup of epilepsy is to aid in the localization of the EZ by documenting deficits related to the epileptic focus. Additional uses include estimating the risk of postoperative cognitive deficits (memory or language) and monitoring cognitive function following surgery. Neuropsychologic deficits seen with generalized epilepsy and epilepsy syndromes are not discussed here.

Temporal lobe epilepsy (TLE) is the most common adult focal epilepsy syndrome that is managed surgically. It is also the most studied epilepsy syndrome utilizing neuropsychological techniques. Since mesial TLE affects the hippocampus, it can cause memory deficits. Left TLE is strongly associated with verbal memory deficits. The association of right TLE with non-verbal (visuospatial) deficits is less well established. One suggested explanation is that non-verbal memory may have more bilateral representation. While the mesial temporal lobe is thought to be responsible for memory consolidation and retrieval, the lateral temporal lobe is associated more with content processing. Thus, impairment in delayed recall is likely to be due to mesial rather than lateral TLE. Progressive memory decline in TLE is thought to be associated with the duration of the disorder. The EZ in TLE may affect an extensive network of extratemporal structures, resulting in deficits such as executive dysfunction due to frontal lobe involvement. **Treatment with ASMs may also affect cognition**. All the first-generation ASMs may cause psychomotor slowing and decreased attention and memory. Among the newer ASMs, topiramate and zonisamide, in particular, have been associated with memory dysfunction.

Functional adequacy and functional reserve models: The **"functional adequacy model"** suggests that lower presurgical levels of ipsilateral memory function will result in less postoperative memory loss (and greater seizure control). On the other hand, the **"functional reserve model"** suggests that the integrity of the contralateral temporal lobe best predicts postoperative function. The former theory has generally found greater acceptance.

Frontal lobe epilepsy (FLE) is associated with evidence of frontal lobe dysfunction, including deficits in executive function, planning, response inhibition, and attention. Reliable lateralized differences of cognitive measures distinguishing left and right FLE have not been well described.

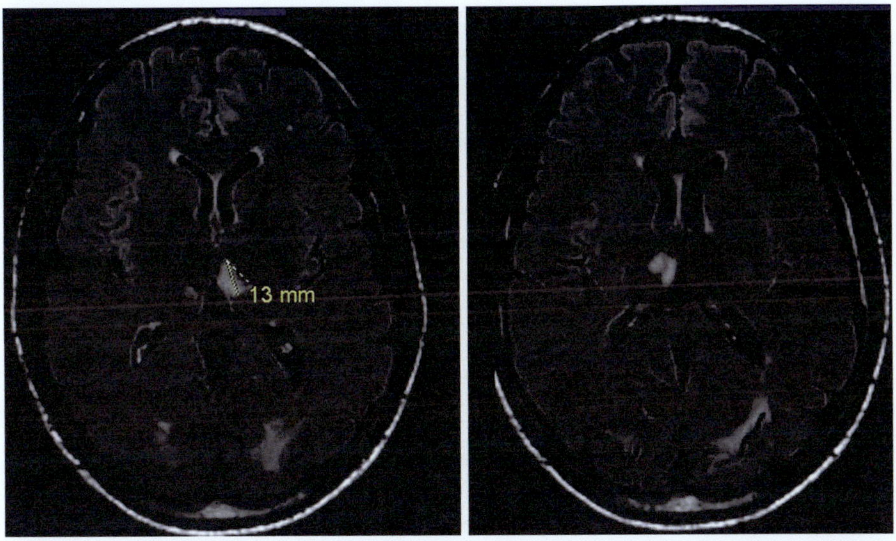

Fig. 21.3 FLAIR MRI showing signal changes in bilateral thalami in a patient following Wada testing (patient-reported new right finger-tip numbness)

21.6 Wada Testing (Intracarotid Amobarbital Procedure) (Fig. 21.3)

The intracarotid amobarbital procedure (IAP) is done to establish language and memory laterality prior to epilepsy surgery (Fig. 21.4). It is eponymously called the Wada test after Juhn Wada, the Japanese scientist who first described it in 1949 and later developed it for cerebral lateralization in epilepsy surgery candidates. The test has considerable variability between epilepsy centers in indication and administration. In some centers, Wada testing is routine prior to epilepsy surgery. Many centers currently reserve this test for surgery on the dominant hemisphere to predict memory performance after surgery. fMRI has replaced Wada testing in some centers, particularly in presurgical testing for right (non-dominant) temporal lobe epilepsy.

Wada Procedure (IAP)

- *The IAP starts with an intracarotid cerebral angiogram to assess cerebral vasculature and cross-filling of the contralateral side. Ipsilateral contrast runoff to the posterior cerebral artery from the internal carotid injection* via *a patent posterior communicating artery is also ensured as the mesial temporal lobe is supplied through the posterior cerebral artery. Abnormal vasculature such as a persistent fetal trigeminal artery supplying the brain stem may preclude the Wada test due to the risk of brain stem suppression.*

(continued)

- *Prior to the injection, the patient is asked to hold out his/her arms and then count (e.g., backward from 100). One cerebral hemisphere is transiently anesthetized with a short-acting anesthetic (sodium amobarbital is traditionally used. Due to drug shortages, others like methohexital and thiopental are also being used but are shorter, limiting the testing time).*
- *As the short-acting anesthetic takes effect (usually within 4–5 s) and the injected cerebral hemisphere "goes to sleep," the contralateral arm becomes weak and simultaneous EEG typically shows ipsilateral slowing. The patient may stop counting and become aphasic, typically when the dominant hemisphere is injected, affecting language function.*
- *The Wada test assesses both language and memory. Effects on language include speech arrest, mutism, and anomia if the injection is in the dominant hemisphere.*
- *To assess memory, familiar objects, symbols, and words on note cards are presented to the patient while the anesthetic is still effective. The continued effect of the anesthetic is assessed by repeatedly testing contralateral arm strength and observing for normalization of the EEG.*
- *As the patient recovers, typically within 5–10 min, memory is tested by asking the patient to recollect the objects, symbols, and words that were presented earlier, either spontaneously or with cues/choices.*
- *At least 30 minutes are allowed for the anesthetic to wear off completely before repeating the test on the contralateral side, if necessary.*
- *Memory/language scores are assigned to each hemisphere to identify the dominant hemisphere.*
- *Wada testing is invasive and can potentially result in complications* (Fig. 21.3).

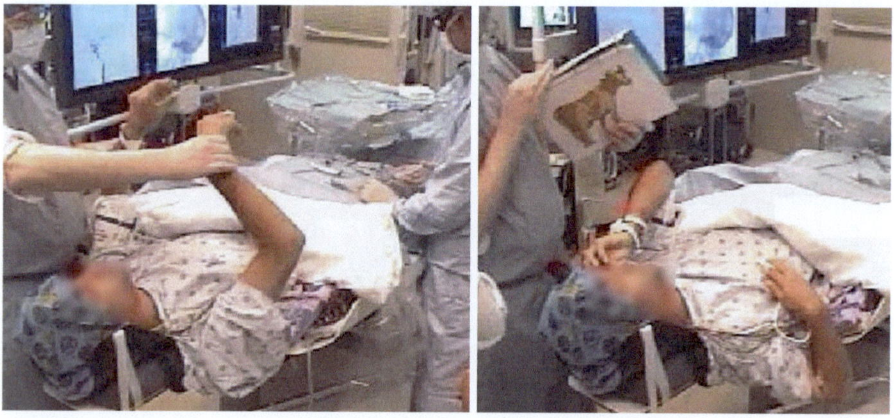

Fig. 21.4 Wada testing: In this example, the injection is on the left side, so the examiner stands on left side to not be in the hemianopic (and hemi-neglected) right side

Although the Wada testing protocol varies much between surgical centers, the Medical College of Georgia protocol has been followed in several research studies and epilepsy centers in the US.

Wada testing may also provide information on the lateralization of the EZ based on memory asymmetry typically seen in TLE. Weaker memory on one side suggests that this is the hemisphere affected by epilepsy. Right TLE may be better lateralized than left TLE using this method.

Wada testing has been criticized as being too cumbersome and stringent, and it may possibly dissuade suitable surgical candidates or unnecessarily deny them surgery. Published case series have reported on patients who did not develop significant postsurgical amnesia after previously having "failed" the Wada test [3].

21.7 Intracranial EEG

Intracranial EEG (icEEG) is the direct cortical recording of EEG from the brain surface or the brain parenchyma and can take different forms (Table 21.1, Figs. 21.5, 21.6, 21.7). The choice between the more commonly used depth and subdural electrodes depends on the characteristics of the lesion that needs to be sampled. It also varies based on the experience of the surgical team at the particular center. In the past, icEEG was performed for most resective surgeries for epilepsy. However, modern neuroimaging has considerably reduced the need for icEEG. Current indications typically include (1) inability to lateralize the EZ using non-invasive testing (MRI, video EEG, functional imaging, neuropsychology), (2) inability to localize the EZ to a particular lobe after the side of onset has been established, and for (3) extra-operative functional mapping of eloquent cortex. Bitemporal epilepsy may be the syndrome where icEEG is most commonly indicated.

21.7.1 Depth Electrodes

Typically a flexible silastic electrode contains 4–10 contacts separated by 1 cm. Neuronavigational techniques are used to safely guide implantation based on a target that needs to be approached. Advantages of depth electrodes include (1) the ability to sample deep structures (e.g., hippocampus, midline cortex, deep heterotopias); (2) utility in the presence of prior surgery (adhesions) when implanting subdural electrodes is technically difficult; (3) avoidance of the need for a craniotomy as these can

Table 21.1 Various approaches of intracranial EEG recording

Typical	(1) Depth electrode (2) Subdural electrode/grid implantation
Rarely, if ever still used	(3) Epidural peg electrodes (4) Foramen ovale electrodes.
Investigational	(5) Intraventricular electrodes (6) Cavernous sinus electrodes

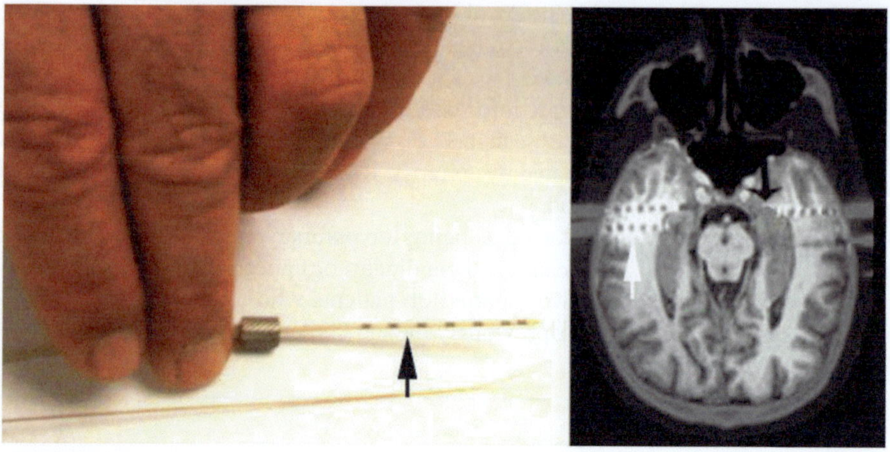

Fig. 21.5 Depth EEG electrode before and after implant

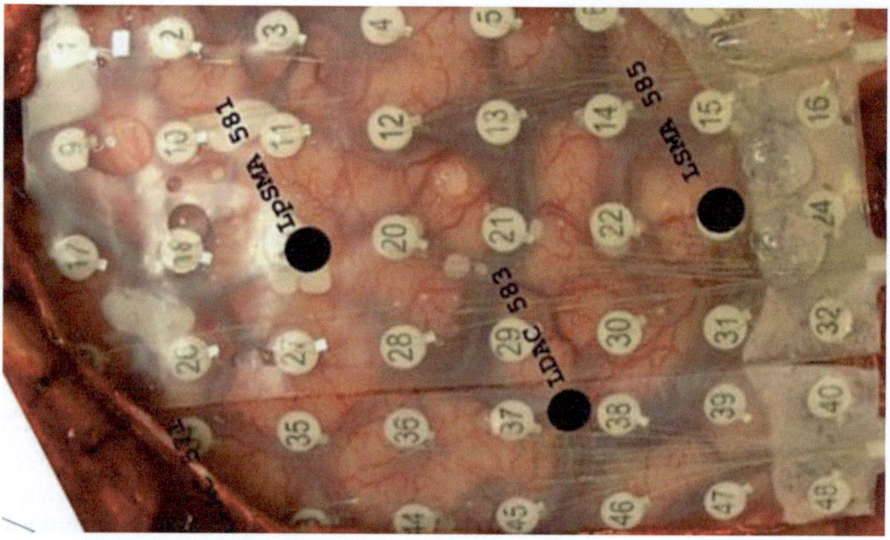

Fig. 21.6 Subdural grid lying on brain surface during surgery

be inserted through drill holes. The complication rate is low (0.5–5%) and includes hemorrhage and infection. Functional deficits related to cortical penetration have not been described. Disadvantages include a very restricted (tunnel vision) sampling.

In the past decade, the invasive technique of stereoelectroencephalography (SEEG) has garnered greater acceptance and adoption as a method of intracranial EEG monitoring with depth electrodes. The SEEG method was popularized in France in the 1950s by Bancaud and Talairach consisting of a multiphase, complex system that utilized a stereotactic frame and a double grid system employed to insert

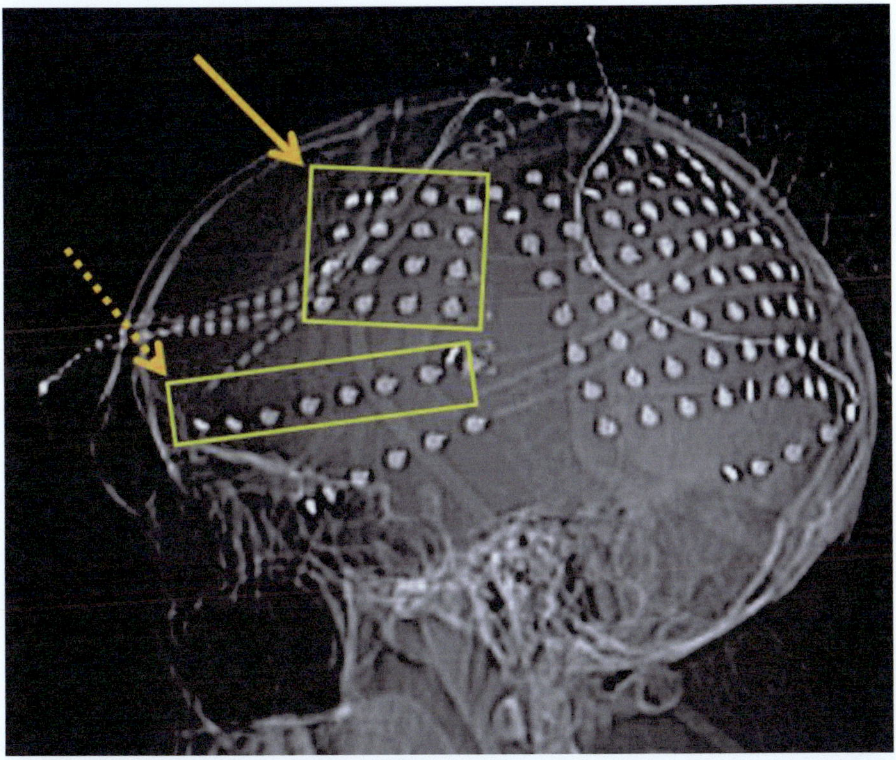

Fig. 21.7 Lateral skull roentgenogram shown demonstrates both strips (dotted arrow) and grids (solid arrow) in a patient undergoing intracranial EEG evaluation

multiple depth electrodes targeting deep brain structures such as the mesial temporal regions for localization of the epileptogenic zone (EZ) [4]. The technique is based on a hypothesis that maps a 3-dimensional network that includes the EZ and its propagation pattern. Depth electrodes used for SEEG have 4–18 contacts spaced 2–10 mm apart that are typically inserted orthogonally through a twist drill hole or a burr hole [5]. Though all current methodologies for SEEG require the integration of stereotactic MRI images with angiography, electrode targeting and placement may be done with a standard frame-based device, a frameless technique, or a robot-assisted device. Generally, SEEG is deemed advantageous to other icEEG methods, given its ability to sample extensive and deep cortical areas with lower surgical risk.

21.7.2 Subdural electrodes (SDE)

SDEs include strips and grid electrodes, which consist of a polyurethane mesh with embedded electrodes placed 1 cm apart. Strips can be inserted via burr holes, while grid electrodes require a craniotomy. Grid electrodes are useful when a large cortical surface needs to be sampled (e.g., frontal lobe) and when functional mapping of the eloquent cortex is desired. Grids have a higher complication rate, including infection, hematoma, and neurological deficits. Strips are also thought to be less precise than depths for assessing mesial TLE [6, 7]. While depths provide true hippocampal sampling, strips often reach only the parahippocampal gyrus. A combination of strips and depths can be used when mesial temporal versus neocortical temporal/extratemporal localization is a concern.

Postoperative hemorrhage and extra-axial collections are more common with subdural electrodes than with depth electrodes, although there may be no difference in clinically significant findings. Subdural grids have more symptomatic extra-axial collecitons [8].

Functional mapping is performed by sequential electrical stimulation of grid electrode contacts. Typically, the stimulation starts at 1 mA and the current is increased step-wise until (1) a functional deficit or symptoms are reported or observed, (2) a predetermined maximum is reached (e.g., 15 mA), or (3) after-discharges or an electrographic seizure are observed. Language is mapped by documenting loss of language function during spontaneous speech, repetition, or comprehension tests. Motor and sensory cortex stimulation leads to positive or negative motor (clonus, weakness) or sensory (tingling, numbness) phenomena. The epileptogenic cortex may also be identified in cortical stimulation by the presence of "delayed responses" (100 ms-1 sec post-stimulation), compared to the normal "early responses" (<100 ms).

Pearls: Surgical Evaluation

Below are some neuropsychological tests and some of the cognitive domains they test

- **Verbal Fluency Test**—semantic knowledge.
- **Boston Naming Test**—dominant temporal lobe function.
- **Trail Making Test** and **Stroop Word-Color Test**—executive function.
- **Wechsler Memory Scale**—long-term and short-term memory.
- **Rey-Osterrieth Complex Figure Test**—visuospatial abilities, memory and executive function.

References

1. Kwan P, Arzimanoglou A, Berg AT, Brodie MJ, Allen Hauser W, Mathern G, Moshé SL, Perucca E, Wiebe S, French J. Definition of drug resistant epilepsy: consensus proposal by the ad hoc task force of the ILAE commission on therapeutic strategies. Epilepsia. 2010;51:1069–77.
2. Kwan P, Brodie MJ. Early identification of refractory epilepsy. N Engl J Med. 2000;342:314–9.
3. Loring DW, Lee GP, Meador KJ, Flanigin HF, Smith JR, Figueroa RE, Martin RC. The intracarotid amobarbital procedure as a predictor of memory failure following unilateral temporal lobectomy. Neurology. 1990;40:605–10.

4. Bancaud J, Angelergues R, Bernouilli C, Bonis A, Bordas-Ferrer M, Bresson M, Buser P, Covello L, Morel P, Szikla G, Takeda A, Talairach J. Functional stereotaxic exploration (SEEG) of epilepsy. Electroencephalogr Clin Neurophysiol. 1970;28:85–6.
5. Jayakar P, Gotman J, Harvey AS, Palmini A, Tassi L, Schomer D, Dubeau F, Bartolomei F, Yu A, Kršek P, Velis D, Kahane P. Diagnostic utility of invasive EEG for epilepsy surgery: indications, modalities, and techniques. Epilepsia. 2016;57:1735–47.
6. Eisenschenk S, Gilmore RL, Cibula JE, Roper SN. Lateralization of temporal lobe foci: depth versus subdural electrodes. Clin Neurophysiol Off J Int Fed Clin Neurophysiol. 2001;112:836–44.
7. Sperling MR, O'Connor MJ. Comparison of depth and subdural electrodes in recording temporal lobe seizures. Neurology. 1989;39:1497–504.
8. Schmidt RF, Wu C, Lang MJ, Soni P, Williams KA Jr, Boorman DW, Evans JJ, Sperling MR, Sharan AD. Complications of subdural and depth electrodes in 269 patients undergoing 317 procedures for invasive monitoring in epilepsy. Epilepsia. 2016;57:1697–708.

Intracranial EEG Monitoring and Electrical Stimulation Mapping

22

Chetan Sateesh Nayak and Jay R. Gavvala

The purpose of an intracranial EEG evaluation in patients with medically intractable epilepsy is to accurately identify the epileptogenic zone (EZ) and delineate the boundaries of resection/ablation with minimal or no permanent functional neurological deficits. Intracranial EEG evaluations can be performed either intra-operatively or extra-operatively depending upon the amount of information that needs to be collected and the level of cooperation required from the patient. Components of an intracranial EEG monitoring either intra-operatively or extra-operatively include the EEG monitoring analysis as well as electrical stimulation mapping.

C. S. Nayak
Intracranial EEG/Epilepsy Clinical Fellow, Texas Comprehensive Epilepsy Program, University of Texas Health Science Center at Houston, Houston, TX, USA

J. R. Gavvala (✉)
Neurology, Texas Comprehensive Epilepsy Program, University of Texas Health Science Center at Houston, Houston, TX, USA
e-mail: jay.r.gavvala@uth.tmc.edu

Z. Haneef (ed.), *Epilepsy Fundamentals*,
https://doi.org/10.1007/978-3-031-77741-7_22

22.1 Intracranial EEG Monitoring

Typical Indications of Invasive Monitoring

- Discordant non-invasive presurgical work-up
- MRI-negative neocortical epilepsy and select cases of mesial temporal epilepsy, such as suspicion for bitemporal epilepsy and where there is a concern for temporal plus epilepsy
- MRI-lesional cases if:
  - Adjacent to eloquent cortex
  - Detailed functional mapping needed
  - Plan to maximally define the epileptic zone for completeness of resection such as in focal cortical dysplasia (FCD)
  - Dual pathology or multifocality (i.e., tuberous sclerosis)
  - If discordance with EEG data (i.e., scalp EEG is non-localizable)

Background In the subset of patients with medically intractable epilepsy who require invasive intracranial evaluation, recording of seizures and localization of seizure onset zone remain to date the best surrogate marker for identifying the epileptogenic tissue [1]. A standard single electrode provides an estimate of the field potential of summation of excitatory and inhibitory post-synaptic evoked potentials roughly from 100 million to 1 billion neurons. Intracranial EEG has the advantage of proximity to the source of electrical activity only separated by highly conductive media and low impedances. Using simultaneous scalp and intracranial recording, cortical spike sources having an area of at least 6 cm^2, although typically at least 10 cm^2 are required to produce scalp-recordable EEG spikes [2]. Intracranial EEG is less susceptible to artifacts and provides a higher signal-to-noise ratio. Additionally, depth electrodes allow for exploring mesial brain structures and deep-seated foci that are not accessible otherwise.

Indication The traditional goal of epilepsy surgery is to resect, ablate or disconnect the epileptogenic zone (EZ), which is the area of the cortex indispensable for seizure generation and whose resection leads to seizure freedom. This classical utilization of an intracranial evaluation seeks to identify the EZ in patients whose standard non-invasive phase I presurgical evaluation is non-localizing or reveals discordant information. With greater advances in technology, intracranial evaluations are increasingly being used to inform neuromodulatory therapies, including in patients where a single EZ cannot be identified and who are not candidates for resective/ablative surgery. The decision about implantation is discussed during a multi-disciplinary surgical conference attended by neurosurgeons, neurologists, neuropsychologists, radiologists, trainees, and nurses, among others. The typical indications are shown in the above Typical Indications of Invasive Monitoring box. The standard approach is to record seizures in the epilepsy monitoring unit.

Intra-operative ECOG There is only limited evidence to support the utilization of pre and post-resection ECOG for localization of epileptic focus. Recording in the intra-operative settings may be adequate in (1) children, especially if younger, (2) lesions with concordant non-invasive evaluations in focal cortical dysplasia, and (3) as an adjunct in multiple subpial transection (MSTs).

Limitations Seizure onset is almost always not recorded. A tunnel vision is related to the limited sampling; thus, successful localization is guided by a strong hypothesis. Anesthesia may limit the analysis of epileptic activity. The ideal agents for intra-operative recording are those with minimal effect on baseline spike frequency. Pharmacological seizure activation is not commonly utilized now but caution must be exercised with inhaled agents that tend to suppress background EEG activity, with reports of enflurane exhibiting an activating effect. Synthetic opiates such as remifentanil and alfentanil may increase the yield of recording epileptiform activity. A few studies have shown that dexmedetomidine has no or little activating effect on epileptiform activity. Propofol, barbiturates, and benzodiazepines increase EEG background frequency and may obscure epileptiform discharges.

22.2 Types of Electrodes

Subdural Electrodes (SDE) These can be implanted over the cortical convexity in the basal neocortex or the inter-hemispheric regions. It allows for broader sampling and detailed mapping, although there is no coverage of the depths of the sulcus. It requires a craniotomy and, consequently, presents a relatively higher risk of surgical complications (Fig. 22.1) [3].

Depth Electrodes These are placed into the substance of the brain using either frame-based or frameless stereotaxic techniques. These electrodes allow targeting deep-seated intracerebral structures at the bottom of the sulcus, insula, and the brain's mesial and basal surfaces. Furthermore, the depth electrodes can also be used to plan minimally invasive stereotactic laser ablation procedures for the EZ. Finally, their placement does not require a craniotomy and outcome studies demonstrate lower rates of surgical complications compared to subdural evaluations (Fig. 22.1).

However, it is important to differentiate the technical approach of placement of stereotactic depth electrodes from the discipline of stereo-electroencephalography (SEEG), which is a methodologic practice based on exploring epileptogenic networks using anatomo-electro-clinical correlation [4]. In the last decade, SEEG is being increasingly adopted as the mainstay of intracranial evaluation in most epilepsy centers in the United states [5].

Decision to Evaluate with SEEG and SDE Over the last 10 years, there has been a paradigm shift with SEEG implantations representing a vast majority of intracranial evaluations in the United States. This was especially seen in cases of repeat intracranial evaluations, evaluation of suspected insular epilepsy, mesial temporal lobe epilepsy and MRI-negative frontal lobe epilepsy. Conversely, SDE evaluation tend to be preferred for patients whose suspected EZ are in close proximity to the eloquent cortex [5].

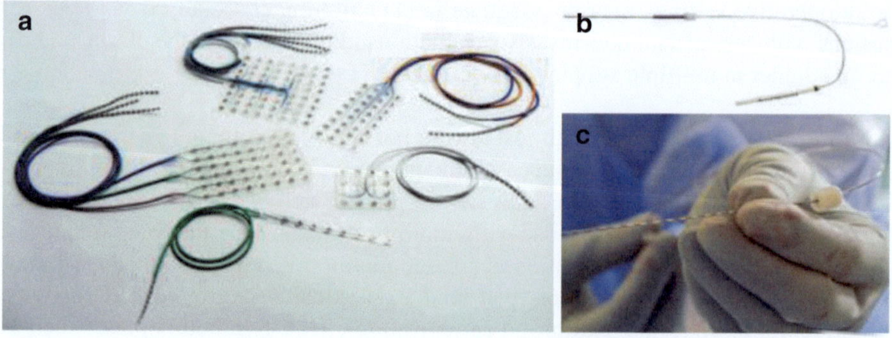

Fig. 22.1 A combination of commercially available (**a**) subdural electrode grids (8 × 8, 6 × 8, 4 × 8, 4 × 4) and contact strip (1 × 8) and 14 contact SEEG depth electrodes (**b**). (**c**) demonstrates a close-up view of the SEEG depth electrode illustrating the 2 mm cylindrical contacts. Images courtesy of PMT Corporation (Chanhassen, MN)

Table 22.1 Cortical zones and methods of assessment in approach to medically intractable epilepsy

Cortical zone	Method of assessment
Ictal onset zone	EEG, ictal SPECT
Irritative zone	EEG, magnetoencephalography, and functional MRI triggered by EEG
Symptomatogenic zone	History and semiology
Epileptogenic lesion	MRI
Eloquent cortex	Cortical stimulation, functional MRI, evoked potentials, magnetoencephalography
Functional deficit zone	Physical examination, neuropsychiatric testing, EEG, PET, SPECT, MRS
Epileptogenic zone	None (theoretical construct)

22.3 Analysis Techniques

Recording of seizures and localization of seizure onset zone remains to date, the best surrogate biomarker of the epileptogenic zone (Table 22.1) [6].

Filter Settings Before reading the intracranial data, the filter and sensitivity settings need to be adjusted. The settings are different from standard scalp EEG, given the high amplitude and frequency of the cortical signals. Commonly used settings include sensitivity: 75 uV/mm; low-frequency filter: 1.6 Hz & high-frequency filter: 500–1000 Hz. The high-frequency cut-off is influenced by the sampling rate per the Nyquist theorem. In addition to the traditional viewing bands of scalp EEG, it is important to recognize the importance of frequency bands on either end of the spectrum. High-frequency oscillations (HFOs) known as ripples (80 to 250 Hz) and fast ripples (250 to 500 Hz) have been recorded from microelectrodes and commercially available intracranial macroelectrodes (subdural and depth) in patients with intractable focal epilepsy. HFOs in the ripple range have been reported spontaneously over broad areas of the neocortex in non-epileptic regions during NREM sleep whereas fast ripples seem to be more tightly linked to the epileptic process. To date, there is no reliable signal parameter to differentiate epileptic from non-epileptic HFOs [7]. However, the co-occurrence of ripples or gamma activity with interictal spikes increases the specificity of both as a marker of epileptogenic brain regions [8]. On the slower end, ictal infraslow activity (commonly but inaccurately referred to as "DC shift") has been shown to have a strong localizing value for the EZ [9].

Bipolar Vs. Referential Montage From a theoretical standpoint, referential montage has the advantage of representing the electrical activity as is, so carefully selected and uninvolved references are needed for that purpose. Certain recording systems enable a so-called active ground reference system, which actively suppresses common electrical noise. Although bipolar montage alters signal content [10], as it functions as a spatial filter canceling common signal in two adjacent electrodes, some experts consider it ideal to visualize and localize fast and infraslow activity <1 Hz.

Interpretation of Intracranial Recordings Particularly relevant to SEEG interpretation is the need to appreciate electrode location with image co-registration techniques. This point cannot be understated, as without an exact localization of each electrode, it is impossible to appreciate the 3D relationship of associated contacts as well as the significance of the activity seen. Interpretation of intracranial EEG involves pattern recognition to differentiate physiological from pathological waveforms. The physiological variants similar to scalp EEG should be recognized, including posterior dominant rhythm and lambda waves with posterior explorations, mu activity with central explorations, and beta oscillations with central and dorsolateral prefrontal explorations [11]. In addition, some additional patterns are exclusive to intracranial EEG from depth electrodes as it can record from the deep-seated cortical structures such as sleep spindles from the thalamus and limbic structures (hippocampus and entorhinal cortex) [11]. During sleep, slow oscillations (<1 Hz) [12], frontally predominant delta (1–4 Hz) [13], K-complexes and vertex waves are generated in the cortex [14]. Physiological high-frequency oscillations (HFOs) including ripples (80–250 Hz) and to a lesser extent fast ripples (>250 Hz) can be seen during awake and sleep periods, especially in eloquent cortices [15].

Interictal epileptiform discharges (IEDs) on intracranial EEG have a higher amplitude, are more frequent and typically have variable polarity in contrast to scalp waveforms (Fig. 22.2a) [16]. These include sharp waves, spikes, spike and wave complexes (Fig. 22.2b), repetitive spikes (Fig. 22.2c) and bursts of fast spikes [17]. IED propagation is important to recognize by comparing the peaks' latency between channels, which can be a reliable marker of the seizure onset zone [18]. IEDs could help define the extent of the epileptogenic network [19]. IEDs are more reliable in localizing to the EZ if they occur as a single population and are associated with HFOs [20] (Fig. 22.2d/e), especially very high-frequency oscillations (VHFOs; 500–2 kHz) [21]. Intracranial IEDs may or may not be visible on scalp EEG

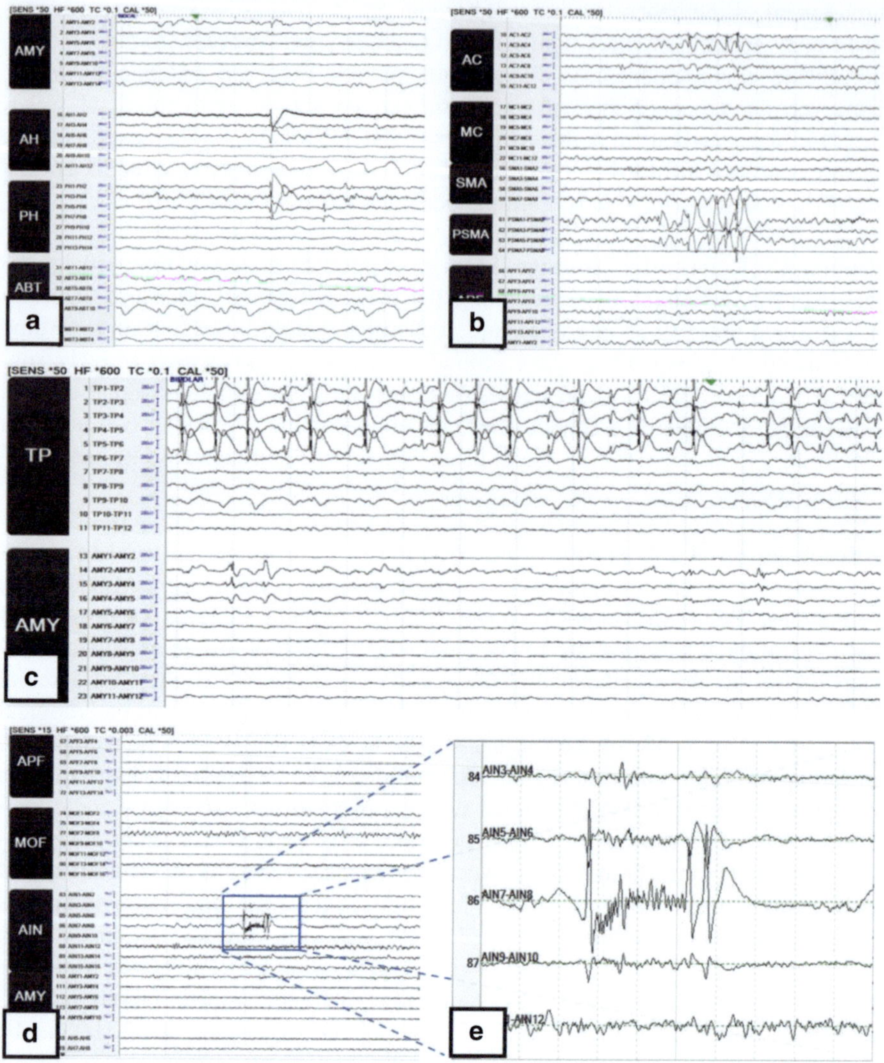

Fig. 22.2 Intracranial interictal epileptiform discharges: (**a**) high amplitude polyphasic spike and wave in the hippocampus (AH & PH) and lower amplitude, triangular appearing discharge in the amygdala (AMY); (**b**) spike and wave complexes in anterior cingulate (AC) & pre-SSMA (PSMA); (**c**) near continuous spike and wave discharges in temporal pole (TP) and (**d/e**) repetitive spikes with superimposed high-frequency oscillations (HFOs) in anterior insula (AIN)

depending on the location, amplitude and orientation of the cortical source [22]. Intracranial IEDs tend to be more frequent and broader during NREM sleep and more localized in REM sleep [23]. IEDs recorded from the hippocampus have a high amplitude and polyphasic appearance and are very frequent, usually occurring as runs. IEDs recorded from the amygdala have a lower amplitude and a triangular appearance [24] (Fig. 22.2a). IEDs arising from FCDs have a typical appearance of near-continuous repetitive spiking with superimposed rhythmic slowing/faster frequencies (Fig. 22.2d). Other characteristic IED patterns include low voltage fast activity in polymicrogyria [17] and small amplitude spikes on a flattened background in periventricular nodular heterotopia [25].

Ictal activity recorded on intracranial EEG has been widely accepted as the most reliable method of determining the EZ. Several characteristic seizure onset patterns (SOP) were observed in SEEG recordings (Fig. 22.3). These include: 1) low voltage fast activity (LVFA); 2) preictal spiking followed by LVFA; 3) bursts of polyspikes followed by LVFA; 4) slow-wave or baseline shift followed by LVFA; 5) rhythmic slow spikes; 6) theta/alpha sharp activity; 7) beta sharp activity; and 8) delta brush [26]. The SOP is strongly linked to histopathology of the epileptogenic lesion. The most common SOP in FCD Type I is LVFA followed by slow-wave or baseline shift followed by LVFA. On the other hand, in FCD Type II, bursts of polyspikes followed by LVFA are the most common followed by LVFA. Similarly, in mesial temporal sclerosis, preictal spiking followed by LVFA was the most common SOP [17]. The SOP has also been related to post-surgical outcomes. LVFA has been associated with the best post-surgical outcomes. Conversely, rhythmic slow spikes, theta/alpha sharp activity, beta sharp activity and delta brush have poor post-surgical outcomes [27, 28]. Despite this fact, there is no validated way to distinguish ictal onset rhythms from propagation patterns, and any frequency category may represent either an onset pattern or spread from another unmeasured location.

Surgical Outcomes Over the past decade, there has been a plethora of literature reporting on long-term outcomes following epilepsy surgery, with chances of long-term and sustained seizure freedom ranging from 27% in frontal lobe epilepsy [29] and up to 80% in lesional mesial temporal lobe epilepsy [30]. However, on longer follow-up of patients with temporal lobectomies, that number goes down to 55% at 5 years and 49% at 10 years [31]. This still compares favorably to a 5%-per-year chance of seizure freedom using anti-seizure medications alone in medically intractable cases [32].

Safety The common surgical complications associated with intracranial evaluation include hemorrhage, infections and hardware-related. Other less common complications include CSF leak, cerebral edema and vascular injury. Overall, the complications with SEEG are significantly lower when compared to SDE [33], presumably due to the more invasive nature of SDE requiring a craniotomy. A meta-analysis from 2016 reported complication rates of 4% hemorrhage and 2.3% infections with subdural electrodes [3], and 1% hemorrhage and 0.8% infections with SEEG [34]. However, a more recent single-center observational study reported a higher total hemorrhage rate of 19.1% with SEEG, although the symptomatic hemorrhage rate was more in line with previously reported data at 2.2% [35]. Duration of implantation of invasive intracranial monitoring correlates with histopathological changes such as microhemorrhages and inflammatory response [36].

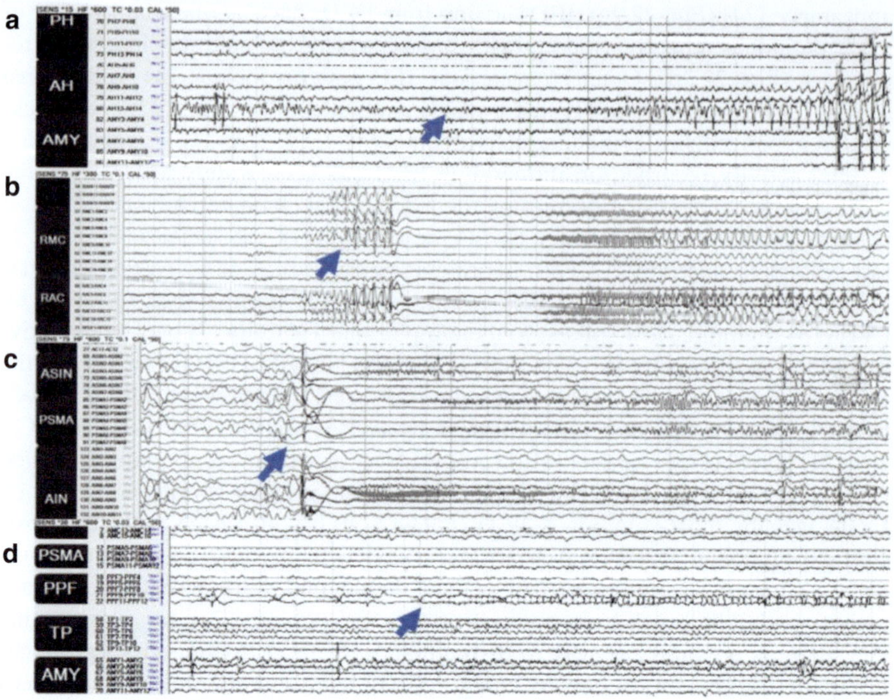

Fig. 22.3 Common SEEG seizure onset patterns: (**a**) Low voltage fast activity (LVFA) in the middle temporal gyrus (AH13–14) (**b**) Pre-ictal spiking with rhythmic spikes of low frequency followed by LVFA in the right anterior and middle cingulate (RAC & RMC), (**c**) Slow wave or baseline shift followed by LVFA in the anterior insula (AIN), anterior superior insula (ASIN) & pre-SSMA (PSMA) and (**d**) theta/alpha sharp activity with progressive increasing amplitude in the posterior prefrontal region (PPF)

22.4 SEEG Implantation Schemes

The SEEG method is a minimally-invasive intracranial monitoring technique that allows spatiotemporal exploration of epileptogenic networks and their relation to semiology based on the anatomo-electro-clinical correlation. However, SEEG comes with a significant sampling bias. Therefore, its successful utilization depends on the presence of a robust, well-formulated hypothesis of the epileptogenic zone generated after going through a comprehensive phase I presurgical evaluation [37]. Although the implantation scheme needs to be personalized according to the findings in each case, a few overarching guidelines can help devise appropriate electrode sampling [38]. Below we describe some common SEEG implantation schemes. The schematic representation was created using the Epilepsy Tracking and Optimized Management Engine (EpiTOMe) [39].

Temporal Lobe Epilepsy (TLE) An important distinction in temporal lobe SEEG evaluations is the subtype of temporal lobe epilepsy of which predominant categories are mesial temporal lobe epilepsy, lateral TLE, temporopolar and mesiolateral TLE. While each has clinical and electrographic distinctions, commonalities in sampling include the mesial temporal structures (hippocampus, amygdala, entorhinal and para-hippocampal gyrus), basal temporal region, temporal pole as well as the lateral temporal regions including superior, middle and inferior temporal gyrus (STG, MTG & ITG) (Fig. 22.4a). In certain cases, consideration for temporal plus epilepsies may also require sampling of the operculo-insular regions, posterior cingulate, orbitofrontal regions and/or the posterior regions (posterior insula, posterior fusiform gyrus, planum temporale, supramarginal gyrus and the temporo-parieto-occipital (TPO) junction) (Fig. 22.4b). Consider contralateral exploration in cases where bitemporal findings are present.

Insular Epilepsy Insular seizures are difficult to assess with subdural grids and are more readily assessed using the SEEG methodology. Implantation of SEEG electrodes may be done using either orthogonal or oblique trajectories. Orthogonal implantations have the advantage of sampling both the insular and opercular regions simultaneously, allowing a distinction between an operculo-insular epilepsy and purely insular or opercular epilepsy. Conversely, oblique trajectories can sample a larger volume of the insular cortex but not the operculum. To appropriately distinguish the nature of opercular or insular involvement, a combination of orthogonal and oblique trajectories is recommended for insular explorations. The insula is highly connected to its neighboring brain regions, particularly the frontal, temporal and parietal cortices, leading to a diverse semiology. Therefore, sampling of these regions is indicated based on the seizure semiology (Fig. 22.4b). Additionally, the insula is also highly connected to the contralateral insula, which may lead to false lateralization of the EZ, and bilateral implantation is recommended when suspecting insular epilepsy.

Frontal Lobe Epilepsy (FLE) The frontal lobes comprise about 1/3 of the cortical surface area, with a significant part located in the mesial and basal regions. The frontal lobes are also highly connected to other brain regions, including intra-lobar connections by means of juxtacortical U fibers, inter-lobar connections such as supper, middle and inferior longitudinal fasciculi, as well as interhemispheric connections using commissural fibers. These factors, including the prominent bilateral connectivity, can make intracranial evaluations in frontal lobe epilepsy more challenging and complicated to interpret. Conceptualization of frontal lobe epilepsies can be delineated into precentral, premotor and prefrontal epilepsies (Fig. 22.4c). Precentral cortex epilepsies demonstrate elementary motor behaviors such as clonic jerks or tonic posturing. It is also important in precentral epilepsies to appreciate the structural and functional connectivity of the peri-rolandic region, and epilepsies presenting with simple motor or somatosensory semiology may arise anywhere in the frontal or parietal regions (Fig. 22.4d). Premotor epilepsies include symptoms such as proximal symmetric or asymmetric tonic posturing, head/eye version, and axial/proximal complex motor behaviors. Prefrontal epilepsies include more complex patterns of gestural motor behaviors, which can have naturalistic appearances

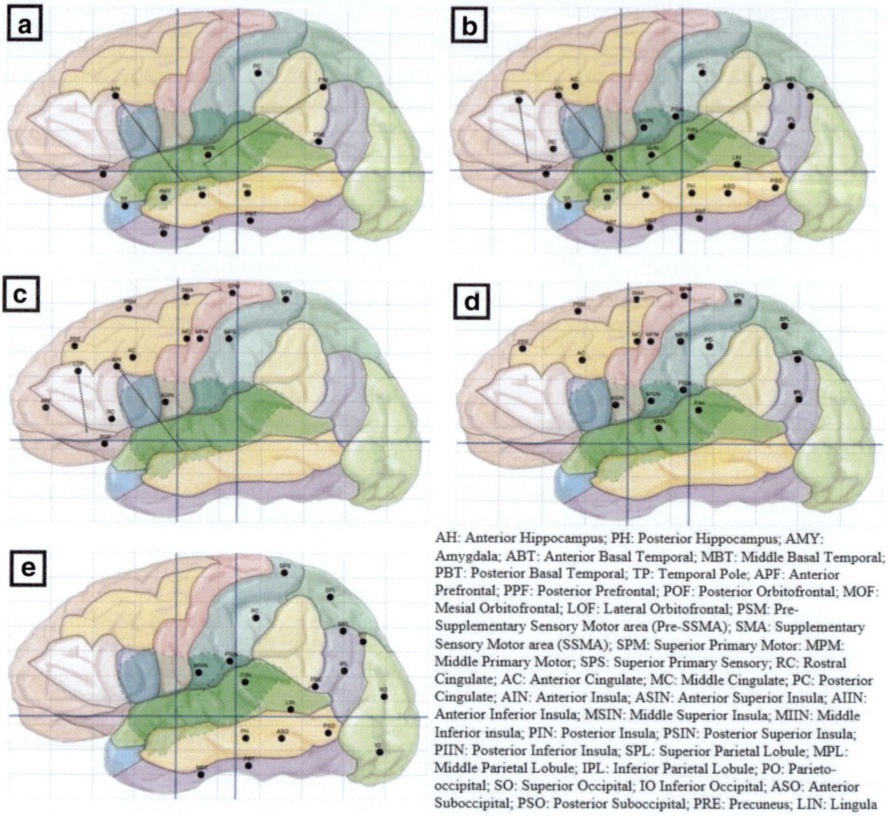

Fig. 22.4 Sample implantation schemes for SEEG for (**a**) left temporal epilepsy, (**b**) left temporal plus epilepsy, (**c**) left frontal epilepsy, (**d**) left peri-rolandic epilepsy and (**e**) left posterior quadrant epilepsy. Note: For this figure, please note that this is an exhaustive scheme of electrodes that could be considered in a temporal plus epilepsy evaluation. The implantation plan for an individual patient needs to be tailored to the EZ hypothesis based on the clinical data and temporal plus subtype suspected (Orbitofrontal, perisylvian, insular, temporo-parieto-occipital)

and emotional features. There commonly is an alteration in awareness with vocalization. More anterior prefrontal structures will also typically include distal > proximal stereotypies. This rostro-caudal gradient of motor movements highlights the crucial role of semiologic evaluation in conceptualizing a FLE [40].

Posterior Quadrant Epilepsy Intracranial explorations for posterior quadrant epilepsy are relatively infrequent. For parietal lobe exploration involving the superior and inferior parietal lobule, it is important to sample the primary somatosensory areas (S1), somatosensory association cortex, parietal operculum, intra-parietal sulcus, inferior parietal cortex, posterior cingulate and posterior insula. Post-central region epilepsies should also include sampling the peri-rolandic cortex and premotor cortex [38]. Based on the clinical hypothesis of seizure propagation, consideration should be given to sampling the central, frontal, occipital and the temporo-parietal junction. At times, contralateral exploration may also be indicated (Fig. 22.4e).

Occipital seizures tend to propagate rostrally by means of either the ventral pathway (occipito-temporal) or the dorsal pathway (occipito-parietal-frontal). For the exploration of the ventral pathway (infra-calcarine seizure onset), it is important to explore the lingual gyrus, cuneus, posterior fusiform gyrus, anterior calcarine fissure and the temporo-occipital junction. For the exploration of the dorsal pathway (supra-calcarine seizure onset), it is important to explore the superior and inferior parietal lobule, posterior cingulate, cuneus, lingual gyrus and the TPO junction (Fig. 22.4e).

22.5 Electrical Stimulation Mapping (ESM)

Electrical stimulation mapping is the application of an electrical stimulus to a population of neurons to activate or deactivate local functional networks, which can be used to localize function and provoke habitual electroclinical seizures. Less frequently, it is also utilized for cortico-cortical evoked potentials (CCEPs).

ESM can be done either intra-operatively (during surgery under local anesthesia in the operating room) or extra-operatively (during video EEG monitoring in the EMU). ESM is best performed when the patient is awake and relaxed. It is important to keep the patient informed about the stimulation's clinical responses, as these may be distressing if unexpected (e.g., forced movements and speech arrest) and on the risk of provoking seizures. For extra-operative ESM, the procedure is typically performed in the later part of the intracranial evaluation usually after the spontaneous seizures have been captured and shortly before the explantation of the intracranial hardware. Typically, ESM is performed after anti-seizure medications have been restarted [41].

ESM for mapping eloquent cortex and seizure provocation is usually done concurrently to better understand if the epileptogenic network coincides with the functional network. In SEEG, cortical stimulation is performed between two contiguous contacts of the electrode using bipolar and biphasic current, whereas in subdural evaluations, stimulation can be performed between two adjacent contacts or between a contact of interest and a "reference electrode." While there are variations in institutional practices, ESM is typically performed via high-frequency and low-frequency stimulation. In the United States, the majority of centers utilize high-frequency stimulation of 50 Hz in trains of 3–8 seconds. This practice is effective for functional mapping, but it may also cause false positives if applied for seizure provocation. In contrast, many centers utilize low-frequency, 1 Hz pulse stimulations over a longer period of time (typically 20–60 seconds) as a tool for seizure provocation. Furthermore, in Europe, many centers utilize this stimulation paradigm for functional mapping of "electrically active" regions such as the hippocampus, Heschl gyrus and peri-rolandic cortex [42]. Typically, current intensities are gradually increased by 0.5–1 mA until the function is detected, side effects are reported, or after-discharges are seen. Longer pulse widths and higher current intensities may be needed in children due to incomplete myelination of the central nervous system (CNS).

High-frequency stimulation is usually preferred for mapping eloquent brain regions. Longer train durations are utilized for testing language and negative motor responses. Low-frequency stimulation is preferred for seizure provocation to reduce the chances of false positives. It has been proposed that stimulation with lower frequencies is effective, but this may reduce the sensitivity of electrical stimulation to detect function, especially over language sites [43]. Conversely, seizures induced by high-frequency stimulation are prone to false positives. Generally the value of seizure provocation is in capturing habitual electroclinical seizures.

After-Discharges After-discharges (ADs) are repetitive epileptiform discharges or rhythmic waves that are induced by cortical stimulation and are generally short and self-limited. The occurrence of ADs invalidates the results of functional mapping and suggests that the localization of the functional response to the stimulated contact is not reliable [44]. Moreover, it is possible that the stimulation of those contacts at higher intensities could progress to a seizure. On the other hand, stimulation after a few-minute breaks and with lower frequencies and currents may reduce the risk of ADs [43]. It is possible to acutely abort ADs and seizures using repeat high-frequency stimulation of the involved contacts [45].

Critical Regions for Mapping Function The primary motor area (M1), where the contralateral side of the body is represented over relatively large brain regions, responds to electrical stimulation, often in clonic movements in adults. The primary sensory areas present with contralateral electrical feeling or tingling, as reported by patients. In the premotor regions, the motor manifestations are often tonic. In supplementary sensory areas, such as SSMA (mesial frontal), S2 (inferior parietal), and insula, there is a bilateral representation of the body in anterior-posterior orientation. The stimulation of the primary auditory cortex in the planum temporale and the primary visual cortex in the para-calcarine cortex generates primitive auditory and visual hallucinations, respectively. There are two negative motor areas where stimulation leads to arrest of tongue movements, and hence speech-over the frontal convexity caudal to Broca's, and in the mesial frontal cortex in the pre-SMA region. It is paramount to test for negative motor areas when evaluating language. The primary motor and sensory thresholds are lower compared to premotor and language sites. The primary language sites include anterior language, also known as Broca's; posterior language, also known as Wernicke's; basal temporal, involved in the processing of visual naming; and temporal pole, involved in the processing of auditory naming. Naming decline has been reported in resections within the dominant left temporal lobe [46].

Mapping of Other Brain Regions The stimulation of insula correlates with complex sensory symptoms involving both sides of the body, pain, visceral motor and gustatory responses. Stimulation of the cingulate may lead to hallucinations of floating, fear, simple or complex motor experiences. Stimulation has also been used to study cognition and experiential phenomena. Some areas of the brain are considered indispensable such as the primary hand motor area and the primary anterior frontal language site. Other areas may be considered for resection if they overlap with the seizure onset zone, as the quality of life following epilepsy surgery correlates with seizure freedom even in the presence of new minor deficits [47].

Limitations Although brain tissue does not perceive pain, stimulation may cause irritation of the dura mater and piercing nerves, which may cause pain and interfere with the procedure. After-discharges and provoked seizures may interfere with and delay mapping. Of note, it is important to recognize the limitations of cortical stimulation in assessing the complexity of human cognition and predicting functions post-surgery. In a survey study, 41% of centers reported at least one case of language decline despite preserving all language identified by electrical cortical stimulation [48].

Safety Electrical stimulation does have the potential for neuronal injury, which is related to charge density and charge per phase. Charge per phase is determined by multiplying the stimulation current by the duration of one phase of the biphasic pulse used and is expressed in units of μC per phase. Charge density is calculated by dividing the charge per phase by the electrode contact surface area. This is an important consideration as the surface areas of SEEG electrodes and subdural electrodes are significantly different, with SEEG electrodes typically having about 25% of the surface area of clinical subdural electrodes. While chronic stimulation in animal models suggests a safety threshold of 30 $\mu C/cm^2$ charge density, limited human stimulation data in subdural cases suggested charge densities over 50 $\mu C/cm^2$ do not cause any pathologic changes consistent with electrical injury [49].

Acknowledgments The authors would like to acknowledge Drs. Rafeed Alkawadri and Paul Van Ness, who previously authored this chapter; some of the information has been adapted into the current chapter.

References

1. Asano E, Juhász C, Shah A, Sood S, Chugani HT. Role of subdural electrocorticography in prediction of long-term seizure outcome in epilepsy surgery. Brain J Neurol. 2009;132:1038–47.
2. Tao JX, Baldwin M, Hawes-Ebersole S, Ebersole JS. Cortical substrates of scalp EEG epileptiform discharges. J. Clin. Neurophysiol. Off. Publ. Am. Electroencephalogr. Soc. 2007;24:96–100.
3. Arya R, Mangano FT, Horn PS, Holland KD, Rose DF, Glauser TA. Adverse events related to extraoperative invasive EEG monitoring with subdural grid electrodes: a systematic review and meta-analysis. Epilepsia. 2013;54:828–39.
4. Lhatoo S, Lacuey N, Ryvlin P. Principles of stereotactic electroencephalography in epilepsy surgery. J Clin Neurophysiol Off Publ Am Electroencephalogr Soc. 2016;33:478–82.
5. Gavvala J, Zafar M, Sinha SR, Kalamangalam G, Schuele S, American SEEG Consortium, supported by The American Clinical Neurophysiology Society. Stereotactic EEG practices: a

survey of United States tertiary referral epilepsy centers. J. Clin. Neurophysiol. Off. Publ. Am. Electroencephalogr. Soc. 2022;39:474–80.
6. Nair DR, Burgess R, McIntyre CC, Lüders H. Chronic subdural electrodes in the management of epilepsy. Clin. Neurophysiol. Off. J. Int. Fed. Clin. Neurophysiol. 2008;119:11–28.
7. Alkawadri R, Gaspard N, Goncharova II, Spencer DD, Gerrard JL, Zaveri H, Duckrow RB, Blumenfeld H, Hirsch LJ. The spatial and signal characteristics of physiologic high frequency oscillations. Epilepsia. 2014;55:1986–95.
8. Ren L, Kucewicz MT, Cimbalnik J, Matsumoto JY, Brinkmann BH, Hu W, Marsh WR, Meyer FB, Stead SM, Worrell GA. Gamma oscillations precede interictal epileptiform spikes in the seizure onset zone. Neurology. 2015;84:602–8.
9. Ikeda A, Taki W, Kunieda T, Terada K, Mikuni N, Nagamine T, Yazawa S, Ohara S, Hori T, Kaji R, Kimura J, Shibasaki H. Focal ictal direct current shifts in human epilepsy as studied by subdural and scalp recording. Brain J Neurol. 1999;122(Pt 5):827–38.
10. Zaveri HP, Duckrow RB, Spencer SS. On the use of bipolar montages for time-series analysis of intracranial electroencephalograms. Clin Neurophysiol Off J Int Fed Clin Neurophysiol. 2006;117:2102–8.
11. Frauscher B, von Ellenrieder N, Zelmann R, Doležalová I, Minotti L, Olivier A, Hall J, Hoffmann D, Nguyen DK, Kahane P, Dubeau F, Gotman J. Atlas of the normal intracranial electroencephalogram: neurophysiological awake activity in different cortical areas. Brain J Neurol. 2018;141:1130–44.
12. Steriade M, Contreras D, Curró Dossi R, Nuñez A. The slow (< 1 Hz) oscillation in reticular thalamic and thalamocortical neurons: scenario of sleep rhythm generation in interacting thalamic and neocortical networks. J Neurosci. 1993;13:3284–99.
13. von Ellenrieder N, et al. How the human brain sleeps: direct cortical recordings of Normal brain activity. Ann Neurol. 2020;87:289–301.
14. Colrain IM, Webster KE, Hirst G, Campbell KB. The roles of vertex sharp waves and K-complexes in the generation of N300 in auditory and respiratory-related evoked potentials during early stage 2 NREM sleep. Sleep. 2000;23:97–106.
15. Frauscher B, et al. High-frequency oscillations in the Normal human brain. Ann Neurol. 2018;84:374–85.
16. Cooper R, Winter AL, Crow HJ, Walter WG. Comparison of subcortical, cortical and scalp activity using chronically indwelling electrodes in man. Electroencephalogr Clin Neurophysiol. 1965;18:217–28.
17. Di Giacomo R, Uribe-San-Martin R, Mai R, Francione S, Nobili L, Sartori I, Gozzo F, Pelliccia V, Onofrj M, Lo Russo G, de Curtis M, Tassi L. Stereo-EEG ictal/interictal patterns and underlying pathologies. Seizure. 2019;72:54–60.
18. Alarcon G, Garcia Seoane JJ, Binnie CD, Martin Miguel MC, Juler J, Polkey CE, Elwes RD, Ortiz Blasco JM. Origin and propagation of interictal discharges in the acute electrocorticogram. Implications for pathophysiology and surgical treatment of temporal lobe epilepsy. Brain J Neurol. 1997;120(Pt 12):2259–82.
19. Bartolomei F, Trébuchon A, Bonini F, Lambert I, Gavaret M, Woodman M, Giusiano B, Wendling F, Bénar C. What is the concordance between the seizure onset zone and the irritative zone? A SEEG quantified study. Clin. Neurophysiol. Off. J. Int. Fed. Clin. Neurophysiol. 2016;127:1157–62.
20. Roehri N, Pizzo F, Lagarde S, Lambert I, Nica A, McGonigal A, Giusiano B, Bartolomei F, Bénar CG. High-frequency oscillations are not better biomarkers of epileptogenic tissues than spikes. Ann Neurol. 2018;83:84–97.
21. Brázdil M, Pail M, Halámek J, Plešinger F, Cimbálník J, Roman R, Klimeš P, Daniel P, Chrastina J, Brichtová E, Rektor I, Worrell GA, Jurák P. Very high-frequency oscillations: novel biomarkers of the epileptogenic zone. Ann Neurol. 2017;82:299–310.
22. Abraham K, Marsan CA. Patterns of cortical discharges and their relation to routine scalp electroencephalography. Electroencephalogr Clin Neurophysiol. 1958;10:447–61.
23. Malow BA, Lin X, Kushwaha R, Aldrich MS. Interictal spiking increases with sleep depth in temporal lobe epilepsy. Epilepsia. 1998;39:1309–16.

24. Kahane P, Debeau F. Intracerebral depth electrode electroencephalography. In: Ebersole JS, Husain AM, Nordli Jr DR, editors. Current practice of clinical electroencephalography. Wolters Kluwer; 2014. p. 394–440.
25. Mirandola L, Mai RF, Francione S, Pelliccia V, Gozzo F, Sartori I, Nobili L, Cardinale F, Cossu M, Meletti S, Tassi L. Stereo-EEG: diagnostic and therapeutic tool for periventricular nodular heterotopia epilepsies. Epilepsia. 2017;58:1962–71.
26. Lagarde S, Bonini F, McGonigal A, Chauvel P, Gavaret M, Scavarda D, Carron R, Régis J, Aubert S, Villeneuve N, Giusiano B, Figarella-Branger D, Trebuchon A, Bartolomei F. Seizure-onset patterns in focal cortical dysplasia and neurodevelopmental tumors: relationship with surgical prognosis and neuropathologic subtypes. Epilepsia. 2016;57:1426–35.
27. Singh S, Sandy S, Wiebe S. Ictal onset on intracranial EEG: do we know it when we see it? State of the evidence. Epilepsia. 2015;56:1629–38.
28. Jiménez-Jiménez D, Nekkare R, Flores L, Chatzidimou K, Bodi I, Honavar M, Mullatti N, Elwes RD, Selway RP, Valentín A, Alarcón G. Prognostic value of intracranial seizure onset patterns for surgical outcome of the treatment of epilepsy. Clin. Neurophysiol. Off. J. Int. Fed. Clin. Neurophysiol. 2015;126:257–67.
29. Khoo A, de Tisi J, Foong J, Bindman D, O'Keeffe AG, Sander JW, Miserocchi A, McEvoy AW, Duncan JS. Long-term seizure, psychiatric and socioeconomic outcomes after frontal lobe epilepsy surgery. Epilepsy Res. 2022;186:106998.
30. Deleo F, Garbelli R, Milesi G, Gozzo F, Bramerio M, Villani F, Cardinale F, Tringali G, Spreafico R, Tassi L. Short- and long-term surgical outcomes of temporal lobe epilepsy associated with hippocampal sclerosis: relationships with neuropathology. Epilepsia. 2016;57:306–15.
31. de Tisi J, Bell GS, Peacock JL, McEvoy AW, Harkness WF, Sander JW, Duncan JS. The long-term outcome of adult epilepsy surgery, patterns of seizure remission, and relapse: a cohort study. Lancet Lond Engl. 2011;378:1388–95.
32. Callaghan B, Schlesinger M, Rodemer W, Pollard J, Hesdorffer D, Allen Hauser W, French J. Remission and relapse in a drug-resistant epilepsy population followed prospectively. Epilepsia. 2011;52:619–26.
33. Tandon N, Tong BA, Friedman ER, Johnson JA, Von Allmen G, Thomas MS, Hope OA, Kalamangalam GP, Slater JD, Thompson SA. Analysis of morbidity and outcomes associated with use of subdural grids vs Stereoelectroencephalography in patients with intractable epilepsy. JAMA Neurol. 2019;76:672–81.
34. Mullin JP, Shriver M, Alomar S, Najm I, Bulacio J, Chauvel P, Gonzalez-Martinez J. Is SEEG safe? A systematic review and meta-analysis of stereo-electroencephalography-related complications. Epilepsia. 2016;57:386–401.
35. McGovern RA, Ruggieri P, Bulacio J, Najm I, Bingaman WE, Gonzalez-Martinez JA. Risk analysis of hemorrhage in stereo-electroencephalography procedures. Epilepsia. 2019;60:571–80.
36. Fong JS, Alexopoulos AV, Bingaman WE, Gonzalez-Martinez J, Prayson RA. Pathologic findings associated with invasive EEG monitoring for medically intractable epilepsy. Am J Clin Pathol. 2012;138:506–10.
37. Khoo HM, Hall JA, Dubeau F, Tani N, Oshino S, Fujita Y, Gotman J, Kishima H. Technical aspects of SEEG and its interpretation in the delineation of the epileptogenic zone. Neurol Med Chir (Tokyo). 2020;60:565–80.
38. Chassoux F, Navarro V, Catenoix H, Valton L, Vignal J-P. Planning and management of SEEG. Neurophysiol Clin Clin Neurophysiol. 2018;48:25–37.
39. Tao S, Lhatoo S, Hampson J, Cui L, Zhang G-Q. A bespoke electronic health record for epilepsy care (EpiToMe): development and qualitative evaluation. J Med Internet Res. 2021;23:e22939.
40. McGonigal A. Frontal lobe seizures: overview and update. J Neurol. 2022;269:3363–71.
41. Arya R, Aungaroon G, Zea Vera A, Horn PS, Byars AW, Greiner HM, Mangano FT, Holland KD. Fosphenytoin pre-medication for pediatric extra-operative electrical stimulation brain mapping. Epilepsy Res. 2018;140:171–6.
42. Trébuchon A, Chauvel P. Electrical stimulation for seizure induction and functional mapping in Stereoelectroencephalography. J. Clin. Neurophysiol. Off. Publ. Am. Electroencephalogr. Soc. 2016;33:511–21.

43. Zangaladze A, Sharan A, Evans J, Wyeth DH, Wyeth EG, Tracy JI, Chervoneva I, Sperling MR. The effectiveness of low-frequency stimulation for mapping cortical function. Epilepsia. 2008;49:481–7.
44. So EL, Alwaki A. A guide for cortical electrical stimulation mapping. J. Clin. Neurophysiol. Off. Publ. Am. Electroencephalogr. Soc. 2018;35:98–105.
45. Lesser RP, Kim SH, Beyderman L, Miglioretti DL, Webber WR, Bare M, Cysyk B, Krauss G, Gordon B. Brief bursts of pulse stimulation terminate afterdischarges caused by cortical stimulation. Neurology. 1999;53:2073–81.
46. Hamberger MJ. Cortical language mapping in epilepsy: a critical review. Neuropsychol Rev. 2007;17:477–89.
47. Spencer SS, Berg AT, Vickrey BG, Sperling MR, Bazil CW, Haut S, Langfitt JT, Walczak TS, Devinsky O, Multicenter Study of Epilepsy Surgery. Health-related quality of life over time since resective epilepsy surgery. Ann Neurol. 2007;62:327–34.
48. Hamberger MJ, Williams AC, Schevon CA. Extraoperative neurostimulation mapping: results from an international survey of epilepsy surgery programs. Epilepsia. 2014;55:933–9.
49. Cogan SF, Ludwig KA, Welle CG, Takmakov P. Tissue damage thresholds during therapeutic electrical stimulation. J Neural Eng. 2016;13:021001.

Surgical Management 23

Garrett Banks and Zulfi Haneef

23.1 Introduction

Epilepsy surgery is the most effective method for treating drug resistant epilepsy that does not respond to anti-seizure medications. Surgical treatment of epilepsy can be divided into (1) resections (e.g., anterior temporal lobectomy), (2) disconnections (e.g., hemispherotomy), (3) ablations (e.g., laser ablation), and (4) neuromodulation (e.g., VNS and RNS). Among these techniques, resection results in the best seizure control, and neuromodulation is the least effective. In pooled analyses, 30% of epilepsy patients are drug resistant, and it has been estimated that approximately half of these patients are candidates for resection or ablation [1, 2]. For the remaining patients, palliative procedures such as disconnection or neuromodulation can be considered, which aim at reducing a patient's seizure burden. This chapter describes some of the commonly performed epilepsy surgeries.

23.2 Resections

Epilepsy resection is a technique where the area of the brain necessary and sufficient for seizure generation, referred to as the *epileptogenic zone* (EZ), is surgically removed. Resections are performed after an extensive multidisciplinary evaluation to identify the EZ using electrophysiology, imaging, neuropsychology, and other

G. Banks (✉)
Neurosurgery, Baylor College of Medicine, Michael E. DeBakey VA Medical Center, Houston, TX, USA
e-mail: Garrett.Banks@bcm.edu

Z. Haneef
Neurology, Baylor College of Medicine, Epilepsy Center of Excellence, Michael E DeBakey VA Medical Center, Houston, TX, USA

Z. Haneef (ed.), *Epilepsy Fundamentals*,
https://doi.org/10.1007/978-3-031-77741-7_23

data. Depending on where the EZ localizes, a standard resection (e.g., anterior temporal lobectomy, amygdalohippocampectomy) or a tailored resection (tailored to the putative EZ) can potentially be performed. In most cases, the goal of resection is complete seizure freedom.

As per data from the National Association of Epilepsy Centers (NAEC), there were 3263 resective epilepsy surgeries (1940 temporal, 1323 extra-temporal) and 817 non-resective intracranial monitoring studies performed in the US during 2021 [3]. Additionally, there were 140 hemispherotomies and 618 laser ablations. Surgical resection has been shown to be cost-effective compared to continued medical treatment in patients with focal drug resistant epilepsy, and also leads to an improvement in their quality of life [4].

Anterior Temporal Lobe Resection (ATLR, Also Called Anterior Temporal Lobectomy) Resection for temporal lobe epilepsy (TLE) is the most common adult epilepsy surgery and may account for 50–60% of non-modulatory surgical procedures. ATLR for TLE leads to some of the best results in epilepsy surgery with 70% of patients attaining seizure freedom at 1 year and 55% continuing to be seizure-free at 5 years. The resection involves the removal of the anterior temporal lobe and mesial structures (Figs. 23.1 and 23.2).

During the ATLR procedure, the anterior part of the middle and lower temporal gyri (and sometimes, part of the superior temporal gyrus, especially in non-dominant side resections) are resected laterally. This allows for the opening of the lateral ventricle and gaining access to the amygdala and hippocampus medially, which are removed along with the temporal pole. The anterior fusiform and parahippocampal gyri are also removed to complete the removal of the anterior temporal lobe. Posterior resection margins are typically more conservative on the left side (dominant) and may be performed with speech mapping to avoid language areas. Notably, the posterior temporal lobe is typically spared for both right and left-sided resections.

Fig. 23.1 "Standard" ATLR margins

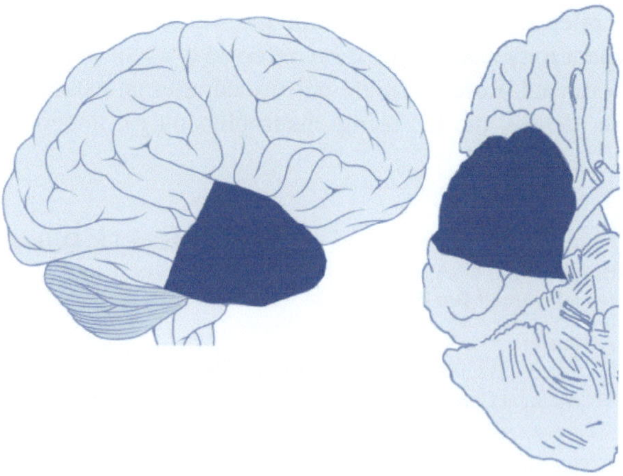

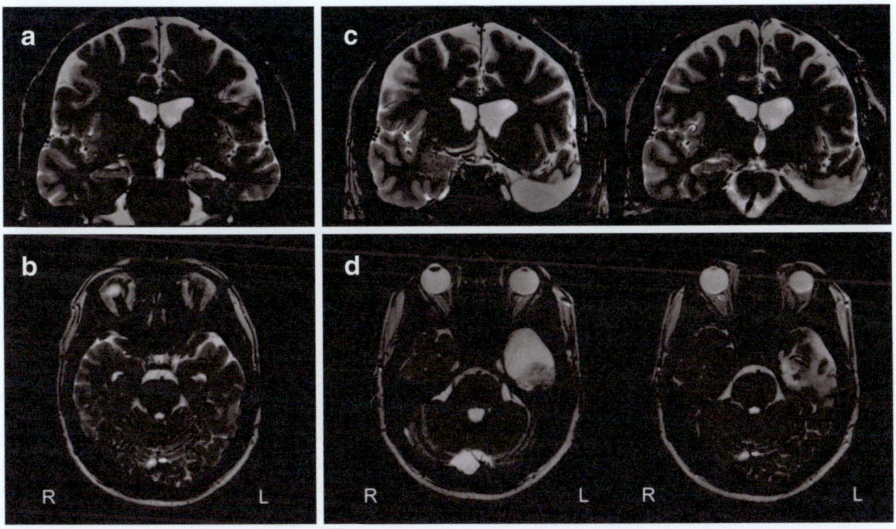

Fig. 23.2 Demonstration of anterior temporal lobectomy for mesial temporal lobe epilepsy. (**a**, **b**) Pre-operative T2-weighted coronal (A) and axial (B) MRI showing increased T2 signal, decreased size, and altered internal architecture of the left hippocampus, consistent with mesial temporal sclerosis. (**c**, **d**) Postoperative T2-weighted coronal (C) and axial (D) MRI demonstrating the resection cavity after left anterior temporal lobectomy. The resection involves the anterior hippocampus to the level of the tectal plate, the amygdala, and approximately 4 cm resection of the anterolateral temporal neocortex, including the middle and inferior temporal gyri. (Reproduced with permission from Chang et al. [5])

Selective Amygdalohippocampectomy (SAH) Due to most people with mesial temporal lobe epilepsy having seizures that originate from mesial-basal temporal lobe structures, including the hippocampus, amygdala and parahippocampus, efforts were made to develop a procedure to solely resect these structures [6]. These efforts led to the development of the selective amygdalohippocampectomy (SAH). In this technique, a neurosurgeon accesses the mesial structures via either the Sylvian fissure, a trans-cortical pathway, or a subtemporal approach (Fig. 23.3). A comparative meta-analysis in 2013 concluded that a standard ATLR leads to a higher chance of achieving freedom from disabling seizures (Engel Class 1 outcome) compared to a SAH, with a number needed to treat of 13 for an additional patient to achieve Class 1 outcome after ATLR. While less efficacious, the proposed benefit of the SAH is superior neuropsychological outcomes due to less temporal cortex being removed during the procedure. Studies comparing the two have shown greater declines in verbal memory after an ATLR compared to SAH for dominant-side surgery. Notably, both procedures cause a significant deficit when performed on the dominant side [7].

Hemispherectomy Hemispherectomy was first described by Dandy in 1928 for glioblastoma resection and by McKenzie in 1938 for epilepsy treatment. This is typically performed for developmental/childhood conditions involving catastrophic hemispheric epilepsy, such as hemimegalencephaly, large porencephaly, Sturge

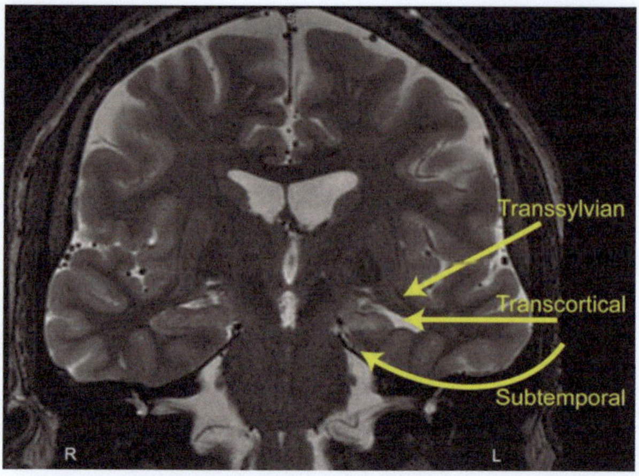

Fig. 23.3 Microsurgical approaches for selective amygdalohippocampectomy. Three main approaches are shown for selective resection of the mesial temporal structures while preserving the lateral temporal neocortex. The trans-sylvian approach includes microsurgical splitting of the sylvian fissure and traversing a portion of the temporal stem. The transcortical approach proceeds through a limited lateral temporal corticectomy, while the subtemporal approach involves gentle elevation of the temporal lobe to identify and enter the collateral sulcus. L: left; R: right. (Reproduced with permission from Chang et al., 2015) [5]

Weber, and Rasmussen syndrome. Engel Class I or II outcomes are seen in up to 80% of patients. In an anatomic hemispherectomy, resection typically involves the frontal, temporal, parietal, and occipital neocortices while preserving subcortical structures (Fig. 23.4a). Complications are typically more common with an anatomical hemispherectomy compared to focal epilepsy resections. Adverse outcomes may include hemiparesis, hemianopia, worsened language function, and hydrocephalus requiring shunting. However, because hemispherectomy is most often performed in children already suffering from marked hemiparesis and hemianopia, the additional deficits incurred through surgery are generally low. Several factors must be taken into account before proceeding with this procedure, including severity of illness, radiographic and electrographic findings, patient age and capacity for neural remodeling, and baseline neurological deficits.

Lesionectomy and Cortical Resection Complete removal of an epileptogenic lesion is called lesionectomy. The underlying pathology can include a variety of lesions, including tumors (e.g., meningioma), vascular malformations (e.g., cavernoma), and developmental abnormalities (e.g., dysplasia) among others. An isolated lesionectomy may be performed or may include additional cortical resection, because the regions around the lesion may be epileptogenic. Intraoperative electrocorticography can be utilized during the operation to look for adjacent epileptogenic tissue.

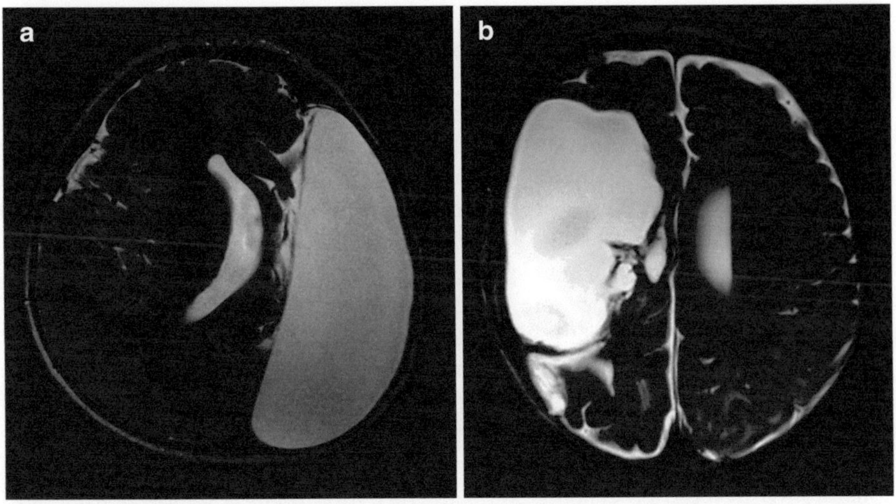

Fig. 23.4 T2-weighed axial MRI images depicting examples of an anatomic (**a**) and functional (**b**) hemispherectomy. (Image courtesy: Joyce Matsumoto, UCLA, California)

23.3 Disconnections

Functional disconnection of pathology may be favored instead of resection in some instances. The goal of these procedures is to disconnect the epileptogenic cortex from the rest of the brain to prevent seizure spread.

Disconnection Syndrome

Due to callosal interruption, communication between language centers in the left hemisphere and the somatosensory system of the right hemisphere is disconnected, leading to the disconnection syndrome. Features of the disconnection syndrome include *left tactile anomia, left apraxia, and left agraphia.*

Left tactile anomia: This refers to the inability to name objects perceived on the left side. This is because stimuli in the right hemisphere (e.g., objects seen in the left visual space or held in the left hand) cannot be named by the language centers in the left hemisphere.

Left-sided ideomotor apraxia: Commands to perform actions interpreted in the left hemisphere language centers can be performed by the left hemisphere (right arm/leg) but not by the disconnected right hemisphere (left arm/leg).

Left-hand agraphia: The patient can write with the right hand but not the left hand, as the language centers in the left are not connected to left-hand movements initiated in the right hemisphere.

Hemispherotomy Hemispherotomy is a procedure in which the epileptogenic hemisphere is disconnected from ipsilateral subcortical structures and the contralateral hemisphere. The earliest version of this technique, pioneered by Rasmussen in 1974, was termed the "functional hemispherectomy." While several anatomic variations of functional hemispherectomy have been described, surgery commonly includes resection of the temporal, posterior frontal and anterior parietal lobe, with disconnection of the corpus callosum and subpial disconnection of the occipital and residual frontal lobe (Fig. 23.4b). When comparing the anatomic hemispherectomy (Fig. 23.4a) to the functional hemispherectomy (Fig. 23.4b), significantly less tissue is removed in the latter procedure. The functional hemispherectomy carries comparatively less morbidity, and lower risk of a condition called superficial cerebral hemosiderosis [8]. Superficial cerebral hemosiderosis refers to chronic progressive neurologic deficits secondary to iron deposition in neurologic tissues from chronic bleeding. Further evolutions of this procedure are the vertical paramedian hemispherotomy, the lateral peri-insular hemispherotomy and the transsylvian hemispherotomy. In these procedures comparatively little brain is removed while still accomplishing the disconnection of the ipsilateral cortex.

Callosotomy In this procedure the corpus callosum is severed to limit interhemispheric spread of epileptic activity. Indications include disabling atonic seizures (or "drop attacks"), which improve by 50–80%, and complex partial seizures with bilateral frontal manifestations. It is commonly utilized for such conditions as Lennox-Gastaut syndrome, tuberous sclerosis and West syndrome. Different techniques are described, but a disconnection of the anterior two-thirds of the corpus callosum is preferred by most experts. While the procedure is often done as the initial primary surgical treatment, one recent paper showed that a corpus callosotomy can be a very effective treatment in patients who have had a previous vagal nerve stimulator and still do not have good control of their seizures [9]. One frequent complication of callosotomy is the **disconnection syndrome**, with unilateral tactile anomia, unilateral apraxia, and alexia.

Multiple Subpial transections (MST) MST are typically performed to minimize functional loss when seizure onset involves eloquent cortex. The procedure involves disrupting the horizontally oriented intracortical fibers (Fig. 23.5) to limit the tangential cortical seizure spread. This preserves function by sparing vertical cortical architecture. However, the utility of MST is a subject of debate among specialists, with seizure freedom rates around 20% [10]. Also, residual neurological deficits can occur in up to 25% of cases.

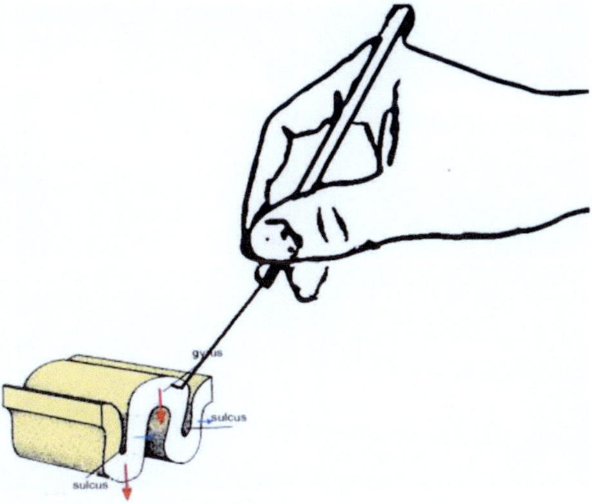

Fig. 23.5 Multiple subpial transections

23.4 Neuroablation

Neuroablation techniques for epilepsy are a set of procedures that aim to destroy epileptogenic tissue using a minimally invasive or noninvasive method. These procedures range from placement of a fiber optic probe to make a controlled ablation (laser ablation) to methods that do not even require an incision, such as focused ultrasound and gamma knife.

MR Guided Laser Ablation, Stereotactic laser **thermoablation**, (e.g., Visualase by Medtronic, Neuroblate by Monteris, Fig. 23.6) or Magnetic Resonance guided laser interstitial thermal therapy (MRgLITT), is a relatively new technique that has been increasing in popularity for the treatment of drug resistant epilepsy. The treatment first involves stereotactic placement of a flexible laser optical fiber within a cooling catheter in the operating room [11]. This fiber is secured with a bolt that is screwed into the skull, and the patient is transported to an MRI. Inside the MRI, the fiber optic cable is connected to a 15 W 980 nm diode laser, and the laser is powered while monitoring the brain using a specialized MRI sequence that can detect temperature changes. Surgeons can make adjustments to the power of the laser while monitoring the temperature in real time to allow for a carefully controlled ablation of the epileptogenic zone.

Seizure freedom rates following MRgLITT average 55.8% at 1 year and 52% at 2 years [12]. While seizure freedom rates vary from study to study, evidence suggests that ablations including more anterior, medial, and inferior temporal lobe structures, involving greater amygdalar ablation, were more likely to be associated with Engel class I outcomes [13]. While seizure freedom rates are lower than ATL, the procedure removes substantially less brain tissue and is suggested by some groups to lead to better postoperative cognitive functional outcomes [14]. Limited

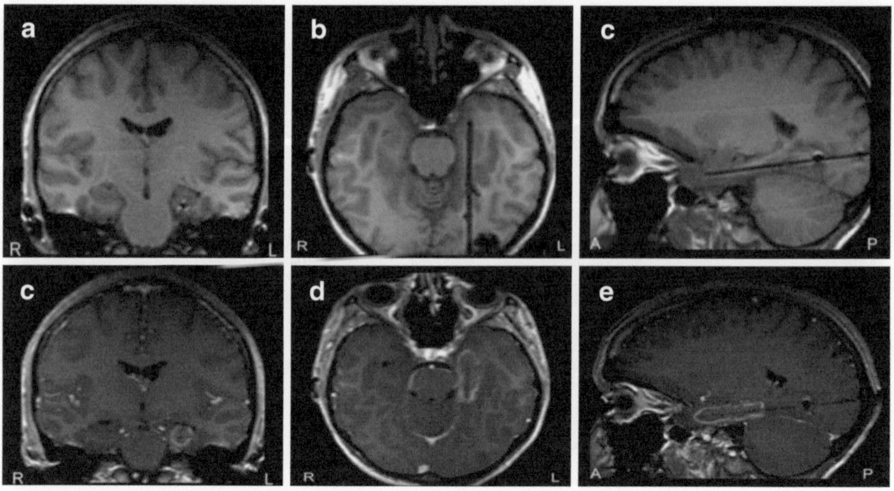

Fig. 23.6 Stereotactic laser thermoablation in mesial temporal lobe epilepsy (**a–c**) T1-weighted peri-procedural MRI coronal (A), axial (B), and sagittal (C) images showing laser probe placement along the axis of the left hippocampus, prior to thermoablation in a patient with mesial temporal lobe epilepsy (**d–f**) Contrast-enhanced T1-weighted MRI coronal (D), axial (E), and sagittal (F) images after thermoablation of mesial temporal lobe structures, with contrast enhancement observed in the region of ablation A: anterior; L: left; P: posterior; R: right. (Reproduced with permission from Chang et al. [5])

studies have shown that laser ablation of the dominant hippocampus leads to less verbal memory decline compared to open resection of the anterior temporal lobe; however, no head-to-head trials have been performed.

One additional indication for MRgLITT is in the treatment of hypothalamic hamartomas. Using the minimally invasive approach of laser ablation, a large case series showed 93% of patients were able to achieve seizure freedom in 1 year [15]. This excellent result was achieved with very minimal morbidity in patients.

MR-Guided Focused Ultrasound Focused ultrasound is a technique that allows for controlled thermal ablation of areas of the brain without needing to make an incision. The intracranial use of focused ultrasound has been FDA-approved for Essential tremor and Parkinson's Disease, but research is ongoing to determine its utility in drug resistant epilepsy. The technique utilizes an array of 1024 ultrasound transducers affixed above the head, with a water bath between the transducers and the scalp. Similar to MRgLITT, the ultrasound is utilized while in an MRI scanner so that a temperature-sensitive MRI sequence can be run to monitor the tissue temperature in real time. The patient can also be kept awake during the procedure to monitor for adverse effects during the sonication.

In one case report a patient with mesial temporal epilepsy had their left hippocampus treated with focused ultrasound [16]. During the procedure they were unable to reach true ablation temperatures, but the patient notably was seizure free at 12 months after the procedure. In another case report, a patient was able to achieve seizure freedom after focused ultrasound treatment of a hypothalamic hamartoma [17]. Lastly, in an open-label phase one trial, two patients underwent a unilateral anterior thalamotomy for medically refractory focal onset epilepsy [18]. One patient was seizure free at 1 year and the other had a significant reduction in their seizure burden. It is important to note that these trials are only for a limited number of patients. Still, they show that there is significant potential for focused ultrasound treatment of medically refractory seizures in the future.

Other Ablative Techniques **Radiofrequency thermocoagulation**: This technique can be performed using a RF generator connected to already implanted stereotactic EEG electrodes after detecting potential sites of epileptogenicity. The procedure is often proceeded by functional mapping of the electrodes to ensure the area is not required for an essential function (eloquent). Both mapping and RF ablation procedure can be performed at the bedside. The procedure's efficacy is lower than MRgLITT, with case series showing seizure freedom in only 18% of patients [19].

- **Stereotactic radiosurgery**: In this procedure focused radiation (e.g., gamma knife) is delivered to lesions with some sparing of surrounding tissue (Fig. 23.7). A latent period of efficacy is usually noted due to radiosurgery requiring time to cause cell death.

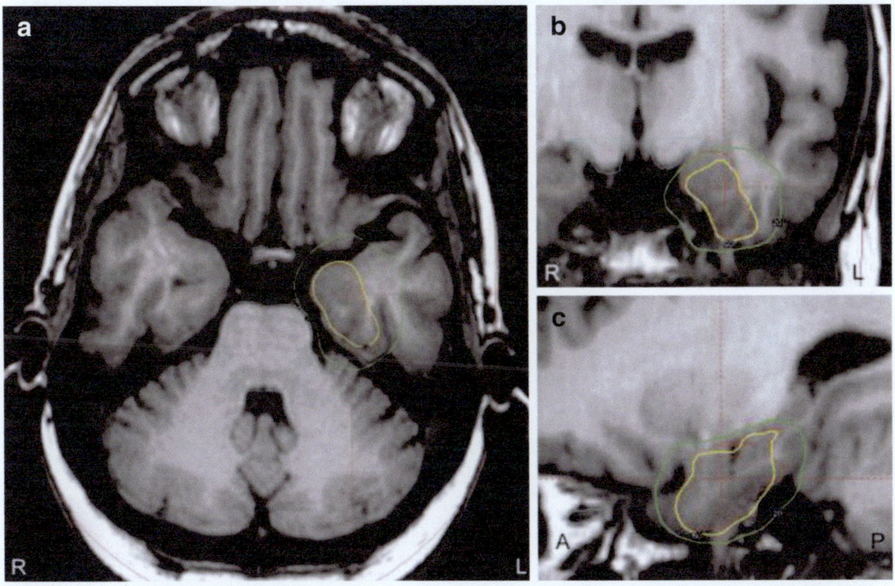

Fig. 23.7 Gamma knife stereotactic radiosurgery planning in a patient with mesial temporal lobe epilepsy On axial (**a**), coronal (**b**), and sagittal (**c**) T1-weighted MRI images, treatment is planned at the mesial temporal structures, with 50% isodose line shown in yellow, 20% isodose line displayed in green, and crosshairs approximating the center of the target. Care is taken to limit the radiation dose to the optic apparatus and brainstem. A: anterior; L: left; P: posterior; R: right. (Reproduced with permission from Chang et al. [5])

23.5 Neuromodulation

Neuromodulation techniques stimulate neural tissue with the aim of reducing seizures without the removal or destruction of brain tissue.

Vagus Nerve stimulation (VNS, Fig. 23.8) VNS is a technique that is usually offered when patients with pharmacoresistant seizures are not candidates for epilepsy surgery with resection/ablation techniques or decline such surgeries. An incision along the anterior left sternomastoid is used to approach and isolate the vagus nerve, where a wire is wrapped around the nerve and connected to a pulse generator implanted in the left upper chest wall.

Repetitive stimuli to the vagus nerve may reduce seizure frequency through an anticonvulsant effect resulting from increased norepinephrine tone [20]. It has also been suggested that ascending stimulation along the vagus nerve into the solitary nucleus alters cortical synchronizability via ascending synapses. The antiepileptic effect increases over months to years and a greater than 50% seizure reduction (or "responder rate") is seen in approximately 50% of patients after 1–2 years of treatment [21]. Seizure freedom may be attained in **5%**. Adverse effects include hoarseness (from recurrent laryngeal stimulation), cough, dyspnea, and neck pain.

In June 2015, the FDA approved a new VNS device (marketed as Aspire SR by Cyberonics, USA.) that uses a customizable cardiac-based algorithm to detect relative heart rate increases during seizures and deliver responsive vagal stimulation to abort them, thus allowing detection and treatment of seizures in real time for selected patients. Preliminary results show that seizure detection based on heart rate changes 20% above baseline was highly sensitive. Due to intrapatient variability of seizure-associated heart rate increases, candidate identification may be challenging. A suggested approach is to identify patients who have at least one seizure with at least

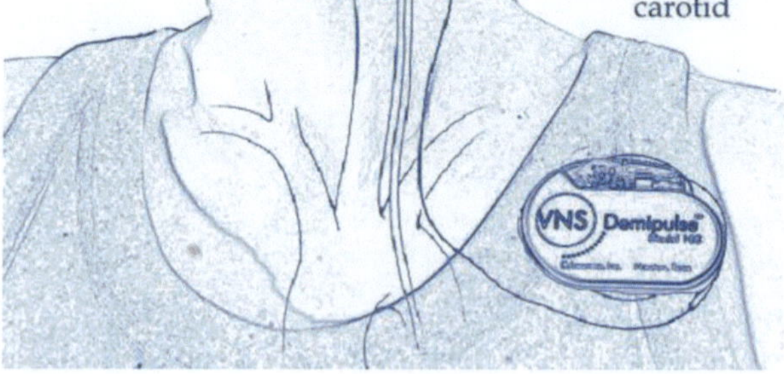

Fig. 23.8 Vagus nerve stimulator (VNS)

20% heart rate during an epilepsy monitoring unit study [22]. Further studies are awaited to assess the long-term benefit of this new device.

Responsive neurostimulation (RNS, Fig. 23.9) This is a closed-loop system where intracranially implanted electrodes (1 or 2 depth/strip electrodes with a maximum of 8 contacts placed according to seizure onset localization) detect epileptiform abnormality using automated algorithms and deliver a current to disrupt seizure activity (responsive neurostimulation). The benefit of RNS is thought to be due to both the responsive component and plasticity resulting from chronic stimulation.

In this procedure the patient undergoes a small craniectomy to remove a section of bone the size of the neurostimulator, and the neurostimulator is affixed inside the skull defect to be flush with the rest of the skull. This technique has FDA approval for the treatment of focal onset epilepsy. In these patients, a median seizure frequency reduction of 53% and a responder rate of 55% with stimulation were reported at 2 years [24]. On average, there was no significant worsening of mood, and 44% of patients reported improvements in quality of life [25].

In addition to focal onset epilepsy, there is an ongoing trial for generalized epilepsy where the responsive neurostimulation leads are placed in the centromedian thalamus. The trial is called NAUTILUS (NCT05147571) and seeks to study the efficacy of RNS in reducing the frequency of general tonic-clonic seizures.

Deep Brain Stimulation Deep brain stimulation is a technique where electrodes are implanted into a subcortical structure and attached to a pulse generator implanted

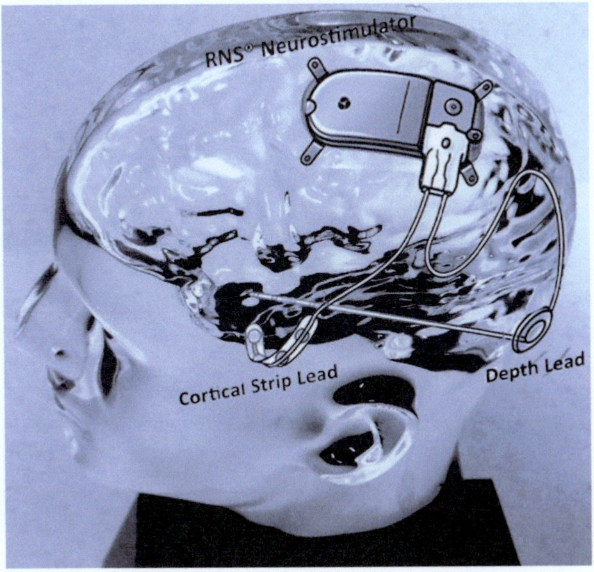

Fig. 23.9 Implanted RNS Neurostimulator, Depth Lead and Cortical Strip Lead. (with permission from Heck et al. [23])

in a subclavicular pocket. The pulse generator uses the electrodes to deliver open-loop stimulation to the subcortical structure. While the technique has been utilized and FDA-approved for decades for the treatment of tremor and Parkinson's disease, recently the technique has been adapted for medically refractory epilepsy as well. In 2017 deep brain stimulation was FDA-approved for anterior thalamic stimulation for partial or secondary generalized seizures after the success of the SANTE trial [26]. Patients were found to have a 41% reduction in seizures at 1 year and 69% at 5 years.

Another area where deep brain stimulation has been utilized is the centromedian thalamus. This stimulation area has shown good efficacy in patients suffering from Lennox-Gastaut syndrome (LGS) in a double-blind randomized study. In half of the stimulation group, there was a more than 50% reduction in seizure frequency [27].

Pearls: Post-Epilepsy Surgery Outcomes

Two grading scales have been used to evaluate outcomes after epilepsy surgery. The first grading scale, the Engel Classification, was published in 1993 and primarily divided outcomes into four categories. The classification also has subcategories for the four classifications but has been primarily criticized for being vague. In 2001 another classification system was proposed, the International League Against Epilepsy Classification, which attempted to be more rigorous regarding which patients should be assigned to each group. While the table below tries to match groups together, the vague nature of what a disabling seizure is, and what worthwhile improvement consists of makes it difficult to match the groups perfectly.

Outcomes after epilepsy surgery			
Engel classification		ILAE classification	
I	Free of disabling seizures	**1**	Completely seizure free, no auras
		2	Only aura, no other seizures
II	Rare disabling seizures	**3**	One to three seizure days per year. (+/−) auras
III	Worthwhile improvement	**4**	Four seizure days per year, up to 50% reduction in seizures. (±) auras
IV	No worthwhile improvement	**5**	Less than 50% reduction in seizures, up to 100% increase in seizures. (±) auras
		6	More than 100% increase of baseline seizure days. (±) auras

Acknowledgments Dario J. Englot for co-authoring a previous version of this chapter.

References

1. Kalilani L, Sun X, Pelgrims B, Noack-Rink M, Villanueva V. The epidemiology of drug-resistant epilepsy: a systematic review and meta-analysis. Epilepsia. 2018;59:2179–93.
2. De Tisi J, Bell GS, Peacock JL, McEvoy AW, Harkness WF, Sander JW, Duncan JS. The long-term outcome of adult epilepsy surgery, patterns of seizure remission, and relapse: a cohort study. Lancet. 2011;378(9800):1388–95. https://doi.org/10.1016/S0140-6736(11)60890-8.
3. NAEC Update – 2021 Center Annual Report Data. Available online at https://naec-epilepsy.org/wp-content/uploads/2024/06/the-2022-Accreditation-Program-and-Epilepsy-Center-Data-Trends-Slides.pdf
4. Sheikh SR, Kattan MW, Steinmetz M, Singer ME, Udeh BL, Jehi L. Cost-effectiveness of surgery for drug-resistant temporal lobe epilepsy in the US. Neurology. 2020;95(10):e1404–16. https://doi.org/10.1212/WNL.0000000000010185.
5. Chang EF, Englot DJ, Vadera S. Minimally invasive surgical approaches for temporal lobe epilepsy. Epilepsy Behav EB. 2015;47:24–33.
6. Spencer D, Burchiel K. Selective amygdalohippocampectomy. Epilepsy Res Treat. 2012;2012:382095.
7. Nascimento FA, Gatto LA, Silvado C, Mäder-Joaquim MJ, Moro MS, Araujo JC. Anterior temporal lobectomy versus selective amygdalohippocampectomy in patients with mesial temporal lobe epilepsy. Arq Neuropsiquiatr. 2016;74(1):35–43. https://doi.org/10.1590/0004-282X20150188.
8. Bahuleyan B, Robinson S, Nair AR, Sivanandapanicker JL, Cohen AR. Anatomic hemispherectomy: historical perspective. World Neurosurg. 2013;80:396–8.
9. Roth J, Bergman L, Weil AG, Brunette-Clement T, Weiner HL, Treiber JM, Shofty B, Cukiert A, Cukiert CM, Tripathi M, Sarat Chandra P, Bollo RJ, Machado HR, Santos MV, Gaillard WD, Oluigbo CO, Ibrahim GM, Jallo GI, Shimony N, O'Neill BR, Budke M, Pérez-Jiménez MÁ, Mangano FT, Iwasaki M, Iijima K, Gonzalez-Martinez J, Kawai K, Ishishita Y, Elbabaa SK, Bello-Espinosa L, Fallah A, Maniquis CAB, Ben-Zvi I, Tisdall M, Panigrahi M, Jayalakshmi S, Blount JP, Dorfmüller G, Bulteau C, Stone SS, Bolton J, Singhal A, Connolly M, Alsowat D, Alotaibi F, Ragheb J, Uliel-Sibony S. Added value of corpus callosotomy following vagus nerve stimulation in children with Lennox-Gastaut syndrome: a multicenter, multinational study. Epilepsia. 2023;64(12):3205–12. https://doi.org/10.1111/epi.17796.
10. Rolston JD, Deng H, Wang DD, Englot DJ, Chang EF. Multiple subpial transections for medically refractory epilepsy: a disaggregated review of patient-level data. Neurosurgery. 2018;82:613–20.
11. Willie JT, Laxpati NG, Drane DL, Gowda A, Appin C, Hao C, Brat DJ, Helmers SL, Saindane A, Nour SG, Gross RE. Real-time magnetic resonance-guided stereotactic laser amygdalohippocampotomy for mesial temporal lobe epilepsy. Neurosurgery. 2014;74(6):569–84; discussion 584-5. https://doi.org/10.1227/NEU.0000000000000343.
12. Youngerman BE, Banu MA, Khan F, McKhann GM, Schevon CA, Jagid JR, Cajigas I, Theodotou CB, Ko A, Buckley R, Ojemann JG, Miller JW, Laxton AW, Couture DE, Popli GS, Buch VP, Halpern CH, Le S, Sharan AD, Sperling MR, Mehta AD, Englot DJ, Neimat JS, Konrad PE, Sheth SA, Neal EG, Vale FL, Holloway KL, Air EL, Schwalb JM, D'Haese PF, Wu C. Long-term outcomes of mesial temporal laser interstitial thermal therapy for drug-resistant epilepsy and subsequent surgery for seizure recurrence: a multi-Centre cohort study. J Neurol Neurosurg Psychiatry. 2023;94(11):879–86. https://doi.org/10.1136/jnnp-2022-330979.
13. Wu C, Jermakowicz WJ, Chakravorti S, Cajigas I, Sharan AD, Jagid JR, Matias CM, Sperling MR, Buckley R, Ko A, Ojemann JG, Miller JW, Youngerman B, Sheth SA, McKhann GM, Laxton AW, Couture DE, Popli GS, Smith A, Mehta AD, Ho AL, Halpern CH, Englot DJ, Neimat JS, Konrad PE, Neal E, Vale FL, Holloway KL, Air EL, Schwalb J, Dawant BM, D'Haese PF. Effects of surgical targeting in laser interstitial thermal therapy for mesial temporal lobe epilepsy: a multicenter study of 234 patients. Epilepsia. 2019;60(6):1171–83. https://doi.org/10.1111/epi.15565.

14. Drane DL, Willie JT, Pedersen NP, Qiu D, Voets NL, Millis SR, Soares BP, Saindane AM, Hu R, Kim MS, Hewitt KC, Hakimian S, Grabowski T, Ojemann JG, Loring DW, Meador KJ, Faught E Jr, Miller JW, Gross RE. Superior verbal memory outcome after stereotactic laser Amygdalohippocampotomy. Front Neurol. 2021;12:779495. https://doi.org/10.3389/fneur.2021.779495.
15. Curry DJ, Raskin J, Ali I, Wilfong AA. MR-guided laser ablation for the treatment of hypothalamic hamartomas. Epilepsy Res. 2018;142:131–4.
16. Abe K, Yamaguchi T, Hori H, Sumi M, Horisawa S, Taira T, Hori T. Magnetic resonance-guided focused ultrasound for mesial temporal lobe epilepsy: a case report. BMC Neurol. 2020;20(1):160. https://doi.org/10.1186/s12883-020-01744-x.
17. Yamaguchi T, Hori T, Hori H, Takasaki M, Abe K, Taira T, Ishii K, Watanabe K. Magnetic resonance-guided focused ultrasound ablation of hypothalamic hamartoma as a disconnection surgery: a case report. Acta Neurochir. 2020;162(10):2513–7. https://doi.org/10.1007/s00701-020-04468-6.
18. Krishna V, Mindel J, Sammartino F, Block C, Dwivedi AK, Van Gompel JJ, Fountain N, Fisher R. A phase 1 open-label trial evaluating focused ultrasound unilateral anterior thalamotomy for focal onset epilepsy. Epilepsia. 2023;64(4):831–42. https://doi.org/10.1111/epi.17535.
19. Cossu M, Fuschillo D, Casaceli G, Pelliccia V, Castana L, Mai R, Francione S, Sartori I, Gozzo F, Nobili L, Tassi L, Cardinale F, Lo RG. Stereoelectroencephalography-guided radiofrequency thermocoagulation in the epileptogenic zone: a retrospective study on 89 cases. J Neurosurg. 2015 Dec;123(6):1358–67. https://doi.org/10.3171/2014.12.JNS141968.
20. Krahl SE, Clark KB, Smith DC, Browning RA. Locus coeruleus lesions suppress the seizure-attenuating effects of vagus nerve stimulation. Epilepsia. 1998;39:709–14.
21. Wasade VS, Schultz L, Mohanarangan K, Gaddam A, Schwalb JM, Spanaki-Varelas M. Long-term seizure and psychosocial outcomes of vagus nerve stimulation for intractable epilepsy. Epilepsy Behav. 2015;53:31–6. https://doi.org/10.1016/j.yebeh.2015.09.031.
22. Boon P, Vonck K, van Rijckevorsel K, El Tahry R, Elger CE, Mullatti N, Schulze-Bonhage A, Wagner L, Diehl B, Hamer H, Reuber M, Kostov H, Legros B, Noachtar S, Weber YG, Coenen VA, Rooijakkers H, Schijns OE, Selway R, Van Roost D, Eggleston KS, Van Grunderbeek W, Jayewardene AK, McGuire RM. A prospective, multicenter study of cardiac-based seizure detection to activate vagus nerve stimulation. Seizure. 2015;32:52–61. https://doi.org/10.1016/j.seizure.2015.08.011.
23. Heck CN, King-Stephens D, Massey AD, Nair DR, Jobst BC, Barkley GL, Salanova V, Cole AJ, Smith MC, Gwinn RP, Skidmore C, Van Ness PC, Bergey GK, Park YD, Miller I, Geller E, Rutecki PA, Zimmerman R, Spencer DC, Goldman A, Edwards JC, Leiphart JW, Wharen RE, Fessler J, Fountain NB, Worrell GA, Gross RE, Eisenschenk S, Duckrow RB, Hirsch LJ, Bazil C, O'Donovan CA, Sun FT, Courtney TA, Seale CG, Morrell MJ. Two-year seizure reduction in adults with medically intractable partial onset epilepsy treated with responsive neurostimulation: final results of the RNS system pivotal trial. Epilepsia. 2014;55(3):432–41. https://doi.org/10.1111/epi.12534.
24. Bergey GK, Morrell MJ, Mizrahi EM, Goldman A, King-Stephens D, Nair D, Srinivasan S, Jobst B, Gross RE, Shields DC, Barkley G, Salanova V, Olejniczak P, Cole A, Cash SS, Noe K, Wharen R, Worrell G, Murro AM, Edwards J, Duchowny M, Spencer D, Smith M, Geller E, Gwinn R, Skidmore C, Eisenschenk S, Berg M, Heck C, Van Ness P, Fountain N, Rutecki P, Massey A, O'Donovan C, Labar D, Duckrow RB, Hirsch LJ, Courtney T, Sun FT, Seale CG. Long-term treatment with responsive brain stimulation in adults with refractory partial seizures. Neurology. 2015;84(8):810–7. https://doi.org/10.1212/WNL.0000000000001280.
25. Meador KJ, Kapur R, Loring DW, Kanner AM, Morrell MJ, RNS® System Pivotal Trial Investigators. Quality of life and mood in patients with medically intractable epilepsy treated with targeted responsive neurostimulation. Epilepsy Behav. 2015;45:242–7. https://doi.org/10.1016/j.yebeh.2015.01.012.
26. Salanova V, Witt T, Worth R, Henry TR, Gross RE, Nazzaro JM, Labar D, Sperling MR, Sharan A, Sandok E, Handforth A, Stern JM, Chung S, Henderson JM, French J, Baltuch G, Rosenfeld WE, Garcia P, Barbaro NM, Fountain NB, Elias WJ, Goodman RR, Pollard

JR, Tröster AI, Irwin CP, Lambrecht K, Graves N, Fisher R, SANTE Study Group. Long-term efficacy and safety of thalamic stimulation for drug-resistant partial epilepsy. Neurology. 2015;84(10):1017–25. https://doi.org/10.1212/WNL.0000000000001334.
27. Dalic LJ, Warren AEL, Bulluss KJ, Thevathasan W, Roten A, Churilov L, Archer JS. DBS of thalamic Centromedian nucleus for Lennox-Gastaut syndrome (ESTEL trial). Ann Neurol. 2022;91(2):253–67. https://doi.org/10.1002/ana.26280. Epub 2021 Dec 28. Erratum in: Ann Neurol. 2022 Aug;92(2):347. doi: 10.1002/ana.26356

Index

Z. Haneef (ed.), *Epilepsy Fundamentals*,
https://doi.org/10.1007/978-3-031-77741-7

MIX
Papier aus verantwortungsvollen Quellen
Paper from responsible sources
FSC® C105338
FSC
www.fsc.org

If you have any concerns about our products,
you can contact us on
ProductSafety@springernature.com

In case Publisher is established outside the EU,
the EU authorized representative is:
Springer Nature Customer Service Center GmbH
Europaplatz 3, 69115 Heidelberg, Germany

Printed by Libri Plureos GmbH
in Hamburg, Germany